Enzymatic and Chemical Synthesis of Nucleic Acid Derivatives

Enzymatic and Chemical Synthesis of Nucleic Acid Derivatives

Edited by
Jesús Fernández-Lucas
María-José Camarasa Rius

WILEY-VCH

Editors

Jesús Fernández-Lucas
European University of Madrid
Dept. of Pharmacy, Biotechn. & Nutrition
Calle Tajo, s/n
28670 Villaviciosa de Odón
Spain

María-José Camarasa Rius
Institutio de Química Médica
Calle Juan de la Cierva, 3
28006 Madrid
Spain

Library of Congress Card No.:
applied for

British Library Cataloguing-in-Publication Data
A catalogue record for this book is available from the British Library.

Bibliographic information published by the Deutsche Nationalbibliothek
The Deutsche Nationalbibliothek lists this publication in the Deutsche Nationalbibliografie; detailed bibliographic data are available on the Internet at <http://dnb.d-nb.de>.

Print ISBN: 978-3-527-34376-8
ePDF ISBN: 978-3-527-81207-3
ePub ISBN: 978-3-527-81209-7
oBook ISBN: 978-3-527-81210-3

Typesetting SPi Global, Chennai, India
Printing and Binding C.O.S. Printers Pte Ltd, Singapore

Printed on acid-free paper

10 9 8 7 6 5 4 3 2 1

Contents

Preface

Nucleosides and nucleotides are key endogenous molecules engaged in many biological events such as RNA and DNA synthesis, metabolism, enzyme regulation, etc. Nucleosides and nucleotides are the major components of nucleic acids. Synthetic nucleos(t)ides have been extensively modified either by chemical or enzymatic procedures to mimic their physiological counterparts and show important antiviral or antiproliferative activities, among others.

The chemical or enzymatic synthesis and modification of nucleos(t)ides is a major research topic in organic, medicinal, and bioorganic chemistry. This book contains a selection of different chapters prepared by highly reputed scientists involved in the field of nucleosides and nucleotides, covering different aspects of their synthesis and/or their applications that provide valuable information to all the scientists working in this field. Each chapter in this volume reflects the own work performed by the respective contributors in the laboratories.

The first seven chapters are involved with the enzymatic or chemoenzymatic synthesis of nucleic acid derivatives. Nowadays, the application of bioprocesses catalyzed by whole cells or enzymes in industrial settings is gaining momentum as compared to traditional chemical synthetic processes. In this context, the enzymatic synthesis of active pharmaceutical ingredients (APIs) shows many advantages, such as one-pot reactions under mild conditions, high stereo- and regioselectivity, and an environmentally friendly technology.

In this context, many different authors are working together to publish a special issue that will focus on the development of new enzymatic or chemoenzymatic strategies for the synthesis of nucleic acid derivatives catalyzed either by whole cells or enzymes, showing the advantages and drawbacks of this methodology versus traditional chemical methods

In the first chapter, Kamel et al. have performed an exhaustive revision about the potential of nucleoside phosphorylases (NPs) as biocatalysts for industrial synthesis of nucleoside analogues, providing detailed information about their substrate specificity and structural properties that influence their activity. Additionally, they describe the potential of mutagenic approaches to widen the substrate spectrum of nucleoside phosphorylases, showing an overview about the upscaling approaches for the production of nucleoside analogues

In Chapter 2, Dr. Daniela Ubiali and Prof. Giovanna Speranza present an overview of the recent work on enzymatic phosphorylation of nucleosides to nucleoside 5′-monophosphate (NMPs) by nonspecific acid phosphatases

(NSAPs) or 2′-deoxyribonucleoside kinases (dNKs). In Chapter 3, Dr. Lewkowicz and Dr. Iribarren summarize the available information regarding the biocatalyzed whole cell synthesis of nucleosides and their analogues referring to relevant examples and recent researches. Furthermore, Dr. Trelles et al. performed an interesting revision about the different cell-immobilization techniques applied to the development of active biocatalysts for the synthesis of nucleic acid derivatives by immobilized cells (Chapter 4). In a similar way, Dr. Arroyo analyzes recent literature concerning the development of novel enzymatic production of nucleic acid derivatives using immobilized enzymes (Chapter 5), including detailed examples and procedures. In Chapter 6, Dr. Ding illustrates the most current relevant examples of *in vivo* and *in vitro* multienzymatic systems as an efficient alternative to traditional multistep chemical procedures used for the synthesis of nucleic acid derivatives.

In a different way, Dr. Eremeeva and Dr. Herdewijn, in Chapter 7, present an exhaustive revision about the enzymatic synthesis of redesigned nucleic acids (xeno-nucleic acids, XNA) using polymerases. In this sense, the enzymatic production of several XNAs, such as base-modified XNAs, sugar-modified XNAs, or the phosphodiester backbone-XNA, is reported.

The last five chapters concentrate on the chemical synthesis and/or applications of these unique molecules. Particularly, Chapters 8–12 deal with the chemical synthesis of nucleosides and oligonucleotides. Thus, in Chapter 8, Dr. Seley-Radke reports a historic perspective of the synthetic approaches followed for the preparation of a unique class of nucleosides, the fleximer nucleosides, designed to understand better how the flexibility in the nucleobase affects the receptor-ligand recognition and function and their potential to overcome the challenges of binding site mutations. Fleximers are flexible nucleosides with several key advantages over natural purine nucleosides.

Chapter 9 by Dr. Eritja et al. focuses on the synthesis of oligonucleotides bearing nucleic acid derivatives of biomedical and structural interest. In this chapter, an overview of valuable and versatile nucleic acid derivatives developed by the group during the past decade is presented together with their emerging applications as well as a highlight of the important recent progress made in the field.

In Chapter 10, Dr. Morales describes the synthesis and applications of carbohydrate-oligonucleotide conjugates. The interest in such conjugates is to take advantage by integrating the properties of both kinds of molecules (carbohydrates and oligonucleotides) in a single molecule. This type of conjugates have attracted a lot of attention to explore their application as potential therapeutic agents, aptamers, ribozymes, biosensors, biochips, diagnostic tools, etc.

Chapter 11 focuses on microarrays; Dr. Damha and colleagues detail the advances in light-directed synthesis of high-density microarrays. This interesting chapter gathers the phosphoramidite chemistry applied to the photolithographic synthesis of microarrays; the recent improvements in the synthesis of DNA microarrays; the synthesis of RNA microarrays; the enzymatic approaches to RNA array synthesis; and the synthesis of 2′F-ANA microarrays.

Finally, in Chapter 12, Dr. Schinazi and coworkers describe how SAMHD1 (a dNTP triphosphohydrolase that degrades dNTPs into 2′-deoxynucleosides and triphosphates) mediates the negative regulation of cellular dNTP levels.

We hope that the readers will find the contents of this book interesting and instructive and challenging enough to be involved in the fantastic field of nucleos(t)ides and nucleic acids. Finally, last but not the least, the editors sincerely thank all the authors who have contributed to this book with very interesting chapters.

Jesús Fernández-Lucas
Universidad Europea de Madrid
María-José Camarasa Rius
Instituto de Química Médica
Consejo Superior de Investigaciones Científicas

1

Enzymatic Synthesis of Nucleoside Analogues by Nucleoside Phosphorylases

Sarah Kamel[1*], *Heba Yehia*[1,2*], *Peter Neubauer*[1], *and Anke Wagner*[1,3]

[1] *Technische Universität Berlin, Department of Bioprocess Engineering, Institute of Biotechnology, Ackerstraße 76, 13355 Berlin, Germany*
[2] *National Research Centre, Department of Chemistry of Natural Products, 12622 Giza, Egypt*
[3] *BioNukleo GmbH, Ackerstraße 76, 13355 Berlin, Germany*

1.1 Introduction

1.1.1 Nucleosides and Nucleoside Analogues

Nucleosides primarily consist of a nitrogenous base (nucleobase), which is either a purine base or a pyrimidine base and a five-carbon sugar (pentose). The base and sugar are covalently linked via an *N*-glycosidic bond (Figure 1.1). The pentose sugar moiety of naturally occurring canonical nucleosides is either ribose or deoxy-ribose whereas the nucleobase might be either a purine (adenine, guanine) or a pyrimidine (cytosine, uracil, thymine). These nucleosides are structural subunits of nucleic acids and are involved in several cellular processes including enzyme regulation and metabolism, DNA and RNA synthesis, and cell signaling [1, 2].

Naturally occurring nucleoside analogues (non-canonical nucleosides) are found in almost all types of RNA especially in tRNAs and they are crucial for RNA processing. Non-canonical analogues are nucleosides with different modifications on the pentose and/or the base [3] (Figure 1.1). There are more than 109 known post-transcriptional modifications in the three phylogenetic domains [4]. Pseudouridine is the most ubiquitous analogue and is sometimes considered as the fifth RNA-related nucleoside [5].

Non-natural nucleoside analogues are synthetic molecules that structurally mimic their physiological counterparts and also act as antimetabolites [2]. Nucleoside analogues access cells through specific nucleoside transporters. Within the cells, they are phosphorylated by nucleoside kinases, which leads to increased levels of di- and tri-phosphorylated nucleoside analogues in virus-infected or cancer cells. The first and the second phosphorylation step can also be catalyzed by viral kinases in cells infected by some DNA viruses. Owing to differences in the substrate spectrum of human and viral kinases, virus-specific drugs can be

*S.K. and H.Y. contributed equally to this work.

Enzymatic and Chemical Synthesis of Nucleic Acid Derivatives, First Edition.
Edited by Jesús Fernández-Lucas and María-José Camarasa Rius.

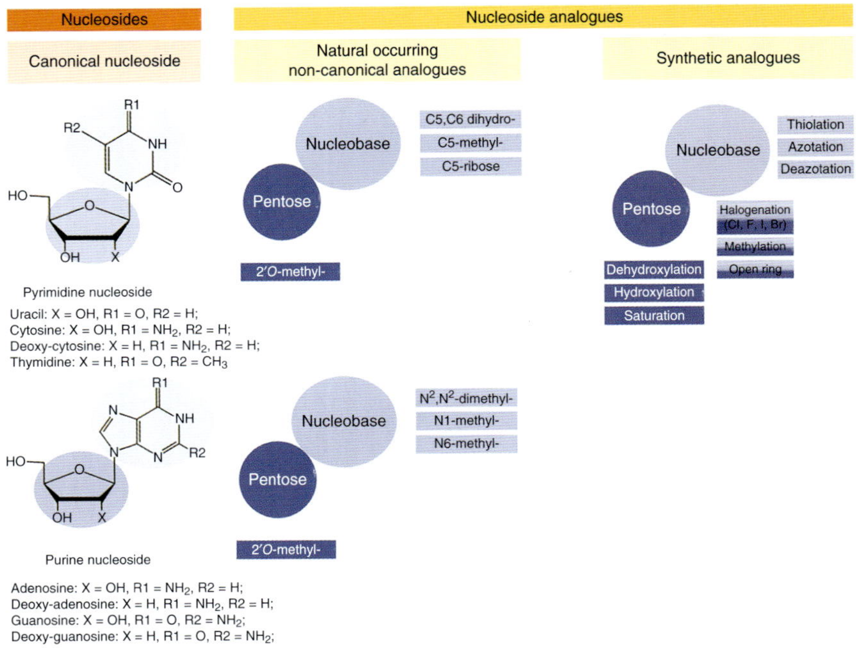

Figure 1.1 Classification of nucleosides and nucleoside analogues. Canonical (unmodified) nucleosides are the building blocks of DNA and RNA. Non-canonical (naturally modified on pentose moiety, base moiety or both) are mainly occurring in RNA. Synthetic nucleosides are used in the treatment of viral and bacterial infections as well as in cancer treatment.

developed. The active forms of nucleoside analogues interfere with intracellular enzymes such as human and viral polymerases, kinases, DNA methyl transferase, ribonucleotide reductase, nucleoside phosphorylases (NPs) or thymidylate synthase [2, 6]. Furthermore, they can be incorporated into newly synthesized DNA and RNA, which may induce termination of the polymerization process, accumulation of mutations in viral progeny, or induction of apoptosis.

For more than 50 years, nucleosides and their analogues have been used as small molecule drugs for the treatment of several viral infections as well as for hematological malignancies and solid tumors. The first FDA approved antiviral nucleoside analogue was idoxuridine, which is used for the treatment of HSV-1 (herpes simplex virus) [7]. In 1969, cytarabine was approved for the treatment of acute myeloid leukemia [2]. Since then, the interest in nucleoside analogues based drugs has tremendously grown. Currently, more than 39 approved nucleoside analogue drugs or drug combinations are approved for the treatment of seven human viral infections, which include HSV, varicella zoster virus (VZV), hepatitis-B virus (HBV), hepatitis-C virus (HCV), human immunodeficiency virus (HIV), respiratory syncytial virus (RSV), and human cytomegalovirus (HCMV) [7]. For treatment of cancer and viral infections, 50% and 20%, respectively, of all approved drugs belong to the class of nucleoside analogues [8]. Additional clinical indications for nucleoside analogues application include chronic hyperuricemia, immune suppression in organ transplant surgeries, and autoimmune disease as well as chronic obstructive pulmonary disease and asthma [2].

Emerging from the significance of nucleoside analogues, there have been continuous attempts to improve and simplify their synthesis processes. With the world moving toward green chemistry approaches, the enzymatic synthesis of nucleoside analogues offers several advantages over chemical methods, which include higher total yields, a higher regio- and stereo-selectivity, and higher product purity. This allows for more biological and clinical trials [9]. Accordingly, enzymatic strategies are considered as a step forward to a more efficient synthesis of nucleosides and their analogues.

1.1.2 Enzymes Involved in the Enzymatic Synthesis of Nucleoside Analogues

Two main classes are employed in the enzymatic synthesis of nucleosides and their analogues: NPs and *N*-deoxyribosyltransferases (NDTs). In this chapter, the focus is on enzymatic approaches using NPs. NPs are of high interest as biocatalysts because of their wide substrate spectrum and abundance in almost all living organisms.

1.2 Nucleoside Phosphorylases

NPs are enzymes belonging to the transferases family (EC 2.4 and EC 2.7.7). NPs catalyze the reversible phosphorolysis of nucleosides into their respective nucleobase and pentofuranose-1-phosphate (Pentose-1P). NPs have been extensively studied since 1911 when Levene and Medigrecenau [10, 11], and Johnes [10–13] observed the enzymatic hydrolysis of nucleosides. Later, Levene et al. isolated an enzyme (nucleosidase) from cattle's spleen, kidney, and pancreas, which catalyzed the hydrolysis of both inosine and adenosine in phosphate buffer, yielding a base and a ribose moiety [14–16]. In 1947, Kalckar demonstrated that the formed ribose was in fact ribose-1-phosphate and that the isolated enzyme was a purine nucleoside phosphorylase (PNP) [17]. Later, it was proven that *Escherichia coli* cells and cell extracts thereof contained enzymes that could phosphorolyze thymidine to thymine and deoxyribose-1-phosphate [18].

1.2.1 Classification and Substrate Spectra of Nucleoside Phosphorylases

Nucleoside phosphorylases are classified based either on their substrate specificity/affinity (Table 1.1) or on their structure [19]. In 2002, Pugmire and Ealick described a structure-based classification of NPs in two distinct families [19, 20]: NP-I and NP-II (Figure 1.2). They demonstrated that members of an NP-I family share the following characteristics [20]: (i) they have a single domain subunit, (ii) they share a common α/β-subunit fold, (iii) their quaternary structure is either trimeric (mammals/higher organisms) or hexameric (bacteria/lower organisms), (iv) they accept both purine nucleosides (bacterial and mammalian PNPs) and

Table 1.1 Nucleoside phosphorylases with acronyms and EC numbers.

Enzyme name	Acronym	EC number
Thymidine phosphorylase	TP	2.4.2.4
Uridine phosphorylase	UP	2.4.2.3
Pyrimidine nucleoside phosphorylase	PyNP	2.4.2.2
Purine nucleoside phosphorylase	PNP	2.4.2.1
5′-Methylthioadenosine phosphorylase	MTAP	2.4.2.28
Guanosine phosphorylase	GP	2.4.2.15

pyrimidine nucleosides (uridine phosphorylase, UP) as substrates, and (v) their substrate-binding sites are similarly arranged. Nevertheless, they are quite different in their quaternary structures, amino acid sequence, and substrate specificity. Additionally, active sites of the hexameric family members are significantly different from those of the trimeric enzymes of higher eukaryotes, which makes them attractive targets for the specific treatment of bacterial or parasitic infections.

Members of the NP-II family are characterized by (i) having two domain subunits: a small α-domain and a large α/β-domain separated by a large cleft, (ii) a dimeric quaternary structure, (iii) being specific to pyrimidine nucleosides, (iv) accepting both thymidine and uridine as substrates in case of lower organisms, and thymidine in higher organisms' members of the family, (v) having more than 30% sequence similarity, and (vi) undergoing high conformational changes and domain movements during catalysis.

1.2.1.1 Nucleoside Phosphorylase-I Family

Members of the NP-I family are subdivided into PNP, uridine nucleoside phosphorylase and 5′-deoxy-5′-methythioadenoside phosphorylase (MTAP) (Figure 1.2). They catalyze reversible phosphorolysis of the *N*-glycosidic bond of purines, uridine, and 5′-deoxy-5′-methythioadenoside (MTA), respectively. PNPs are the most widely studied group of nucleoside phosphorylases due to their importance as drug targets.

Purine Nucleoside Phosphorylases (PNPs) PNPs are further classified into two main groups, trimeric PNPs and hexameric PNPs (Figure 1.2). Trimeric PNPs also named as low molecular mass PNPs (approx. 31 kDa per subunit) are specific for 6-oxopurines (guanine and hypoxanthine) and their nucleosides. The hexameric PNPs (also known as high molecular mass PNPs) consist of subunits with a size of approximately 26 kDa [20] and have a broader substrate spectrum. They accept both, 6-oxopurines and 6-aminopurines (adenine) and their nucleosides. Initially, trimeric forms were described to be specific for mammals and hexameric forms for bacteria. However, later it was shown that *E. coli*, *Bacillus subtilis*, and *Bacillus stearothermophilus* possess both forms [20–24].

Several reports showed that PNPs accept a wide range of substrate, thereby showing the wide substrate spectrum of these enzymes. Generally, the substrate

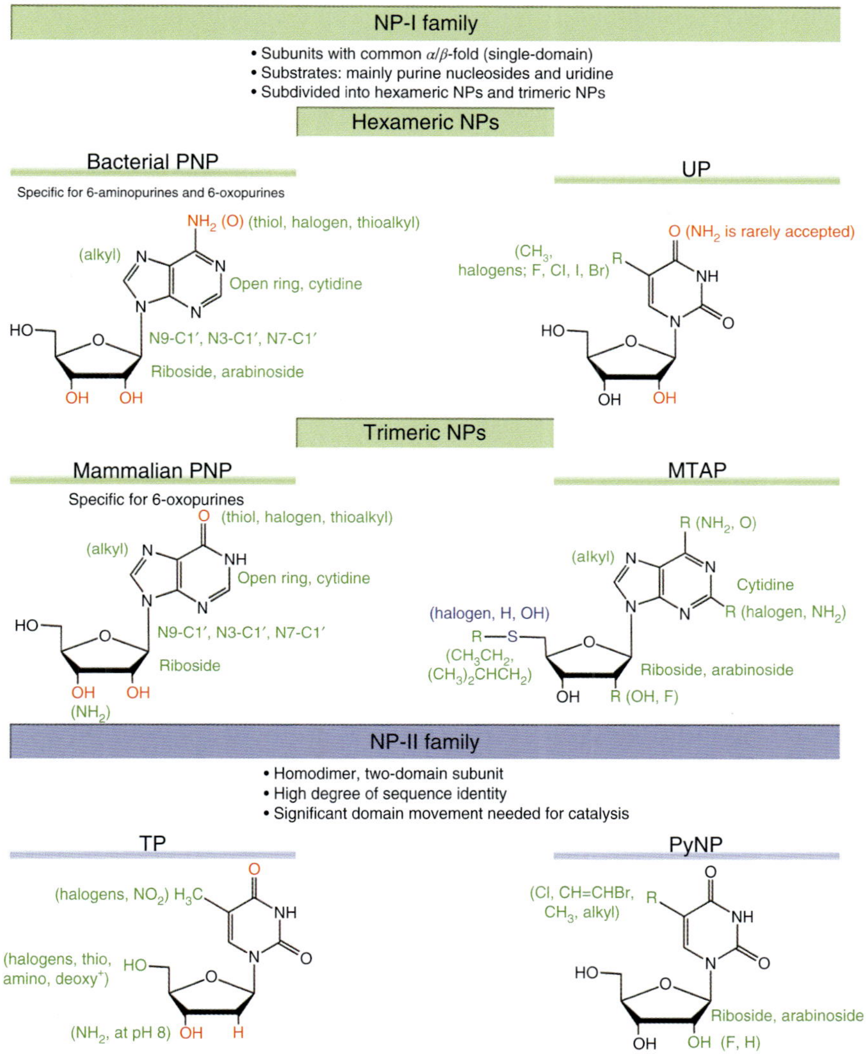

Figure 1.2 Classification of nucleoside phosphorylases (NPs) and their substrate affinities. NPs are classified into two main families: NP-I and NP-II. NP-I family is further subdivided into hexameric and Trimeric NPs. NP-II family is subdivided into two main classes: TP and PyNP. Residues labeled in red are crucial for enzyme–substrate interaction; green and blue labeled substitutions are those accepted by enzymes.

spectrum of hexameric PNPs is larger than that of the trimeric ones [19]. As it will be mentioned in detail later, modifications on the nucleobase and the sugar moiety highly influence the substrate recognition by the enzyme, and both the binding to and releasing from the active sites (Figure 1.2).

PNPs mostly recognize and cleave the N9-C1′ glycosidic bond of purines. However, bonds between N3-C1′ and N7-C1′ are cleaved as well by PNPs from different origins including *E. coli* and bovine and human PNPs [25–27]. Base substitutions with electronegative groups (thiol, halogen, and thioalkyl) at C6 and small alkyl substitution at N7 are well accepted by some members of the PNP family (Figure 1.2). Unexpectedly, open ring purine nucleoside analogues (as ribavirin) bearing structure similarities to inosine were found to be cleaved by some PNPs from various organisms including *E. coli, B. subtilis, Brevibacterium acetylicum*, calf, and human [27–29].

Maintaining the electronegativity at C1′ is crucial for substrate recognition by PNPs. Hence, ribo- and arabinonucleosides are well accepted by PNPs of *E. coli* and a number of thermostable PNPs [30–32]. The loss of 2′-OH and or 3′-OH reduced the phosphorylation efficiency of human PNP. The replacement of 3′-OH by an amino group improved the activity compared to 3′-deoxy-nucleosides due to the retained electronegativity [33]. Nonetheless, there are some exceptions that also accept 3′-deoxynucleosides, such as PNPs isolated from *B. acetylicum, Plasmodium carotovorum*, and *Geobacillus thermoglucosidasius* [30, 34, 35].

An interesting feature of PNPs is their ability to accept pyrimidine nucleosides as substrate. Thus, they may be interesting catalysts for the synthesis of cytidine and deoxycytidine that are not utilized by many pyrimidine nucleoside phosphorylases (PyNPs) [30].

5′-Deoxy-5′-methylthioadenoside Phosphorylase (MTAP) An interesting member of the NP-I family is MTAP, which was primarily characterized in rats to work on MTA. MTAPs show a wide substrate spectrum. In addition to MTA some MTAPs as those isolated from *Pyrococcus furiosus* and *Sulfolobus solfataricus* accept 6-oxopurine nucleosides (guanosine and inosine). For these enzymes, higher sequence similarities to PNPs compared to the known MTAPs were shown [36–38]. Substitutions with halogen or amino groups at the C2 position of the purine ring are very well tolerated [39–43] (Figure 1.2). MTAP isolated from *Aeropyrum pernix* has some unique characteristics: on one hand, it is active toward some pyrimidines including cytidine and deoxycytidine, and on the other hand, unlike most reported MTAPs, it phosphorolyzes 2′-fluoro-modified arabinoside [30], which is an interesting activity when considering the enzymatic synthesis of nucleoside analogues that can be used as drugs.

Uridine Nucleoside Phosphorylase Different prokaryotic and eukaryotic organisms including *E. coli, Lactobacillus casei, Enterobacter aerogenes, Salmonella typhimurium, Saccharomyces cerevisiae, Schistosoma mansoni, Mus musculus, Rattus norvegicus*, and *Homo sapiens* were reported to possess a UP [19].

Although UPs of different origins vary in their substrate spectrum, they show some common features: none of the reported UPs catalyze cytidine phosphorolysis except the UP from *S. cerevisiae*. However, all accept 2′-deoxy nucleosides, although with much lower affinity compared to uridine. Nucleosides that are halogenated or methylated at the C5 position are generally well accepted and tolerated [19] (Figure 1.2).

1.2.1.2 Nucleoside Phosphorylase-II Family

The NP-II family comprises phosphorylases that catalyze the reversible cleavage of the glycosidic bond of thymidine (thymidine phosphorylase [TP]) and pyrimidines (PyNPs) in the presence of phosphate. Although UP catalyzes a similar reaction, it has been classified under the NP-I family due to its distinct structural features. As mentioned earlier, both TP and PyNP share two-domain subunits with a high percentage (≈40%) of sequence and structural similarities [19, 20, 44]. However, one major catalytic difference between TP and PyNP exists: TP has a high specificity toward C2′ deoxy nucleosides compared to PyNPs [45]. Hence, PyNPs possess a wider substrate spectrum than TP.

Thymidine Phosphorylase Structure–function relationship studies have provided some insight into the acceptance of base and sugar modifications by TP. Regarding the sugar moiety, the absence of a hydroxyl group (—OH) at the C2′ position of the nucleoside is crucial for TP binding. Both, *E. coli* and human TP, do not accept uridine as a substrate. In fact, uridine and 2′-methyl uridine were shown to inhibit the catalytic action of *E. coli* TP [46–49]. On the other hand, the presence of the 3′-OH group of the nucleoside is crucial for interactions between the substrate and the enzyme. This is reinforced by the fact that 3′-amino-3′-deoxythymidine was phosphorolyzed by TP only at pH 8 and not at pH 6.5, because at pH 6.5 both the 3′-amino group of the nucleoside and the amino group of the enzyme are positively charged [48]. The 5′-position of the nucleoside does not strongly influence the catalytic reaction. Various 5′ modifications were tolerated by *E. coli* TP including halogens, thio, amino, and deoxy substitutions [48–50] (Figure 1.2). Modifications in the C5 position of the base are generally well tolerated by TP, while substitutions at C6 have a big impact on the catalytic potential. Substitutions that increase the aromaticity of the nucleobase hinder the base to leave the catalytic pocket of the enzyme [49].

Pyrimidine Nucleoside Phosphorylases PyNPs have been isolated and characterized from several organisms including *B. stearothermophilus, B. subtilis, G. thermoglucosidasius, Thermus thermophilus, Hameophilus influenza,* and *Mycoplasma hyorhinis* [45, 51–55]. PyNPs are of increasing interest due to their broader substrate affinity and tolerability compared to those of TP or UP. Thus, a wider application of these enzymes to produce synthetic nucleoside analogues is envisioned.

PyNPs well tolerate modification at the C5 position of the base. PyNPs from various organisms are able to catalyze the glyosidic bond cleavage of 2′-deoxy-5-halogenated nucleoside analogues (Figure 1.2). Additionally, a bromovinyl (CH=CHBr) group at C5 was tolerated by *E. coli* TP and *B. subtilis* PyNP. Several other modifications at C5 including alkyl groups are also well accepted by PyNPs from various sources [45, 56–58] (Figure 1.2). Interestingly, compounds with modifications at the C2′ position of the nucleoside are also accepted as substrates by PyNPs. Arabinosides as well as 2′-deoxy-2′-fluororibosides and 2′-deoxy-2′-fluoroarabinosides are substrates for some PyNPs, predominantly for those isolated from thermophilic organisms such as *T. thermophilus* [54] (Figure 1.2).

1.2.2 Limitations in the Current Classification

Although the classification of PNPs into two main categories as proposed by Pugmire and Ealick [20] is applicable for most of the known PNPs, there are still some exceptions that do not fit into this categories.

The low molecular mass inosine-guanosine phosphorylase extracted from the vegetative state and the spores of *Bacillus cereus* was shown to either have a tetrameric quaternary structure or to equilibrate between the dimeric and the tetrameric form [24]. *B. stearothermophilus* (JTS-859 strain) as mentioned earlier contains both forms of PNPs: a low molecular mass and a high molecular mass PNP. It was reported that the low molecular mass enzyme forms a dimer rather than a trimer whereas the high molecular mass form is a tetramer rather than a hexamer [24].

E. coli hexameric PNP-II (product of *xapA* gene) has an amino acid sequence, promotor structure, and substrate specificity similar to the low molecular mass PNPs. As observed for low molecular mass PNPs, *E. coli* PNP-II does not accept adenosine or adenosine derivatives as substrates [19, 24, 59].

1.2.3 Reaction Mechanism

Nucleoside phosphorylases reversibly cleave *N*-glycosidic bonds of β-nucleosides with configurational inversion producing α-pentofuranose-1-phosphate and a free nucleobase. The catalytic mechanism of the phosphorolytic cleavage has been reported for several members of the NP-I family. Most of the reported NPs were shown to have a similar catalytic mechanism (Figure 1.3). NP-II members are expected to follow the same pattern.

The mechanism described for human PNP illustrated that the general reaction follows an S_N1 mechanism [20, 60]. The nucleoside binds to the enzyme in a high energy conformation producing a steric strain on the glycosidic bond, thus favoring bond cleavage. An oxocarbenium ion is formed during the transition state by the flow of an electron from O4′ to the purine base rendering the base with a negative charge. The formed oxocarbenium ion is stabilized by phosphate ions. Additionally, negative charges on the base moiety are stabilized by interactions of the enzymes active site to the N-atom of the base, thus facilitating glycosidic bond cleavage [19, 20, 60, 61].

The mechanism of phosphorolytic cleavage became clear by the availability of crystal structures; however, the exact mechanism of nucleoside synthesis is still not clear and difficult to elaborate due to the difficulty of crystallizing pentose-1-phospahte alone with PNP. Yet, it is suggested that the presence of the

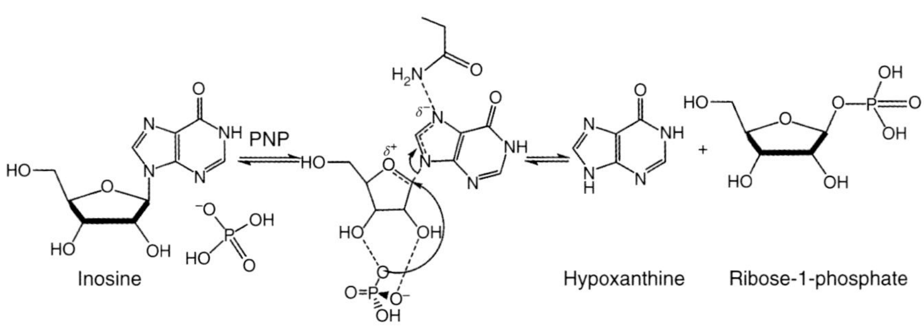

Figure 1.3 Catalytic mechanism described for PNP.

base favors the binding of pentose-1-phosphate to the catalytic site. Otherwise, the latter would cause steric hindrance of base binding [24, 60, 61].

PNP reactions follow non-Michaelis Menten kinetics. However, an exact kinetic mechanistic model was not agreed upon so far [24, 60]. Crystal structure data suggest the formation of ternary complexes of enzyme/nucleoside/phosphate and enzyme/base/pentose-1-phosphate. Whether these complexes are formed in a sequentially defined order or by random binding was not yet confirmed. Nevertheless, while there is no evidence for the occurrence of a ping–pong mechanism, a sequential order of the reaction is assumed by most authors [24].

1.2.4 Domain Structure and Active Site Residues of Nucleoside Phosphorylases

1.2.4.1 NP-I Family Members

As mentioned earlier, members of the NP-I family share a single domain structure that is highly conserved [62]. Nevertheless, amino acid residues in the active sites are not necessarily conserved among different members of the NP-I family.

One interesting difference between the four subgroups (trimeric PNP, MTAP, hexameric PNP, and UP) of the NP-I family is the observed sequence motifs. Although sequence motifs do not necessarily reflect functional differences, they help to show differences and similarities between subgroups of enzymes. There are nine identified sequence motifs in NP-I family (Figure 1.4). These structural regions are denoted with letters A–I. Their exact position in the primary amino acid sequence was reviewed by Pugmire and Ealick in 2002 [20]. Numbering codes were assigned based on structure similarities, for example B_1, and B_2 are similar regarding their quaternary structure's position; however, their amino acid sequence differs significantly. Not all members of the subclasses have the same number of motifs, but all of them encompass motif A_1. Noteworthy, MTAP (E_3, H_2) and UP (H_3, I_1) have two unique motifs that are involved in the active site and inter-subunit interactions [20].

From nine motifs identified in the members of the NP-I family, seven are involved in the formation of the active site and six are involved in subunit interaction. The observed structural similarities between trimeric PNP and MTAP as well as between hexameric PNP and UP are also reflected in the motif structure (Figure 1.4). Amino acid residues involved in binding of phosphate,

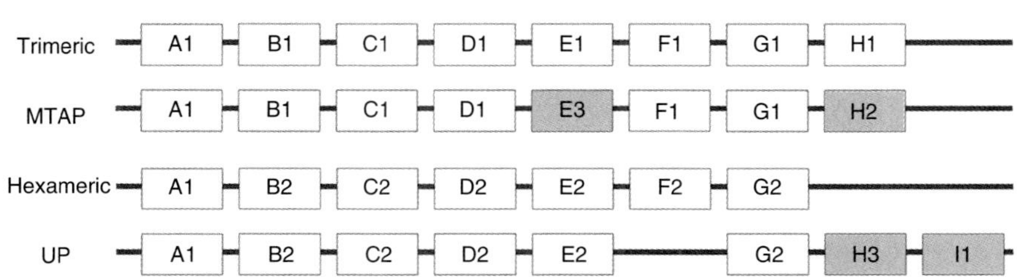

Figure 1.4 Structural motifs described for members of the NP-I family. Highlighted in grey are domains that are unique for either MTAP or UP. Trimeric- trimeric PNP, Hexameric-hexameric PNP.

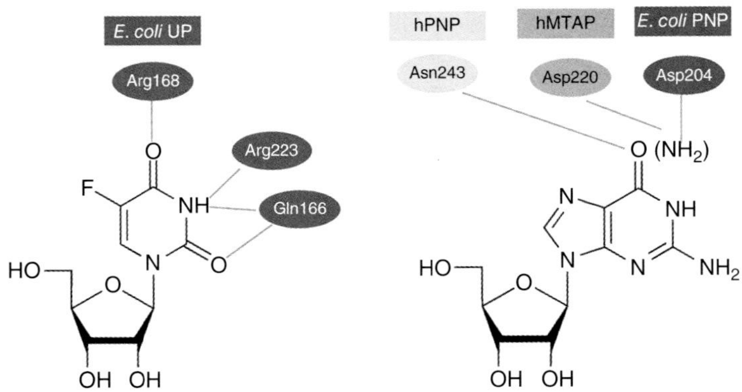

Figure 1.5 Schematic presentation of amino acid residues that define substrate specificity of different nucleoside phosphorylases.

nucleobase, and pentose sugar differ between hexameric and trimeric members of the NP-I family [19, 20]. Detailed information on amino acids involved in substrate binding has been recently reviewed [19].

Functionally, the four subclasses of the NP-I family showed different specificities toward substrates. Trimeric PNPs are specific to 6-oxopurines and MTAPs are described to be specific for 5′deoxy-5′-methylthioadenosinde, whereas hexameric PNPs accept both 6-oxopurines and 6-aminopurines (adenosine). UP is strictly specific to pyrimidine derivatives with no distinct specificity toward the C2′ position. These functional divergences are also reflected by differences in the amino acid residues found in the active sites. hMTAP (Asp220) and *E. coli* PNP (Asp204) form hydrogen bonds with the 6-amino group of the purine base. The replacement of an aspartic acid residue (Asp) with asparagine (Asn) as in the case of human and bovine PNP (Asn243) restricted the activity of human and bovine PNP toward 6-oxopurines (Figure 1.5). As another example, *E. coli* and human UP are conserved in residues Gln166, Arg168, and Arg223, which are described to be responsible for the specificity of UP [19].

1.2.4.2 NP-II Family Members

TP and PyNP are the main two enzyme groups within the NP-II family. Enzymes from both classes share a high degree of sequence similarity (33–67%). Furthermore, they share two main domain subunits, a large α/β domain and a smaller α domain, that are separated from each other by a large cleft. The amino acid residues of the α domain are more conserved. Detailed information on amino acids involved in substrate binding has been recently summarized [19].

Active site residues are highly conserved among TP and PyNP with few exceptions that might be responsible for differences in substrate acceptance. While TP is highly specific for 2′-deoxy nucleosides, PyNP catalyzes the cleavage of a number of 2′-modified nucleosides. A possible explanation that is revealed by structural studies is the substitution of lysine (Lys108) of PyNP with methionine in both *E. coli* and human TP [19]. This replacement alters the hydrogen bonding organization between the 2′-hydroxyl group of the nucleoside and the oxygen atom of the phosphate. An alternative explanation is a different positioning of 2′-modified nucleosides in the active site of TP compared to 2′-deoxy nucleosides [63].

1.3 Enzymatic Approaches to Produce Nucleoside Analogues Using Nucleoside Phosphorylases

Despite the huge advances achieved in the chemical synthesis of nucleosides and their analogues, the existing drawbacks led to a continuous interest in developing enzymatic approaches. In 1954, M. Friedkin and D. Roberts [64, 65] reported the first attempts to enzymatically synthesize pyrimidine nucleoside analogues using TP isolated from horse liver. Since then several studies have been conducted on enzymes to be used in the synthesis of nucleoside analogues.

As mentioned earlier, nucleoside phosphorylases catalyze the reversible cleavage of *N*-glycosidic bond of nucleosides. Thus, the condensation of a nucleobase and pentose-1-phosphate can be used for the synthesis of nucleosides or their analogues. Different synthesis strategies were developed based on this mechanism of action.

1.3.1 One-pot Two-Step Transglycosylation Reaction

Enzymatic transglycosylation has been extensively studied [9, 66, 67]. It is the transfer of a pentofuranose moiety from a chemically synthesized or a naturally occurring nucleoside to a base of interest. Pentose-1P is formed as an intermediate in the reaction (Figure 1.6a). Sugar donor is a pyrimidine nucleoside (cleavage reaction) and sugar acceptor is a purine base (synthesis reaction) or vice versa. This reaction takes place in the presence of phosphate. As noted earlier, reactions catalyzed by NPs are reversible, thus both cleavage and synthesis take place simultaneously and the reaction applies to equilibrium rules and dynamics that might hinder high product yields.

In case of producing a purine nucleoside via transglycosylation, it is generally assumed that the reaction equilibrium is on the side of the starting nucleoside and that the reaction of the PyNP is faster compared to the PNP due to the modifications in the purine base. In other words, in a transglycosylation reaction the formed intermediate (Pentose-1P) may easily undergo the backward reaction to the starting nucleoside than to the final product. High phosphate concentration enhances nucleoside cleavage [19], but hinders the synthesis of the nucleoside of interest. Therefore, adjusting the phosphate concentration in transglycosylation reactions is a critical factor. Although the transglycosylation reaction was proven to be an efficient way for the synthesis of mainly riboside and deoxy-ribosides nucleoside analogues, the synthesis of some arabinosides and fluoro-modified-ribosides/arabinosides is more challenging [66–68]. Hence, further enzymatic approaches were investigated.

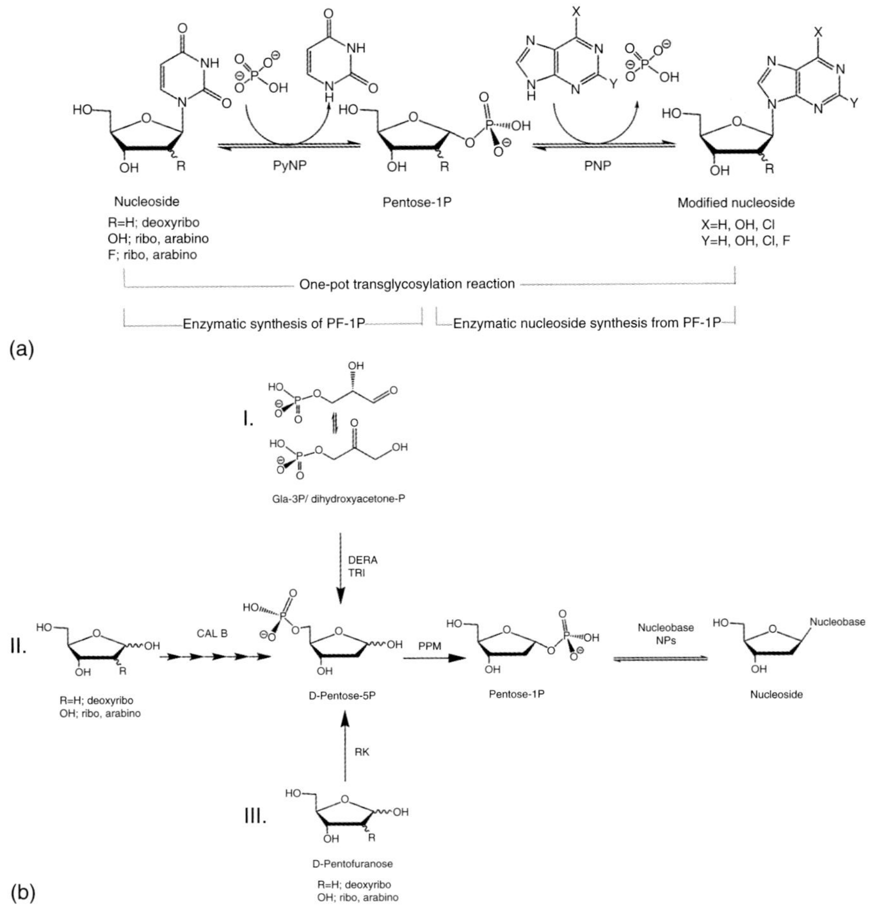

Figure 1.6 Approaches for the synthesis of nucleosides and nucleoside analogues. (a) One-pot transglycosylation reaction. (b(I)) Enzymatic synthesis starting from glyceraldehyde-3-phosphate. (b(II)) Chemo-enzymatic synthesis starting from deoxyribose, ribose, or arabinose. (b(III)) Enzymatic synthesis starting from D-pentoses (deoxyribose, ribose, or arabinose). PF-1P: Pentose-1P

1.3.2 Pentofuranose-1-phosphate as Universal Glycosylating Substrate for Nucleoside Phosphorylase (NP)

1.3.2.1 Nucleoside Synthesis from Chemically Synthesized Pentose-1P

The chemical synthesis of glycosyl phosphates has a long history and has been thoroughly studied and reported [69–73]. Although details of the chemical synthesis will not be discussed in this chapter, it is worth mentioning that the most efficient and currently used method for the synthesis of pyranose-1-phosphate is that proposed by MacDonald in 1962 [70]. Later, this method was also used for the production of α-D-pentofuranose-1-phosphate (Pentose-1P) [68, 74]. Tediousness of the chemical synthesis of Pentose-1P as well as low yields, and the difficulty of purifying the wanted α-anomeric isoform have so far restricted the use of Pentose-1P as a substrate for NPs; generally, the availability of Pentose-1P is a serious bottleneck for the synthesis of modified nucleoside analogues. Nevertheless, the chemical synthesis of Pentose-1P followed by the enzymatic condensation with a nucleobase using NPs is still considered as a valuable chemo-enzymatic route for nucleoside synthesis.

1.3.2.2 Nucleosides Synthesis from D-Glyceraldehyde-3-phosphate

2′-deoxyribonucleoside synthesis from glyceraldehyde-3-phosphate was first reported as a potential enzymatic approach for nucleoside synthesis in the beginning of this century [75, 76]. The authors described the transformation of D-glyceraldehyde-3-phosphate or dihydroxyacetone monophosphate to nucleoside analogues through three main consecutive steps. The transformation was performed as a one-pot reaction without purification of the intermediates. In the first step, a condensation of D-glyceraldehyde-3-phosphate or dihydroxyacetone-phosphate with acetaldehyde is catalyzed by D-2-deoxyribose-5-phosphate aldolase (DERA) and triose phosphate isomerase (TRI). The observed product is D-ribofuranose-5-phosphate. In the second step the stereospecific transformation of D-ribofuranose-5-phosphate into α-D-ribofuranose-1-phosphate (Rib-1P) by phosphopentomutase (PPM) is performed. Finally, the enzymatic condensation of Rib-1P with a heterocyclic nucleobase is catalyzed by NPs (Figure 1.6b(I)).

1.3.2.3 Nucleoside Synthesis from D-Pentose

In 2008 [77], the chemo-enzymatic synthesis of β-D-ribofuranosyl, β-D-deoxyribofuranosyl and β-D-arabinofuranosyl nucleosides from 5-phosphate of D-ribose, D-arabinose and 2-deoxy-D-ribose was reported. The pentose-5-phosphate was chemo-enzymatically prepared through different selective steps of acetylation and deacetylation in the presence of lipase B from *Candida antartica*. The formed pentofuranose-5-phosphate (Pentose-5P) was transformed into Pentose-1P using PPM. Pentose-1P was then used by nucleoside phosphorylase for the enzymatic condensation reaction with nucleobase yielding nucleosides and their analogues (Figure 1.6b(II)). Inosine, 6-mercaptopurine riboside and ribavirin were successfully synthesized by the above method [77].

More recently [78], a different and more efficient strategy was proposed based on transforming D-pentose to Pentose-5P using ribokinase (RK). Pentose-5P was later converted to Pentose-1P by PPM. Pentose-1P was used as a substrate for NP to synthesize nucleoside analogues in the presence of heterocyclic nucleobases like uracil, thymine, and hypoxanthine (Figure 1.6b(III)).

1.3.2.4 Nucleoside Synthesis from Enzymatically Produced Pentose-1P

We have recently proposed a two-pot enzymatic approach to produce nucleoside analogues as an alternative to the one-pot transglycosylation [79]. This approach is based on the ability of nucleoside phosphorylase to cleave the glycosidic bond of nucleoside yielding Pentose-1P, which is then purified as barium salt. It is applied in a separate step as a substrate for the synthesis of modified nucleosides using NPs (Figure 1.6a). This approach allows for easier optimization of each of the reactions, which leads to increased yields of the desired nucleoside analogue.

1.4 Approaches to Produce Nucleoside Analogues

Since nucleoside phosphorylases were first described by Kalckar [80], many research projects were conducted to test whether definite nucleoside analogues are used as substrates for nucleoside phosphorylases. Different forms of the enzyme (e.g. free or immobilized) were applied for the production of known drugs. One common objective of various studies was the optimization of already existing methods for large-scale industrial production with the aim to overcome some drawbacks and limitations such as (i) product inhibition, (ii) poor solubility of substrates, and/or (iii) consumption of large amounts of enzymes that might be hard to express. Such scale-up activities focused mainly on compounds with biological activity or of commercial value. In the coming sections, we try to summarize these approaches.

1.4.1 Whole Cell Catalysis

The application of whole cells can be regarded as an easy, time-saving, and economic option to produce nucleoside analogues. Expensive and time-consuming steps like cells lysis, enzyme extraction, and purification are avoided. Additionally, the retrieval and reuse of cells via centrifugation or ultrafiltration is possible. Best yields and conversion rates were achieved by cells harvested at the stationary phase where the salvage pathway for nucleoside production is strongly activated, which correlate with large amounts of nucleoside phosphorylases [81]. Ribavirin, dideoxyinosine, adenine arabinoside, and 2′-amino-2′-deoxyadenosine were synthesized using whole cells and different pentofuranosyl donors [81–86]. One requirement for the successful application of whole cell biocatalysts is that both the sugar donor and acceptor are well soluble. Using whole cell catalysts, side reactions that are catalyzed by other enzymes acting on the same class of substrates have to be considered, for example deaminases or lipases [87, 88].

Complex media were proven to be better than defined media for producing NPs in *Enterobacter* species [89]. Many nucleobases, nucleosides, Rib-1P, and nucleotide monophosphates as inducing agents for potentiating the expression of NPs were tested. Conversion by some eukaryotic cancer cell lines was also reported [90, 91].

Another biologically important molecule is benzimidazole, whose nucleoside exhibits antitumor, antifungal, antiparasitic, analgesics, antiviral, and antihistamine activities, and is being used in the treatment of cardiovascular diseases and in therapies related to neurology, endocrinology, and ophthalmology. Bentancor and coworkers [92] published a screening study for producing benzimidazole riboside and 2′-deoxyriboside using whole cells catalysts. Uridine was applied as a ribose donor and thymidine as the 2′-deoxyribose donor, respectively. Optimal temperature for the reaction was 60 °C for ribosides and 45 °C for deoxyribosides, respectively, suggesting that the higher temperature inhibits the action of thymidine phosphorylase [93, 94]. However, recent data show that α-deoxyribose-1-phosphate is unstable and degrades at higher temperature [95].

Mainly, free cells were applied to produce nucleoside analogues, but immobilized cells were also used. Immobilized whole cells can be either alive or dead depending on (i) whether the enzyme is intra- or extracellular, (ii) the molecular weight and the size of the enzyme, (iii) substrate and product, and (iv) the cell's permeability to them. As an example, *Aeromonas hydrophila* cells were entrapped in agarose beads and afforded the same kinetics and yields as those by the free cells. The entrapped cells, however, were used three times before any decrease in activity was detected [81]. A mutant strain of *E. coli* was cross-linked to glutaraldehyde and successfully used to produce ribavirin and the anticancer drug cladribine [96].

The widespread application of whole cell extracts is restricted because broadly used bacterial whole cells that are used as biocatalysts for the production of modified nucleosides co-express a number of enzymes that can (i) consume substrates; (ii) catalyze undesired transformations of substrates or formed nucleosides; and (iii) secrete desired products in the medium, which makes purification challenging [97].

1.4.2 Crude Enzyme Extract

Compared to purified enzyme, the application of crude cell extract is a less laborious and a more economical approach. It is suitable to carry out preliminary screening assays to determine the catalytic potential of enzymes especially in the cases of overexpression where the target enzyme is available in high concentration. A drawback of crude enzyme extracts is a shorter half-life compared to that of purified enzymes [53, 56, 82, 98–100].

The application of crude enzyme extracts is principally advantageous in case of thermophilic enzymes. If the enzymes are heterologously expressed in a recombinant host, proteolysis and interference from the host proteins can be inhibited through an additional heat shock step [101]. This was implemented in the preliminary tests conducted for the characterization of nucleoside phosphorylases of *T. thermophilus*. The extract was heated at 80 °C for one hour and then centrifuged. Afterward, the supernatant was successfully tested for its phosphorolysis activity with a variety of nucleosides [31].

As compared to whole cell catalysis, cell lysate or enzyme extracts result in a more reproducible and robust reaction as the Pentose-1P intermediate is not consumed as an energy source, which might happen under certain conditions using bacterial cells. It can finally end up with the reaction being totally reserved in the backward direction [102].

1.4.3 Application of Purified Enzymes

The use of purified enzyme preparations has the advantage that side-reactions are avoided. To date, NPs from different mesophiles, like *E. coli* [103–105] or *B. subtilis* [106], were applied in the synthesis of pharmacologically active compounds [103, 104].

Thermostable NPs from thermophilic or hyperthermophilic microorganisms catalyze enzymatic reactions with the same high regio- and stereo-selectivity compared to mesophilic NPs, but have several advantages for large-scale applications: (i) thermophilic enzymes can be easily purified with high protein recovery after heterologous production and costs are low as a heat treatment usually is sufficient as a purification step [31]; (ii) a high pH tolerance is observed, which reduces the need to finely regulate the pH of the reaction; (iii) it was shown that PNPs of *T. thermophilus* show a remarkable resistance to aggressive organic solvents such as dimethylformamide or dimethyl sulfoxide, which are required in the synthesis of compounds with low water solubility; (iv) using thermophilic catalysts, reactions can be performed at elevated temperatures, which are frequently required to reduce the viscosity of the medium or to increase the solubility and concentration of some substrates [107]; (v) it was observed that an increased thermal stability correlates with a broader substrate spectrum [30, 31, 54, 108].

For industrial applications of purified enzymes, it is highly desirable to use either immobilized catalysts or continuous reactors to be able to recover the enzyme from the reaction mixture. This simplifies downstream processing and facilitates biocatalyst recycling. As immobilization might also increase the stability of nucleoside phosphorylases, diverse methods have been exploited [57, 101, 109, 110]. Approaches for both immobilized enzymes and continuous reactors are described in more detail in the following paragraphs.

1.4.3.1 Immobilized Enzymes

Biocatalyst immobilization denotes binding an enzyme to an inert support (natural, synthetic, or inorganic in nature) where it maintains its activity without blocking or disrupting the active sites. The ideal support has to be stable, endures sheering force of the reaction, can be regenerated, and has no effect on the type of reaction [111, 112]. Usually, enzyme immobilization has a positive effect regarding thermal stability, pH resistance, and longevity (e.g. storage time and number of operational cycles).

Immobilization of purified enzymes is an important tool for large-scale commercial application of biocatalysts. Considering the cost load of the enzyme in the reaction, one economically favorable feature is the ability to reuse the enzyme as long as it retains its activity. The general immobilization principles for enzymes are (i) adsorption to the support's surface, (ii) cross-linking (enzyme molecules bind to each other) in the presence of spacers (e.g. polyethylene glycol) to reduce steric hindrance, (iii) covalent binding to a support, (iv) entrapment within a polymer, (v) encapsulation within a surrounding matrix, and (vi) affinity binding [113, 114] (Figure 1.7). The most described immobilization method for NPs is entrapment in different matrices like agar, agarose, and alginates [81, 86, 116].

Immobilization of pure enzyme to produce nucleoside analogues in a transglycosylation reaction might be a complicated process as it involves the co-binding of a PNP and PyNP. Multimeric enzymes are known to be quite difficult to immobilize as it is necessary to consider the correct assembly and multisubunit attachment, least desorption, and minimal deactivation [117–119]. However, feasibility was proven as immobilized PyNPs from *B. subtilis* and TP from *E. coli* were applied for the synthesis of 5-halogenated-pyrimidine-2′-deoxyribonucleosides by transglycosylation in fully aqueous medium [57]. For both enzymes, immobilization was shown to strongly enhance the stability upon incubation in acetonitrile. Both PyNP and TP could be re-used for at least five times with no loss of productivity. In another approach, immobilized UP of *Clostridium perfringens*

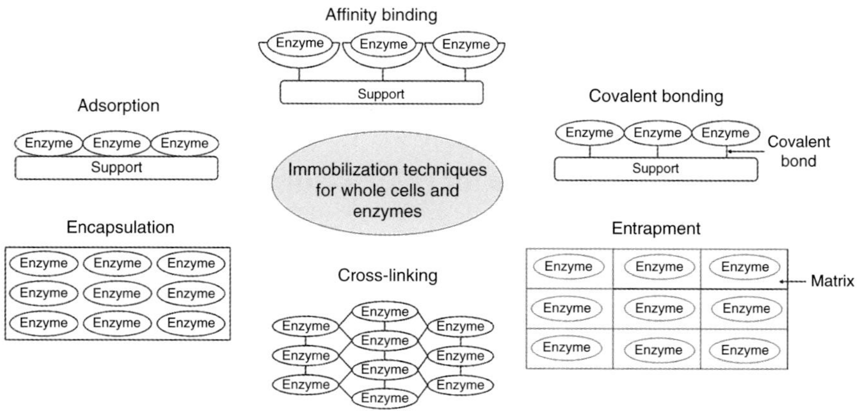

Figure 1.7 Immobilization techniques applied for whole cells and purified enzymes. Source: Adapted from Sirisha et al. 2016 [111] and Zhao et al. 2015 [115].

and PNP of *A. hydrophila* were used as biocatalysts for the production of vidarabine in a transglycosylation reaction [120].

Different matrices and techniques were compared by Rocchietti and coworkers for UP and PNP from *B. subtilis* [117]. While covalent bonding distorted the protein structure and resulted in decreased activity and stability, UP was best immobilized via ionic bonding to Sepabeads and post-immobilization treatment with polyaldehyde was necessary to cross-link the support and protein. PNP was best immobilized on glycoxyl-agarose using the surfactant Triton X-100 that prevents agglomeration of the protein and hence, allows diffusion into the support. The combined use of these two enzymes allowed the reaction to take place at a higher optimal temperature (45 °C) and pH (pH = 10) compared to the free enzymes. Using these conditions, guanine conversion proceeded, which is usually limited due to low solubility.

1.4.3.2 Enzyme Reactors

Few trials were recorded in literature using different NPs in enzyme reactors for the production of nucleoside analogues. Preparative scale nucleoside synthesis was conducted in column reactors by some research groups. Hori and colleagues operated their experiment for 17 days at 60 °C to produce 5-methyl uridine from a thymine and inosine mixture [101]. Immobilized crude enzyme extract from the thermophilic bacterium *B. stearothermophilus* was bound to an ion-exchange column. The reaction mixture was fed in a pre-optimized rate with the aim to overcome the equilibrium constant and to guarantee that the hypoxanthine solubility limit is not exceeded [101].

In two other approaches an enzyme column reactor was connected to a HPLC [121, 122]. The purified PNP from *A. hydrophila* was immobilized on silica particles of definite particle size. The system could run over the analytical column or bypass it via a six-port switching valve and it was used to test the substrate affinity of the enzyme and compare the reaction kinetics of free and immobilized enzyme [121]. Later it was shown that the system was stable for 10 months [122].

The described system was used for a faster routine activity screening. Based on a "design of experiment" optimized study, they produced different modified nucleosides including a number of halogenated analogues.

1.5 Upscaling Approaches for the Production of Nucleoside Analogues

Despite the high cost of soluble enzymes that limited the wide-spread application of nucleoside phosphorylases in industrial processes, some successful upscaling approaches were reported. These are almost all dedicated to the synthesis of molecules with known biological activity. For example, many research studies were conducted to execute large-scale industrial enzymatic or chemo-enzymatic processes to produce the antiviral drug ribavirin from 1,2,4-triazole-3-carboxamide (TCA) using purified enzymes or bacterial cell cultures.

Chen et al. performed a fed-batch fermentation process for a recombinant *Bacillus* strain using glucose as the sole carbon source in an attempt to produce ribavirin in a preparative scale [123]. The fermentation was carried out in 3 l volume and TCA was added in the middle of the log phase. Tween 80 was added to increase the permeability of the cell membrane for the enzyme and thereby increase TCA conversion to ribavirin.

A method carried out in 7.5 l volume was described using a genetically engineered *Bacillus* strain that secretes PNP extracellularly. TCA was used as the substrate for the production of ribavirin [124]. In another upscaling experiment, the chemo-enzymatic synthesis of ribavirin in a 20 l reaction volume was described [125]. A conversion rate of 97% was observed in the enzymatic reaction using TCA and guanosine as substrates and a heterologously produced *E. coli* PNP as biocatalyst.

1.6 Production of Pharmaceutically Active Compounds by Nucleoside Phosphorylases

The chemo-enzymatic synthesis of nucleosides with biological relevance was well described in a review by Mikhailopulo in 2007 [67]. The application of nucleoside phosphorylases in the production of drugs like cladribine or mizoribine was proven (Table 1.2). Both are useful drugs for the treatment of cancer. In 2003, the enzymatic synthesis of clofarabine, fludarabine, or vidarabine with *E. coli* RK, PPM, and PNP [136] was shown.

The phosphorolytic cleavage of a number of other approved drugs like ribavirin or didanosin was also shown [34] (Table 1.2). Using PNP of *B. acetylicum* the relative activity was 22% for both compounds compared to inosine. Recently, the phosphorolytic cleavage of nucleoside analogue drugs listed in the European Pharmacopoeia by thermophilic PyNPs, PNPs, and MTAPs was tested [19]. Both cytarabine and zidovudine were accepted as substrates by PyNP, while

Table 1.2 Drugs synthesized by different nucleoside phosphorylases.

<table>
<tr><th colspan="3">Purine-based drugs</th><th colspan="3">Pyrimidine-based drugs</th></tr>
<tr><td colspan="2">Purine nucleoside phosphorylase (PNP)</td><td>*S*-methylthio-adenine phosphorylase (MTAP)</td><td>Pyrimidine phosphorylase (PyNP)</td><td>Thymidine phosphorylase (TP)</td><td>Uridine phosphorylase (UP)</td></tr>
<tr><td colspan="3">6-Methyl purine [32, 39]</td><td colspan="3" rowspan="2">Floxuridine [57, 126, 127]</td></tr>
<tr><td colspan="3">Cordycepin [34, 108]</td></tr>
<tr><td>Fludarabine [32]</td><td>Clofarabine [128]</td><td rowspan="5">Cladribine [41]</td><td colspan="2" rowspan="2">Idoxuridine [129, 130]</td><td rowspan="5"></td></tr>
<tr><td>Vidarabine [31, 120]</td><td>6-Phenyl purine riboside [32]</td></tr>
<tr><td>Nelarabine [131]</td><td>Ribavirin [28, 29]</td><td rowspan="3"></td><td>Tegafur [132]</td></tr>
<tr><td>Didanosine [34, 35]</td><td>Purinethol riboside [116]</td><td rowspan="2">Stavudine [133]</td></tr>
<tr><td>Thioguanine riboside [134]</td><td>5-Aza-7-deazaguanine [9, 135]</td></tr>
</table>

gemcitabine and lamivudine were not cleaved. PNPs and MTAPs of thermophilic microorganisms phosphorolyzed ribavirin and didanosin. Nucleoside analogues with open sugar ring structures like aciclovir, ganciclovir, and tenofovir were not accepted as substrate. This is in accordance with studies performed with three PNPs of *T. thermophilus* or *B. acetylicum* and aciclovir as substrate. No phosphorolytic activity was observed [31, 137]. For ganciclovir it was even shown that it is an inhibitor for PNP [24].

1.7 Outlook for the Application of Nucleoside Phosphorylase in the Production of Nucleoside Analogues

Nucleoside phosphorylases play a crucial role in nucleoside metabolism. The substrate spectra of naturally available NPs have been studied intensively during the past decades. This knowledge was the basis for the application of nucleoside phosphorylases in the production of nucleoside based drugs or precursors for modified oligonucleotides (e.g. aptamers). A number of different approaches have been developed to increase product yields and drastically reduce costs, which opens the way to use NPs for industrial-scale production. The limitations of available nucleoside phosphorylases will be overcome in the near future by optimizing the biocatalysts using genetic engineering.

References

1 Savić, D., Stanković, T., Lavrnja, I. et al. (2015). Purine nucleoside analogs in the therapy of cancer and neuroinflammation. *Mol. Inhib. Target Ther.* 1 (1): 3–14.

2 Jordheim, L.P., Durantel, D., Zoulim, F., and Dumontet, C. (2013). Advances in the development of nucleoside and nucleotide analogues for cancer and viral diseases. *Nat. Rev. Drug Discovery* 12 (6): 447–464.

3 Agris, P.F. (2015). The importance of being modified: an unrealized code to RNA structure and function. *RNA* 21 (4): 552–554.

4 Rozenski, J., Crain, P.F., and McCloskey, J.A. (1999). The RNA modification database: 1999 update. *Nucleic Acids Res.* 27 (1): 196–197.

5 Charette, M. and Gray, M.W. (2000). Pseudouridine in RNA: what, where, how, and why. *IUBMB Life (International Union Biochem Mol Biol Life)* 49 (5): 341–351.

6 Galmarini, C.M., Mackey, J.R., and Dumontet, C. (2002). Nucleoside analogues and nucleobases in cancer treatment. *Lancet Oncol.* 3 (7): 415–424.

7 De Clercq, E. (2016). Approved antiviral drugs over the past 50 years. *Clin. Microbiol. Rev.* 29 (3): 695–747.

8 Vande Voorde, J., Balzarini, J., and Liekens, S. (2014). An emerging understanding of the Janus face of the human microbiome: enhancement versus impairment of cancer therapy. *J. Antimicrob. Chemother.* 69 (10): 2878–2880.

9 Mikhailopulo, I.A. and Miroshnikov, A.I. (2013). Some recent findings in the biotechnology of biologically important nucleosides. *Biochem. Biotechnol. Mod. Med.* 328–353.

10 Levene, P.A. and Medigrecenau, F. (1911). On nucleases. *J. Biol. Chem.* 9 (5): 389–402.

11 Levene, P.A. and Medigrecenau, F. (1911). On nucleases. *J. Biol. Chem.* 9 (1): 65–83.

12 Jones, W. (1911). Concerning nucleases. *J. Biol. Chem.* 9 (2): 129–138.

13 Jones, W. (1911). On the physiological agents which are concerned in the nuclein fermentation, with special reference to four independent desamidases. *J. Biol. Chem.* 9 (2): 160–180.

14 Levene, P.A., Yamagawa, M., and Weber, I. (1924). On nucleosidases. I. General properties. *J. Biol. Chem.* 60 (3): 693–706.

15 Levene, P.A. and Weber, I. (1924). On nucleosidases. II. Purification of the enzyme. *J. Biol. Chem.* 60 (3): 707–715.

16 Levene, P.A. and Weber, I. (1924). On nucleosidases. III. The degree of specificity of nucleosidase and the distribution of it in various organs and in various species. *J. Biol. Chem.* 60 (3): 717–720.

17 Kalckar, H.M. (1947). The enzymatic synthesis of purine riboside. *J. Biol. Chem.* 167 (2): 477–486.

18 Manson, L.A. and Lampen, J.O. (1951). The metabolism of deoxyribose nucleosides in *Escherichia coli*. *J. Biol. Chem.* 193 (2): 539–547.

19 Yehia, H., Kamel, S., Paulick, K. et al. (2017). Substrate Spectra of Nucleoside Phosphorylases and their Potential in the Production of Pharmaceutically Active Compounds. *Curr. Pharm. Des.* 23 (45): 6913–6935.

20 Pugmire, M.J. and Ealick, S.E. (2002). Structural analyses reveal two distinct families of nucleoside phosphorylases. *Biochem. J* 361: 1–25.

21 Hori, N., Watanabe, M., Yamazaki, Y., and M. (1989). Purification and characterization of second thermostable purine nucleoside phosphorylase in *Bacillus stearothermophilus* JTS 859. *Agric. Biol. Chem.* 53 (12): 3219–3224.

22 Seeger, C., Poulsen, C., and Dandanell, G. (1995). Identification and characterization of genes (xapA, xapB, and xapR) involved in xanthosine catabolism in *Escherichia coli. J. Bacteriol.* 177 (19): 5506–5516.

23 Senesi, S., Falcone, G., Mura, U. et al. (1976). A specific adenosine phosphorylase, distinct from purine nucleoside phosphorylase. *FEBS Lett.* 64 (2): 1715–1722.

24 Bzowska, A., Kulikowska, E., and Shugar, D. (2000). Purine nucleoside phosphorylases: properties, functions, and clinical aspects. *Pharmacol. Ther.* 88 (3): 349–425.

25 Bzowska, A., Ananiev, A.V., Ramzaeva, N. et al. (1994). Purine nucleoside phosphorylase: inhibition by purine N(7)- and N(9)-acyclonucleosides; and substrate properties of 7-β-D-ribofuranosylguanine and 7-β-D-ribofuranosylhypoxanthine. *Biochem. Pharmacol.* 48 (5): 937–947.

26 Bzowska, A., Kulikowska, E., Poopeiko, N.E., and Shugar, D. (1996). Kinetics of phosphorolysis of 3-(β-d-ribofuranosyl)adenine and 3-(β-d-ribofuranosyl)hypoxanthine, non-conventional substrates of purine-nucleoside phosphorylase. *Eur. J. Biochem.* 239 (1): 229–234.

27 Wielgus-Kutrowska, B., Kulikowska, E., Wierzchowski, J. et al. (1997). Nicotinamide riboside, an unusual, non-typical, substrate of purified purine-nucleoside phosphorylases. *Eur. J. Biochem.* 243 (1–2): 408–414.

28 Furihata, T., Kishida, S., Sugiura, H. et al. (2014). Functional analysis of purine nucleoside phosphorylase as a key enzyme in ribavirin metabolism. *Drug Metab. Pharmacokinet.* 29 (2): 211–214.

29 Xie, X.X., Xia, J.G., He, K.F. et al. (2011). Low-molecular-mass purine nucleoside phosphorylase: characterization and application in enzymatic synthesis of nucleoside antiviral drugs. *Biotechnol. Lett* 33 (6): 1107–1112.

30 Zhou, X., Szeker, K., Janocha, B. et al. (2013). Recombinant purine nucleoside phosphorylases from thermophiles: preparation, properties and activity towards purine and pyrimidine nucleosides. *FEBS J* 280 (6): 1475–1490.

31 Almendros, M., Berenguer, J., and Sinisterra, J.-V. (2012). *Thermus thermophilus* nucleoside phosphorylases active in the synthesis of nucleoside analogues. *Appl. Environ. Microbiol.* 78 (9): 3128–3135.

32 Hassan, A.E., Abou-Elkhair, R.A., Riordan, J.M. et al. (2012). Synthesis and evaluation of the substrate activity of C-6 substituted purine ribosides with *E. coli* purine nucleoside phosphorylase: palladium mediated cross-coupling of organozinc halides with 6-chloropurine nucleosides. *Eur. J. Med. Chem.* 47 (1): 167–174.

33 Stoeckler, J.D., Cambor, C., and Parks, R.E. (1980). Human erythrocytic purine nucleoside phosphorylase: reaction with sugar-modified nucleoside substrates. *Biochemistry* 19 (1): 102–107.

34 Shirae, H. and Yokozeki, K. (1991). Purifications and properties of orotidine-phosphorolyzing enzyme and purine nucleoside phosphorylase from *Erwinia carotovora* AJ 2992. *Agric. Biol. Chem.* 55 (7): 1849–1857.

35 Nannemann, D.P., Kaufmann, K.W., Meiler, J., and Bachmann, B.O. (2010). Design and directed evolution of a dideoxy purine nucleoside phosphorylase. *Protein Eng. Des. Sel.* 23 (8): 607–616.

36 Cacciapuoti, G., Bertoldo, C., Brio, A. et al. (2003). Purification and characterization of 5′-methylthioadenosine phosphorylase from the hyperthermophilic archaeon *Pyrococcus furiosus*: substrate specificity and primary structure analysis. *Extremophiles* 7 (2): 159–168.

37 Cacciapuoti, G., Porcelli, M., Bertoldo, C. et al. (1994). Purification and characterization of extremely thermophilic and thermostable 5′-methylthioadenosine phosphorylase from the archaeon *Sulfolobus solfataricus*. Purine nucleoside phosphorylase activity and evidence for intersubunit disulfide bonds. *J. Biol. Chem.* 269 (40): 24762–24769.

38 Appleby, T.C., Mathews, I.I., Porcelli, M. et al. (2001). Three-dimensional structure of a hyperthermophilic 5′-deoxy-5′-methylthioadenosine phosphorylase from *Sulfolobus solfataricus. J. Biol. Chem.* 276 (42): 39232–39242.

39 Hassan, A.E., Abou-Elkhair, R.A., Parker, W.B. et al. (2016). 6-Methylpurine derived sugar modified nucleosides: synthesis and evaluation of their substrate activity with purine nucleoside phosphorylases. *Bioorg. Chem.* 65: 9–16.

40 Toorchen, D. and Miller, R.L. (1991). Purification and characterization of 5′-deoxy-5′-methylthioadenosine (MTA) phosphorylase from human liver. *Biochem. Pharmacol.* 41 (12): 2023–2030.

41 Fabianowska-Majewska, K., Duley, J., Fairbanks, L. et al. (1994). Substrate specificity of methylthioadenosine phosphorylase from human liver. *Acta Biochim. Pol.* 41 (4): 391–395.

42 Ghoda, L.Y., Savarese, T.M., Northup, C.H. et al. (1988). Substrate specificities of 5′-deoxy-5′-methylthioadenosine phosphorylase from Trypanosoma brucei brucei and mammalian cells. *Mol. Biochem. Parasitol.* 27 (2–3): 109–118.

43 Bacchi, C.J., Goldberg, B., Rattendi, D. et al. (1999). Metabolic effects of a methylthioadenosine phosphorylase substrate analog on *African trypanosomes. Biochem. Pharmacol.* 57 (1): 89–96.

44 Zhou, M., Pugmire, M.J., Vuong, B.Q., and Ealick, S.E. (1999). Cloning, expression and crystallization of pyrimidine nucleoside phosphorylase from *Bacillus stearothermophilus. Acta Crystallogr. D Biol. Crystallogr.* 55 (Pt 1: 287–290.

45 Saunders, P.P. and Wilson, B.A. (1969). Purification and comparative properties of a pyrimidine nucleoside phosphorylase from *Bacillus stearothermophilus. J. Biol. Chem.* 244 (13): 3691–3697.

46 Kono, A., Hara, Y., Sugata, S. et al. (1984). Substrate specificity of a thymidine phosphorylase in human liver tumor. *Chem. Pharm. Bull.* 32 (5): 1919–1921.

47 Schinazi, R.F., Peck, A., and Sommadossi, J.-P. (1992). Substrate specificity of *Escherichia coli* thymidine phosphorylase for pyrimidine nucleosides with anti-human immunodeficiency virus activity. *Biochem. Pharmacol.* 44 (2): 199–204.

48 Panova, N.G., Alexeev, C.S., Kuzmichov, A.S. et al. (2007). Substrate specificity of *Escherichia coli* thymidine phosphorylase. *Biochemistry (Mosc)* 72 (1): 21–28.

49 Hatano, A., Harano, A., Takigawa, Y. et al. (2008). Kinetic parameters and recognition of thymidine analogues with varying functional groups by thymidine phosphorylase. *Bioorg. Med. Chem.* 16: 3866–3870.

50 Panova, N., Kosiova, I., Petrova, M. et al. (2008). Nucleoside phosphonic acids in thymidine phosphorylase inhibition: structure–activity relationship. *Nucleic Acids Symp. Ser.* 52 (1): 665–666.

51 Avraham, Y., Grossowicz, N., and Yashphe, J. (1990). Purification and characterization of uridine and thymidine phosphorylase from Lactobacillus casei. *Biochim. Biophys. Acta* 1040: 287–293.

52 Gao, X.F., Huang, X.R., and Sun, C.C. (2006). Role of each residue in catalysis in the active site of pyrimidine nucleoside phosphorylase from *Bacillus subtilis*: a hybrid QM/MM study. *J. Struct. Biol.* 154 (1): 20–26.

53 Leer, J.C., Hammer-Jespersen, K., and Schwartz, M. (1977). Uridine phosphorylase from *Escherichia coli*. Physical and chemical characterization. *Eur. J. Biochem.* 75: 217–224.

54 Szeker, K., Zhou, X., Schwab, T. et al. (2012). Comparative investigations on thermostable pyrimidine nucleoside phosphorylases from *Geobacillus thermoglucosidasius* and *Thermus thermophilus*. *J. Mol. Catal. B: Enzym.* 84: 27–34.

55 Voorde, J.V., Gago, F., Vrancken, K. et al. (2012). Characterization of pyrimidine nucleoside phosphorylase of *Mycoplasma hyorhinis*: implications for the clinical efficacy of nucleoside analogues. *Biochem. J* 445 (1): 113–123.

56 Scocca, J.J. (1971). Purification and substrate from specificity of pyrimidine nucleoside phosphorylase from *Haemophilus influenzae*. *J. Biol. Chem.* 246 (21): 6606–6610.

57 Serra, I., Bavaro, T., Cecchini, D.A. et al. (2013). A comparison between immobilized pyrimidine nucleoside phosphorylase from *Bacillus subtilis* and thymidine phosphorylase from *Escherichia coli* in the synthesis of 5-substituted pyrimidine 2-deoxyribonucleosides. *J. Mol. Catal. B: Enzym.* 95: 16–22.

58 Hamamoto, T., Noguchi, T., and Midorikawa, Y. (1996). Purification and characterization of purine nucleoside phosphorylase and pyrimidine nucleoside phosphorylase from *Bacillus stearothermophilus* TH 6-2. *Biosci. Biotechnol., Biochem.* 60 (7): 1179–1180.

59 Dandanell, G., Szczepanowski, R.H., Kierdaszuk, B. et al. (2005). *Escherichia coli* purine nucleoside phosphorylase II, the product of the xapA gene. *J. Mol. Biol.* 348 (1): 113–125.

60 Erion, M.D., Stoeckler, J.D., Guida, W.C. et al. (1997). Purine nucleoside phosphorylase. 2. Catalytic mechanism. *Biochemistry* 36 (39): 11735–11748.

61 Mao, C., Cook, W.J., Zhou, M. et al. (1998). Calf spleen purine nucleoside phosphorylase complexed with substrates and substrate analogues. *Biochemistry* 37 (20): 7135–7146.

62 Larson, E.T., Mudeppa, D.G., Gillespie, J.R. et al. (2010). The crystal structure and activity of a putative trypanosomal nucleoside phosphorylase reveal it to be a homodimeric uridine phosphorylase. *J. Mol. Biol.* 396: 1244–1259.

63 Balaev, V.V., Lashkov, A.A., Gabdulkhakov, A.G. et al. (2016). Structural investigation of the thymidine phosphorylase from *Salmonella typhimurium* in the unliganded state and its complexes with thymidine and uridine. *Acta Crystallogr. F Struct. Biol. Commun.* 72 (3): 224–233.

64 Friedkin, M. and Roberts, D. (1954). The enzymatic synthesis of nucleosides. I. Thymidine phosphorylase in mammalian tissue. *J. Biol. Chem.* 207 (1): 245–256.

65 Friedkin, M. and Roberts, D. (1954). The enzymatic synthesis of nucleosides. II. Thymidine and related pyrimidine nucleosides. *J. Biol. Chem.* 207 (1): 257–266.

66 Mikhailopulo, I.a. and Miroshnikov, A.I. (2011). Biologically important nucleosides: modern trends in biotechnology and application. *Mendeleev Commun.* 21 (2): 57–68.

67 Mikhailopulo, I.A. (2007). Biotechnology of nucleic acid constituents-state of the art and perspectives. *Curr. Org. Chem.* 11 (4): 317–335.

68 Mikhailopulo, I.A. and Miroshnikov, A.I. (2010). New trends in nucleoside biotechnology. *Acta Naturae* 2 (2): 36–59.

69 MacDonald, D. (1986). Separation of the anomers of glycosyl phosphates. *Carbohydr. Res.* 6: 376–381.

70 MacDonald, D. (1962). A new route to glycosyl phosphates. *J. Org. Chem.* 27 (3): 1107–1109.

71 MacDonald, D.L. (1966). Preparation of glycosyl phosphates: beta-D-glucopyranosyl phosphate. *Carbohydr. Res.* 3: 117–120.

72 Yamada, K., Matsumoto, N., and Hayakawa, H. (2009). Stereoselective synthesis of 2-deoxy-2-fluoroarabinofuranosyl-α-1-phosphate and its application to the synthesis of 2′-deoxy-2′-fluoroarabinofuranosyl purine nucleosides by a chemo-enzymatic method. *Nucleosides Nucleotides Nucleic Acids* 28 (11–12): 1117–1130.

73 Timmons, S.C. and Jakeman, D.L. (2008). Stereospecific synthesis of sugar-1-phosphates and their conversion to sugar nucleotides. *Carbohydr. Res.* 343 (5): 865–874.

74 Mikhailopulo, I.A. and Cramer, F. (1982). Conformation of N(9)-(β-D-arabinofuranosyl) adenine 5′-monophosphate (ara-AMP) in anhydrous dimethylsulphoxide monitored by 13C NMR. *FEBS Lett.* 144 (1): 109–111.

75 Ouwerkerk, N., Steenweg, M., De Ruijter, M. et al. (2002). One-pot two-step enzymatic coupling of pyrimidine bases to 2-deoxy-D-ribose-5-phosphate. A new strategy in the synthesis of stable isotope labeled deoxynucleosides. *J. Org. Chem.* 67 (5): 1480–1489.

76 Ouwerkerk, N., van Boom, J.H., Lugtenburg, J., and Raap, J. (2000). Chemo-enzymatic synthesis of thymidine 13C-labelled in the 2′-deoxyribose moiety. *Eur. J. Org. Chem.* 861–866.

77 Taverna-Porro, M., Bouvier, L.A., Pereira, C.A. et al. (2008). Chemoenzymatic preparation of nucleosides from furanoses. *Tetrahedron Lett.* 49 (16): 2642–2645.

78 Miroshnikov, A.I., Esipov, R.S., Muravyova, T.I. et al. (2010). A new strategy for the synthesis of nucleosides: one-pot enzymatic transformation of D-pentoses into nucleosides. *Open Conf. Proc. J.* 1 (1): 98–102.

79 Kamel, S., Weiss, M., Klare, H.F.T. et al. (2018). Chemo-enzymatic production of α-D-pentofuranose-1-phosphates using thermostable pyrimidine nucleoside phosphorylases. *Mol. Catal.* 458: 52–59.

80 Kalckar, H.M. (1945). Enzymatic synthesis of a nucleoside. *J. Biol. Chem.* 158 (3): 723–724.

81 Nobile, M., Terreni, M., Lewkowicz, E., and Irabarren, A.M.I. (2010). Aeromonas hydrophila strains as biocatalysts for transglycosylation. *Biocatal. Biotransform.* 28 (5–6): 395–402.

82 Bestetti, G., Cali', S., Ghisotti, D., et al. (2000). Recombinant bacterial strains for the production of natural nucleosides and modified analogues thereof. WO2000039307 A2, filed 23 December 1999 and issued 6 July 2000.

83 Rogert, M.C., Trelles, J.A., Porro, S. et al. (2002). Microbial synthesis of antiviral nucleosides using *Escherichia coli* BL21 as biocatalyst. *Biocatal. Biotransform.* 20 (5): 347–351.

84 Utagawa, T. (1999). Enzymatic preparation of nucleoside antibiotics. *J. Mol. Catal. B: Enzym.* 6 (3): 215–222.

85 Trelles, J.A., Fernández, M., Lewkowicz, E.S. et al. (2003). Purine nucleoside synthesis from uridine using immobilised *Enterobacter gergoviae* CECT 875 whole cells. *Tetrahedron Lett.* 44 (12): 2605–2609.

86 Luo, W., Liu, Y., Zhu, X. et al. (2011). Cloning and characterization of purine nucleoside phosphorylase in *Escherichia coli* and subsequent ribavirin biosynthesis using immobilized recombinant cells. *Enzyme Microb. Technol.* 48 (6–7): 438–444.

87 Carballeira, J.D., Fernandez-Lucas, J., Quezada, M.A. et al. (2009). Biotransformations. In: *Encyclopedia of Microbiology*, 212–251. Elsevier.

88 Pal, S. and Nair, V. (1997). Enzymatic synthesis of thymidine using bacterial whole cells and isolated purine nucleoside phosphorylase. *Biocatal. Biotransform.* 15 (2): 147–158.

89 Wei, X., Ding, Q., Zhang, L. et al. (2008). Induction of nucleoside phosphorylase in *Enterobacter aerogenes* and enzymatic synthesis of adenine arabinoside. *J. Zhejiang Univ. Sci. B* 9 (7): 520–526.

90 Lee, S.H. and Sartorelli, A.C. (1981). Conversion of 6-thioguanine to the nucleoside level by purine nucleoside phosphorylase of sarcoma 180 and sarcoma 180/TG ascites cells. *Cancer Res* 41 (3): 1086–1090.

91 Singh, P.P., Joshi, S., Russell, P.J. et al. (2011). Purine nucleoside phosphorylase mediated molecular chemotherapy and conventional chemotherapy: a tangible union against chemoresistant cancer. *BMC Cancer* 11 (1): 368.

92 Bentancor, L., Trelles, J.A., Nóbile, M. et al. (2004). Benzimidazole as deazapurine analogue for microbial transglycosylation. *J. Mol. Catal. B: Enzym.* 29 (1): 3–7.

93 Quintiliani, M., Balzarini, J., and McGuigan, C. (2013). Design, synthesis, and biological evaluation of C1-phosphonamidate analogues of 2-deoxy-D-ribose-1-phosphate. *Tetrahedron* 69 (43): 9111–9119.

94 Rubira, M.-J., Jimeno, M.-L., Balzarini, J. et al. (1998). Synthesis and NMR conformational studies of stable analogues of 2-deoxy-alfa-D-ribose-1-phosphate. *Tetrahedron* 54: 8223–8240.

95 Kamel, S., Weiss, M., Klare, H.F.T. et al. (2018). Chemo-enzymatic production of α-D-pentofuranose-1-phosphates using thermostable pyrimidine nucleoside phosphorylases. *Mol. Catal.* 458: 52–59.

96 Barai, V.N., Zinchenko, A.I., Eroshevskaya, L.A. et al. (2002). A universal biocatalyst for the preparation of base- and sugar-modified nucleosides via an enzymatic transglycosylation. *Helv. Chim. Acta* 85 (7): 1901–1908.

97 Zhou, X., Szeker, K., Jiao, L.-Y. et al. (2015). Synthesis of 2,6-dihalogenated purine nucleosides by thermostable nucleoside phosphorylases. *Adv. Synth. Catal.* 357 (6): 1237–1244.

98 Hanrahan, J.R. and Hutchinson, D.W. (1992). The enzymatic synthesis of antiviral agents. *J. Biotechnol.* 23 (2): 193–210.

99 Tono, H. and Cohen, S.S. (1962). The activity of nucleoside phosphorylase on 1-β-D-arabinosyluracil within *Escherichia coli*. *J. Biol. Chem.* 237 (4): 1271–1282.

100 Esipov, R.S., Gurevich, A.I., Chuvikovsky, D.V. et al. (2002). Overexpression of *Escherichia coli* genes encoding nucleoside phosphorylases in the pET/Bl21 (DE3) system yields active recombinant enzymes. *Protein Expr. Purif.* 24 (1): 56–60.

101 Hori, N., Watanabe, M., Sunagawa, K. et al. (1991). Production of 5-methyluridine by immobilized thermostable purine nucleoside phosphorylase and pyrimidine nucleoside phosphorylase from *Bacillus stearothermophilus* JTS 859. *J. Biotechnol.* 17 (2): 121–131.

102 Lewkowicz, E.S., Martínez, N., Rogert, M.C. et al. (2000). An improved microbial synthesis of purine nucleosides. *Biotechnol. Lett* 22: 1277–1280.

103 Trelles, J.A., Bentancor, L., Schoijet, A. et al. (2004). Immobilized *Escherichia coli* BL21 as a catalyst for the synthesis of adenine and hypoxanthine nucleosides. *Chem. Biodivers.* 1 (2): 280–288.

104 Krenitsky, T.A., Rideout, J.L., Koszalka, G.W. et al. (1982). Pyrazolo[3,4-d]pyrimidine ribonucleosides as anticoccidials. 1. Synthesis and activity of some nucleosides of purines and 4-(alkylthio)pyrazolo[3,4-d]pyrimidines. *J. Med. Chem.* 25 (1): 32–35.

105 Krenitsky, T.A., Koszalka, G.W., and Tuttle, J.V. (1981). Purine nucleoside synthesis: an efficient method employing nucleoside phosphorylases. *Biochemistry* 20 (12): 3615–3621.

106 Condezo, L.A., Fernandez-Lucas, J., Garcia-Burgos, C.A. et al. (2006). Enzymatic synthesis of modified nucleosides. In: *Biocatalysis in the Pharmaceutical and Biotechnology Industries* (ed. R. Patel), 401–424. CRC Press.

107 Vieille, C. and Zeikus, G.J. (2001). Hyperthermophilic enzymes: sources, uses, and molecular mechanisms for thermostability. *Microbiol. Mol. Biol. Rev.* 65 (1): 1–43.

108 Utagawa, T., Morisawa, H., Yamanaka, S. et al. (1985). Properties of nucleoside phosphorylase from *Enterobacter aerogenes*. *Agric. Biol. Chem.* 49 (11): 3239–3246.

109 Zuffi, G., Ghisotti, D., Oliva, I. et al. (2004). Immobilized biocatalysts for the production of nucleosides and nucleoside analogues by enzymatic transglycosylation reactions. *Biocatal. Biotransform.* 22 (1): 25–33.

110 Zhou, X., Mikhailopulo, I.A., Cruz Bournazou, M.N., and Neubauer, P. (2015). Immobilization of thermostable nucleoside phosphorylases on MagReSyn epoxide microspheres and their application for the synthesis of 2,6-dihalogenated purine nucleosides. *J. Mol. Catal. B: Enzym.* 115: 119–127.

111 Sirisha, V.L., Jain, A., and Jain, A. (2016). Enzyme immobilization: an overview on methods, support material, and applications of immobilized enzymes. *Adv. Food Nutr. Res.* 79: 179–211.

112 Brady, D. and Jordaan, J. (2009). Advances in enzyme immobilisation. *Biotechnol. Lett* 31 (11): 1639–1650.

113 Bickerstaff, G.F. (1997). Immobilization of enzymes and cells: some practical considerations. In: *Immobilization of Enzymes and Cell* (ed. G.F. Bickerstaff), 1–11. Totowa, NJ: Humana Press.

114 Mazid, M.A. (1993). Biocatalysis and immobilized enzyme/cell bioreactors. *Nat. Biotechnol.* 11 (6): 690–695.

115 Zhao, X., Qi, F., Yuan, C. et al. (2015). Lipase-catalyzed process for biodiesel production: enzyme immobilization, process simulation and optimization. *Renewable Sustainable Energy Rev.* 44 (Supplement C): 182–197.

116 Trelles, J.A., Fernández-Lucas, J., Condezo, L.A., and Sinisterra, J.V. (2004). Nucleoside synthesis by immobilised bacterial whole cells. *J. Mol. Catal. B: Enzym.* 30 (5–6): 219–227.

117 Rocchietti, S., Ubiali, D., Terreni, M. et al. (2004). Immobilization and stabilization of recombinant multimeric uridine and purine nucleoside phosphorylases from *Bacillus subtilis*. *Biomacromolecules* 5 (6): 2195–2200.

118 Fernández-Lafuente, R., Hernández-Jústiz, O., Mateo, C. et al. (2001). Biotransformations catalyzed by multimeric enzymes: stabilization of tetrameric ampicillin acylase permits the optimization of ampicillin synthesis under dissociation conditions. *Biomacromolecules* 2 (1): 95–104.

119 Serra, I., Serra, C.D., Rocchietti, S. et al. (2011). Stabilization of thymidine phosphorylase from *Escherichia coli* by immobilization and post immobilization techniques. *Enzyme Microb. Technol.* 49 (1): 52–58.

120 Serra, I., Daly, S., Alcantara, A.R. et al. (2015). Redesigning the synthesis of vidarabine via a multienzymatic reaction catalyzed by immobilized nucleoside phosphorylases. *RSC Adv.* 5 (30): 23569–23577.

121 Calleri, E., Ubiali, D., Serra, I. et al. (2014). Immobilized purine nucleoside phosphorylase from *Aeromonas hydrophila* as an on-line enzyme reactor for biocatalytic applications. *J. Chromatogr. B* 968: 79–86.

122 Calleri, E., Cattaneo, G., Rabuffetti, M. et al. (2015). Flow-Synthesis of nucleosides catalyzed by an immobilized purine nucleoside phosphorylase from *Aeromonas hydrophila*: integrated systems of reaction control and product purification. *Adv. Synth. Catal.* 357 (11): 2520–2528.

123 Chen, N., Xing, C.G., Xie, X.X., and Xu, Q.Y. (2009). Optimization of technical conditions of producing ribavirin by *Bacillus subtilis*. *Ann. Microbiol.* 59 (3): 525–530.

124 Ma, Y., Yang, S., Liu, L. et al. (2014). Fermentative production of ribavirin by overexpressing purine nucleoside phosphorylase in Bacillus. *J. Chem. Pharm. Res.* 6 (5): 1377–1384.
125 Sakharov, V., Baykov, S., Konstantinova, I. et al. (2015). An efficient chemoenzymatic process for preparation of ribavirin. *Int. J. Chem. Eng.* 2015: 5, 734851.
126 el Kouni, M.H., el Kouni, M.M., and Naguib, F.N.M. (1993). Differences in activities and substrate specificity of human and murine pyrimidine nucleoside phosphorylases: implications for chemotherapy with 5-fluoropyrimidines. *Cancer Res.* 53 (16): 3687–3693.
127 Desgranges, C., Razaka, G., Rabaud, M., and Bricaud, H. (1981). Catabolism of thymidine in human blood platelets purification and properties of thymidine phosphorylase. *Biochem. Biophys. Acta* 654: 211–218.
128 Fateev, I.V., Antonov, K.V., Konstantinova, I.D. et al. (2014). The chemoenzymatic synthesis of clofarabine and related 2′-deoxyfluoroarabinosyl nucleosides: the electronic and stereochemical factors determining substrate recognition by *E. coli* nucleoside phosphorylases. *Beilstein J. Org. Chem.* 10 (1): 1657–1669.
129 Blank, J.G. and Hoffee, P.A. (1975). Purification and properties of thymidine phosphorylase from *Salmonella typhimurium*. *Arch. Biochem. Biophys.* 265: 259–265.
130 Hoffee, P.A. and Blank, J. (1978). Thymidine phosphoryalse from *Salmonella typhimurium*. *Methods Enzymol.* 437–442.
131 Averett, D.R., Koszalka, G.W., Fyfe, J.A. et al. (1991). 6-Methoxypurine arabinoside as a selective and potent inhibitor of varicella-zoster virus. *Antimicrob. Agents Chemother.* 35 (5): 851–857.
132 Focher, F. and Spadari, S. (2001). Thymidine phosphorylase: a two-face Janus in anticancer chemotherapy. *Curr. Cancer Drug Targets* 141–153.
133 Panova, N.G., Alexeev, C.S., Polyakov, K.M. et al. (2008). Substrate specificity of thymidine phosphorylase of *E. coli*: role of hydroxyl groups. *Nucleosides Nucleotides Nucleic Acids* 27 (12): 1211–1214.
134 Vande Voorde, J., Liekens, S., and Balzarini, J. (2013). *Mycoplasma hyorhinis*–encoded purine nucleoside phosphorylase: kinetic properties and its effect on the cytostatic potential of purine-based anticancer drugs. *Mol. Pharmacol.* 84 (6): 865–875.
135 Fateev, I.V., Kharitonova, M.I., Antonov, K.V. et al. (2015). Recognition of artificial nucleobases by *E. coli* purine nucleoside phosphorylase versus its Ser90Ala mutant in the synthesis of base-modified nucleosides. *Chem. Eur. J.* 21 (38): 13401–13419.
136 Mikhailopulo, A. and Miroshnikov, A.I. (2013). Some recent findings in the biotechnology of biologically important nucleosides. *Biotechnol. Acta* 6 (4): 63–77.
137 Shirae, H. and Yokozeki, K. (1991). Purification and properties of purine nucleoside phosphorylase from *Brevibacterium acetylicum* ATCC 954. *Agric. Biol. Chem.* 55 (2): 493–499.

2

Enzymatic Phosphorylation of Nucleosides

Daniela Ubiali[1] *and Giovanna Speranza*[2]

[1] *University of Pavia, Department of Drug Sciences, Viale Torquato Taramelli 12, 27100 Pavia, Italy*
[2] *University of Milan, Department of Chemistry, Via Camillo Golgi 19, 20133 Milano, Italy*

2.1 Introduction

Nucleosides and nucleotides are endogenous compounds that are related to the regulation and modulation of many physiological processes. Nucleoside and nucleotide analogues (NAs) are synthetic, chemically modified compounds that mimic their physiological counterparts. They exploit cellular metabolism in order to be incorporated into DNA and RNA, inhibiting cellular division and viral replication; they are used, indeed, as therapeutic agents, mainly in the control of cancer cell growth and viral infections [1]. Nucleotides are also frequently used as prodrugs of their parent nucleosides because the ionic feature of the phosphate group significantly enhances the solubility of poorly water-soluble nucleosides [2].

In addition, some purine nucleotides are of significant economical interest from an industrial viewpoint because they are currently used as foodstuff additives [3]. Specifically, inosine 5′-monophosphate (5′-IMP) and guanosine 5′-monophosphate (5′-GMP) elicit an umami taste in humans and synergistically increase the intensity of the umami taste of monosodium glutamate (MSG). For this reason, they are widely used as flavoring additives in the food industry [4, 5]. The market size for umami seasoning nucleotides increased steadily over decades and the global annual 5′-IMP and 5′-GMP production has reached c. 34 000 tons in 2012 of which Ajinomoto Co. takes a large share [6]. Further functions attributed to nucleotides are immune-, lipidic-, and digestive-related. It has been demonstrated, indeed, that supplementation of pediatric formulas with nucleotides is of benefit in neonatal nutrition [7, 8]. As a result, the current demand for nucleotides in the market is increasing along with the search for novel scalable manufacturing processes [9, 10]. Given the importance of nucleotides, great effort has been devoted to the development of nucleoside phosphorylation methods, both chemical and enzymatic.

The synthesis of nucleosides and nucleotides is not trivial at all. *In vivo*, these molecules can be synthesized either by the de novo or the salvage pathways of nucleosides through complex multi-enzymatic processes. From the chemist's

Enzymatic and Chemical Synthesis of Nucleic Acid Derivatives, First Edition.
Edited by Jesús Fernández-Lucas and María-José Camarasa Rius.

viewpoint, there exist several general procedures to synthesize modified nucleosides and nucleotides. This knowledge still mostly derives from the boost in nucleic acid chemistry registered in the 1950–1960s, just when the implication of DNA and RNA structures in the replicative processes was becoming clear [11].

Concerning phosphorylation, over the years, different protocols for the selective formation of 5′-phosphate esters of nucleosides have been described or patented [12–17]. Most of them, however, rely on the use of $POCl_3$ which is a compound difficult to handle because it is corrosive and highly moisture-sensitive, and hydrolyzes rapidly under storage or standard reaction conditions. Moreover, direct phosphorylation of nucleosides with $POCl_3$ is mostly unfeasible due to the lack of regioselectivity, which results in mixtures of mono- and dinucleotides. Therefore, $POCl_3$-based methods usually require an excess of the phosphorylating agent (and quenching agents, thereof), and protection/deprotection steps often become necessary. Although the use of trialkylphosphates (such as $(Et)_3PO_4$ or $(Me)_3PO_4$) as solvents dramatically improved the regioselectivity for the 5′-position of nucleosides, the nature of the substrate still plays a crucial role in affecting the reaction outcome (yield and purity) [13, 15]. Alternative "fully chemical" methods to carry out 5′-phosphorylation of nucleosides as well as a comprehensive description of the "evolution" in this field from the early synthetic efforts have been nicely reviewed by Vaghefi in 2005 [18].

In cells, the phosphate group is introduced by phosphorylating enzymes, the phosphorylases or phosphotransferases (kinases) that use ATP as an energy-rich phosphate donor. In addition, some of the enzymes that hydrolyze phosphorylated compounds are also able to perform the opposite reaction (transphosphorylation reaction). For this reason, they can be used in synthetic mode (i.e. as biocatalysts for nucleotide synthesis).

The advantage of using an enzymatic route relies on the evidence that phosphorylating enzymes can be stereo- and regioselective making the phosphorylation of polyhydroxy compounds, in principle, much more viable. In addition, enzymes operate under mild conditions and, in general, produce less waste. Thus, as opposed to established chemical "conventional" approaches, enzymatic phosphorylation is a mild reaction falling within the "green chemistry" approach [19–21].

An overview of the recent work on enzymatic phosphorylation of nucleosides to nucleoside 5′-monophosphates is described in this chapter.

2.2 Nonspecific Acid Phosphatases (NSAPs)

Phosphatases form a large class of structurally and mechanistically diverse enzymes that catalyze the hydrolysis of phosphate monoesters [22]. Biochemical and biophysical properties, such as optimum pH value (acid, neutral, or alkaline), substrate specificity (nonspecific versus specific for certain substrates), molecular size (high versus low molecular weight), identity of the enzyme residue that is transiently phosphorylated during catalysis (histidine, serine, and cysteine) and, more recently, amino acid sequence homology have been used over time to classify phosphatases into families [22–25].

The term nonspecific acid phosphatases (NSAPs) refers to a group of enzymes that display optimal catalytic activity at acidic to neutral pH values, which do not exhibit a marked substrate specificity being usually able to hydrolyze a broad range of organic phosphoesters. NSAPs are widely distributed among enteric bacteria as soluble proteins in the periplasmic space or membrane-bound lipoproteins [24, 25].

On the basis of amino acid sequence similarity, three different classes of NSAPs have been identified, designated as class A, B, and C. Members of each class have some common biophysical and functional characteristics, but also exhibit functional differences. Class A NSAPs are further classified into three subclasses (A1–A3). Prototype enzymes for A1, A2, and A3 classes are PhoC of *Morganella morganii* (PhoC-Mm), PhoN of *Salmonella typhimurium* (PhoN-St), and Apy of *Shigella flexneri* (Apy-Sf), respectively [24–26].

Class A acid phosphatases contain a polypeptide component with an Mw of 25–27 kDa possessing sequence motifs, KX_6RP-(X_{12-54})-PSGH-(X_{31-54})-SRX_5HX_3D, shared and conserved by several lipid phosphatases and the mammalian glucose-6-phosphatases [23, 25, 27], which play a very important role in catalysis. Mostly important for the synthesis of nucleotides, class A1 NSAPs exhibit 5′-regioselective PP_i-nucleoside phosphotransferase activity. In spite of the high degree of homology at the amino acid sequence level, the K_m for inosine, which is always considered as the reference substrate, can be significantly different among enzymes belonging to the A1 class, thus suggesting that local sequence differences in the binding site affect the PP_i-nucleoside phosphotransferase reaction [26].

The occurrence in a large variety of plant and animal tissues of enzymes which catalyze the transfer of phosphoric acid from low-energy organic phosphates to nucleosides was first reported by Brawermann and Chargaff [28, 29], who named such enzymes nucleoside phosphotransferases owing to their substrate specificity.

Crude enzyme preparations from plants were mostly investigated. For example, Marutzky et al. described the synthesis of nucleoside-5′-monophosphates in gram scale using a phosphotransferase from carrots and phenylphosphate as the phosphate donor [30].

The nucleoside phosphotransferase reaction of bacteria was mainly studied by Mitsugi and coworkers using *p*-nitrophenylphosphate as the phosphate donor [31, 32]. It was found that bacteria belonging to such genera as *Pseudomonas, Flavobacterium, Serratia, Alcaligenes, Achromobacter*, and *Staphylococcus* have a nucleoside phosphotransferase activity regioselective for the C5′-position, whereas bacteria belonging to genera *Aeromonas, Escherichia, Aerobacter, Proteus*, and *Salmonella* catalyze the phosphorylation in the C3′(2′)-position [33].

A dramatic advancement in the study of the phosphorylation process was derived from the research of Asano et al. who found that the phosphotransferase activity specific for the 5′-position was largely distributed among the bacteria belonging to the family of *Enterobacteriaceae* such as *M. morganii*, which was selected for further studies aimed at 5′-IMP production [34, 35]. A selective nucleoside phosphorylating enzyme was thus purified from *M. morganii* NCIMB10466 crude extract. The enzyme phosphorylated various nucleosides at the 5′-position to produce nucleoside-5′-monophosphates using either the food additive PP_i as the phosphate source, or other energy-rich compounds,

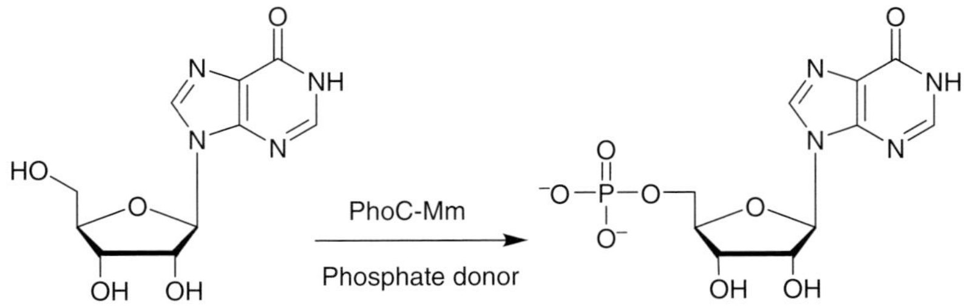

PhoC-Mm: NSAP from *Morganella morganii*;
phosphate donor: PPi, carbamylphosphate [35], acetylphosphate [35].

Scheme 2.1 Regioselective phosphorylation of inosine to 5′-IMP.

such as carbamylphosphate and acetylphosphate (Scheme 2.1) [35]. The enzyme also exhibited phosphatase activity, and dephosphorylated various phosphate esters, thus indicating to be an acid phosphatase. From an applicative synthetic standpoint, the dephosphorylating activity of the enzyme affected the final yield of the synthesized 5′-IMP due to the hydrolysis of the phosphate ester to inosine [35]. In order to control the phosphatase activity and increase the efficiency of the transphosphorylation reaction, a mutated acid phosphatase with an increased phosphotransferase reaction yield was obtained (101 g l^{-1} of 5′-IMP was synthesized from inosine in an 88% molar yield). This improvement was achieved with two mutations, Gly to Asp at position 92 and Ile to Thr at position 171 [36].

As for the mechanism, this phosphorylation reaction is in fact a transphosphorylation, in which a phosphate group is transferred from a phosphate donor, e.g. PP_i, to the alcohol group of an acceptor. As depicted in Scheme 2.2, the enzymatic reaction entails the formation of an activated phosphorylated enzyme intermediate that can either regioselectively transfer its phosphate group to an appropriate substrate acceptor or react with water resulting in hydrolysis and releasing of free phosphate. A further issue of the phosphatase-catalyzed transphosphorylations is the reversibility of the phosphate transfer, i.e. the concurrent hydrolytic activity

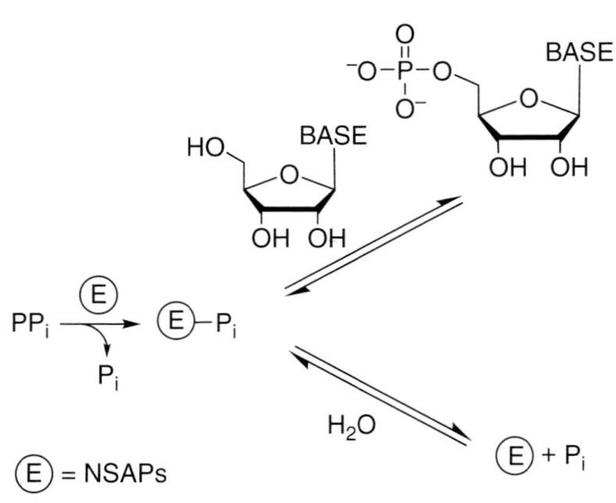

Scheme 2.2 Mechanism of NSAP-catalyzed transphosphorylation and hydrolysis.

of these biocatalysts that results in a progressive dephosphorylation of the formed phosphate ester. Key factors for an effective phophorylation are therefore both a high affinity of the phospho-enzyme intermediate for the substrate and the inhibition of the product hydrolysis [21, 37, 38].

To further improve the production of 5′-IMP, phosphatases from several bacteria (*Providencia stuartii, Enterobacter aerogenes, Escherichia blattae,* and *Klebsiella planticola*) were investigated confirming the pivotal role played by a lower K_m value for inosine in order to achieve an increase in nucleotide productivity. On this ground, by crossing mutational data available for *M. morganii* [36] and structural information of *E. blattae* [27], a rational site-directed mutagenesis study was carried out on this latter enzyme, resulting in a triple mutant able to produce 140 g l^{-1} of 5′-IMP [26, 39–41]. This evidence corroborated that a slight structural change caused by an amino acid replacement around the inosine binding site could significantly reduce the K_m value.

The studies of Asano and coworkers have clearly showed that recombinant acid phosphatases with regioselective phosphotransferase activity (AP/PTase, EC 3.1.3) can be used for the large-scale production of nucleotides [9, 42, 43]. From an industrial viewpoint, the use of PP_i as a phosphate donor has a number of advantages as it can be easily prepared from phosphate at low cost and, not less important, it is safe as it is routinely used as a food additive due to its GRAS status ("generally recognized as safe") acknowledged by the FDA [21].

Recently, an NSAP from *S. flexneri* (PhoN-Sf) [44], an enzyme belonging to the same class A1 acid phosphatase as *M. morganii* PhoC, was covalently immobilized on epoxy-activated supports (Immobeads-150 and Sepabeads™ EC-EP) with high efficiency (70% retained activity). The immobilized enzyme was stable for many months and was used to produce 5′-IMP from inosine and PP_i on gram scale in a packed-bed continuous reaction [43]. This approach resulted to be very useful to circumvent one of the main limitations of phosphatase-catalyzed synthesis, that is competing dephosphorylation of the product due to the natural hydrolytic activity of these biocatalysts. Coupling of immobilization and flow mode reaction allows the physical separation of the phosphatase from the phosphorylated product, thus preventing its hydrolysis. Reaction or protein engineering [9, 37, 45] has been also developed toward this aim.

2.3 Deoxyribonucleoside Kinases (dNKs)

In contrast to phosphatases, deoxyribonucleoside kinases (dNKs, E.C. 2.7.1) are designed by Nature to phosphorylate natural deoxyribonucleosides (dNs) to their corresponding monophosphates (dNMPs) by using ribonucleoside triphosphates, generally ATP, as energy-rich phosphate donors (Scheme 2.3). dNKs catalyze the first and most often rate-limiting step in the salvage pathway of nucleic acids to DNA synthesis. Once inside the cell, the dNMPs are phosphorylated by the nucleoside monophosphate kinases (NMPKs) to diphosphates and again by nucleoside diphosphate kinases (NDPKs) to deoxynucleoside triphosphates (dNTPs). The dNK-catalyzed phosphorylation is irreversible, whereas the phosphotransferase reactions catalyzed by the NMPKs and NDPKs

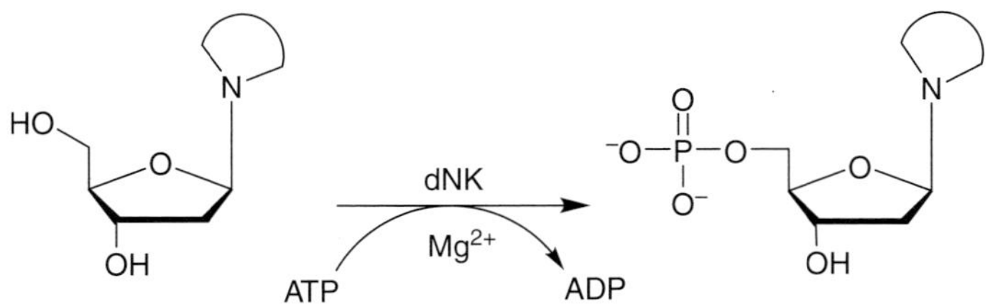

dNK: deoxyribonucleoside kinases.

Scheme 2.3 Regioselective phosphorylation of 2′-deoxyribonucleosides.

are reversible [46]. These reactions are also responsible for the *in vivo* activation of NAs used as therapeutic agents. NAs, in fact, are prodrugs that need to be converted into the corresponding phosphates to exert their pharmacological activity. Upon *in vivo* activation by phosphorylation, NAs are either incorporated in the viral/cellular nucleic acid or are able to inhibit specific enzymes of nucleic acid biosynthesis. They may act, indeed, as antimetabolites to impair DNA and/or RNA processing. Polymerases, reverse transcriptase, and thymidylate synthase are only few examples of enzymes targeted by NAs [1]. The recognition of NAs as substrates by dNKs has been responsible for exploiting these enzymes for both medical and biocatalytic applications.

The use of NAs as anticancer and antiviral drugs is well established. Nevertheless, the emergence of resistance mechanisms (often due to an insufficient intracellular metabolism) that make cancer cells refractory to therapy as well as the high toxicity to non-cancer cells (due to the lack of selectivity of NAs) represent a crucial and challenging limitation to cancer/infection treatments [47]. In addition to the constant search for new molecules, many efforts have been thus paid to improve the activity of molecules already available. Within gene therapy applications, transfer of a specific dNK gene into cancer cells has been proposed as a strategy to both increase the activity of cytotoxic NAs against solid tumors and to minimize the toxicity toward normal cells. The rationale of "suicide-gene therapy," also referred to as "gene-directed enzyme/prodrug therapy" (G-DEPT), relies in fact on the transfection of tumor cells with a foreign gene encoding an enzyme that is capable of activating *in situ* a nontoxic prodrug (i.e. a nucleoside) to an advanced cytotoxic intermediate (i.e. a nucleotide). In other words, upon administration, the prodrug would be only activated in the tumor cells where the "suicide gene" is expressed [48].

The first G-DEPT system aimed at enhancing the susceptibility of cancer cells to NAs was developed by using the thymidine kinase gene of the *Herpes Simplex virus* (HSV-TK) in combination with the prodrug Ganciclovir (GCV) to produce GCV-monophosphate [48, 49]. Successively, other dNKs were assayed toward this aim, including the tomato thymidine kinase (TK, in combination with 3′-azido-3′-deoxythymidine (AZT)) [50, 51] and the dNK from fruitfly *Drosophila melanogaster* (in combination with different NAs, due to its multisubstrate character) [52, 53]. Although positive results were observed both *in vitro* and in animal models in terms of decrease in tumor size, major limitations that emerged from these studies, such as inefficient transfection of cancer cells and poor prodrug activation by the suicide enzymes, have probably discouraged following trials in humans. The successful outcome of a "suicide/prodrug system" depends on a striking difference in expression levels of the suicide gene between cancer and normal cells, which is the essential requirement for enhancing a selective localization of the prodrug activation. Needless to say, the prodrug should not be activated by any other enzyme before reaching the target in order to avoid any undesired effect to normal cells [54].

Substrate promiscuity is a crucial issue also in biocatalysis. In this context, a relaxed substrate specificity is generally a desirable property for enzymes intended for synthetic applications, in order to have flexible and versatile biocatalysts. Substrate specificity, catalytic efficiency, and subcellular localization of different dNKs from various organisms have been very nicely reviewed by Slot Christiansen et al. in 2015 [46]. With the exception of fungi, that appear not to possess any kinase, all organisms studied to date have one or more dNKs displaying a high diversity. Mammalian dNKs were among the first enzymes to be cloned and fully characterized. Four deoxyribonucleoside-specific kinases have been found in mammalian cells and tissues: the cytoplasmic enzymes TK1 and deoxycytidine kinase (dCK), and the mitochondrial enzymes TK2 and deoxyguanosine kinase (dGK). These kinases have distinct but overlapping specificities [55]. The research on mammalian dNKs has been mostly related to the role played by these enzymes in DNA synthesis (also including medical implications arising from defects in mitochondrial DNA) [56], and in the activation of nucleoside-based drugs. Interestingly, the research carried out by Munch-Petersen and coworkers from the late 1990s have highlighted that, in contrast to mammals, insects have only one dNK that possesses all four dNK activities [57–61]. *D. melanogaster* (*Dm*dNK), *Bombyx mori* (*Bm*dNK) [60], and *Anopheles gambiae* (*Ag*dNK) [61] have been investigated in this context, but only *Dm*dNK has found practical application in biocatalysis to date [62, 63]. *Dm*dNK is able to phosphorylate all natural deoxyribonucleosides with high efficiency [57] and also displays a similar broad substrate promiscuity toward NAs used in chemotherapy that has brought about its application in the "suicide gene therapy," as explained above [55]. Thus, *Dm*dNK has been exploited as a biocatalyst for the phosphorylation of both naturally occurring nucleosides and NAs, also on an industrial scale [62, 63]. Specifically, Serra et al. [63] reported the synthesis of arabinosyladenine 5′-monophosphate (araA-MP) and its 2-fluorinated congener fludarabine 5′-monophosphate (FaraA-MP). Both these NAs, which are used against systemic herpetic infections [2] or specific hematological malignancies [64], respectively, are preferably administered as soluble monophosphate prodrugs [2]. *Dm*dNK was immobilized by adsorption on an amino-activated epoxy support and then cross-linked by treatment with a polyaldehyde (c. 40% retained activity). Upon immobilization, the productivity and rate of synthesis of *Dm*dNK in the phosphorylation of arabinosyladenine were dramatically enhanced with respect to the non-immobilized enzyme, affording the title compound in 95% yield and 96.5% purity [63].

A decade after the identification of *Dm*dNK, Piškur and coworkers [65] subcloned three different genes encoding dNKs from *Dictyostelium discoideum*, a social amoeba: a TK, a deoxyadenosine kinase (*Dd*dAK), and a dGK. *Dd*dAK was found to efficiently phosphorylate only 2′-deoxyadenosine (dAdo), its native substrate, and FaraA. Kinetic parameters of purified recombinant *Dd*dAK measured in the phosphorylation of FaraA showed that this enzyme was only seven times less specific for FaraA as for dAdo, and phosphorylated FaraA and dAdo with almost equally high turnover. Hence, *Dd*dAK was immobilized and used as a biocatalyst in the enzymatic synthesis of the above nucleotides [66]. As for *Dm*dNK, biotransformations were carried out in fully aqueous medium and under mild conditions (pH 8, 37 °C) in almost quantitative yield. Moreover, immobilization further increased the stability of the biocatalyst also allowing its recycle and repeated use.

Phosphorylations performed by dNKs need the presence of Mg^{2+}ions. The activity of these kinases is in fact dependent on Mg^{2+} that acts by binding the γ- and β-phosphates of the phosphate donor upon the catalysis, thus "preparing" the ATP to be in the correct conformation before it is bound by the enzyme. The generally accepted mechanism for dNKs is that activation of the 5′-OH of the deoxyribose is required in order to make a nucleophilic attack on the γ-phosphate of the phosphate donor.

To date, examples of dNK-based phosphorylations on a preparative scale are very few, although the enzymatic approach has been demonstrated to be a convenient alternative to $POCl_3$-based phosphorylation. One of the reasons can be ascribed to the high cost of phosphate donors (typically ATP) that have to be used in stoichiometric amounts, unless an efficient ATP-regeneration system is coupled to the main reaction (i.e. phosphorylation). Different ATP-recycling systems have been developed to regenerate the required cofactor that can be used, indeed, in catalytic amounts [20, 67].

Basically, nucleoside triphosphates (NTPs) can be regenerated from nucleoside diphosphates (NDPs) by using four enzymatic systems: (i) pyruvate kinase (PK) coupled with phosphoenolpyruvate (PEP) as the phosphate donor, (ii) acetate kinase (AK)-acetylphosphate, (iii) creatine kinase (CK)-creatine phosphate, and (iv) polyphosphate kinase (PPK)-polyphosphate (polyP). Although these enzymes have been mostly employed in the regeneration of ATP, they can also be used in the regeneration of other NTPs, such as guanosine triphosphate (GTP), uridine 5′-triphosphate (UTP), and cytidine 5′-triphosphate (CTP). In the specific case of ATP and dNKs, it is worth underlining that ATP recycling may be economically beneficial for the process, but also avoids possible enzyme inhibition phenomena caused by adenosine 5′-diphosphate (ADP) accumulation over time, and streamlines the reaction impurity profile (adenosine 5′-monophosphate (AMP) can be formed from ADP, too) [20, 67]. To date, the most economical methodology for NTP regeneration from their corresponding NDP utilizes PPK and polyP [67]. However, one should evaluate if ATP recycling might be beneficial for the specific process under investigation, by considering both the scale of the bioconversion and the value of the target product [63, 66].

2.4 Conclusion

Whether NSAPs are superior to dNKs (or vice versa) as biocatalysts for the enzymatic phosphorylation of nucleosides to 5′-monophosphates is debatable.

On one hand, NSAPs have the advantage of using cheap, easily affordable PP_i as the phosphate donor, whereas dNKs rely on more expensive cofactors that would need to be recycled through an auxiliary enzymatic system. On the other hand, NSAPs are hydrolases rather than transferases; therefore, their use may require a laborious reaction/enzyme optimization in order to enhance phosphorylation over dephosphorylation, whereas dNKs have exclusive phosphotransferase activity. Accumulated research on NSAPs in biocatalysis has provided new enzymes as promising candidates to be implemented for nucleotide synthesis on a preparative scale. As for dNKs, a high number of enzymes from different sources have been also investigated; nevertheless, diversity of this enzyme family and comparative studies between different organisms have been mostly pursued to assist the comprehension of the biological role of dNK as well as evolutionary aspects. The interest in dNKs as biocatalysts is quite recent and coincides with the "shift" of the research to non-mammalian organisms. In this context, multisubstrate dNKs from insects such as silkworm (*Bm*dNK) and mosquito (*Ag*dNK), besides the well investigated *Dm*dNK, appear to be promising candidates for biocatalytic applications. For example, *Ag*dNK [61] has been shown to phosphorylate purine more efficiently than pyrimidine deoxyribonucleosides as compared to *Dm*dNK. In addition, it has been suggested that *Ag*dNK could also phosphorylate some NAs such as stavudine, 2-chloro-deoxyadenosine, and 5-bromo-vinyl-deoxyuridine.

Last but not least, NSAPs and dNKs possess slightly different substrate specificity as NSAPs display higher affinities for ribonucleosides than for 2′-deoxyribonucleosides. This feature complements the specificity of dNKs, mostly directed toward 2′-deoxyribonucleosides, thus implementing the toolset of phosphorylating enzymes.

References

1 Jordheim, L.P., Durantel, D., Zoulim, F., and Dumontet, C. (2013). Advances in the development of nucleoside and nucleotide analogues for cancer and viral diseases. *Nat. Rev. Drug Discovery* 12: 447–464.

2 Superti, F., Ammendolia, M.G., and Marchetti, M. (2008). New advances in anti-HSV chemotherapy. *Curr. Med. Chem.* 15: 900–911.

3 Ledesma-Amaro, R., Jiménez, A., Santos, M.A., and Revuelta, J.L. (2013). Biotechnological production of feed nucleotides by microbial strain improvement. *Process Biochem.* 48: 1263–1270.

4 Cairoli, P., Pieraccini, S., Sironi, M. et al. (2008). Studies on umami taste. Synthesis of new guanosine 5′-phosphate derivatives and their synergistic effect with monosodium glutamate. *J. Agric. Food. Chem.* 56: 1043–1050.

5 Behrens, M., Meyerhof, W., Hellfritsch, C., and Hofmann, T. (2011). Sweet and umami taste: natural products, their chemosensory targets, and beyond. *Angew. Chem. Int. Ed.* 50: 2220–2242.

6 Wendisch, V.F., Eberhardt, D., Herbst, M., and Jensen, J.V.K. (2016). Biotechnological production of amino acids and nucleotides. In: *Biotechnological Production of Natural Ingredients for Food Industry*, 1e (ed. J.L. Bicas, M.R. Maróstica and G.M. Pastor), 60–163. Bentham Science Publishers.

7 Mateos-Vivas, M., Rodríguez-Gonzalo, E., Domínguez-Álvarez, J. et al. (2016). Determination of nucleosides and nucleotides in baby foods by hydrophilic interaction chromatography coupled to tandem mass spectrometry in the presence of hydrophilic ion-pairing reagents. *Food Chem.* 211: 827–835.

8 Del Arco, J., Cejudo-Sanches, J., Esteban, I. et al. (2017). Enzymatic production of dietary nucleotides from low-soluble purine bases by an efficient, thermostable and alkali-tolerant biocatalyst. *Food Chem.* 237: 605–611.

9 Suzuki, E., Ishikawa, K., Mihara, Y. et al. (2007). Structural-based engineering for transferases to improve the industrial production of 5′-nucleotides. *Bull. Chem. Soc. Jpn.* 80: 276–286.

10 Kakehi, M., Usuda, Y., Tabira, Y., and Sugimoto, S. (2009). Method for producing nucleotide by fermentation. US Patent 7,575,901 B2, filed 12 March 2004 and issued 18 August 2009.

11 Vorbrueggen, H. and Ruh-Pohlenz, C. (2000). Synthesis of nucleosides. In: *Organic Reactions*, vol. 55 (ed. L.A. Paquette et al.), 1–630. New York, NY: Wiley.

12 Yoshikawa, M., Kato, T., and Takenishi, T. (1967). A novel method for phosphorylation of nucleosides to 5′-nucleotides. *Tetrahedron Lett.* (50): 5065–5068.

13 Yoshikawa, M., Kato, T., and Takenishi, T. (1969). Studies of phosphorylation. III. Selective phosphorylation of unprotected nucleosides. *Bull. Chem. Soc. Jpn.* 42: 3505–3508.

14 Blumbergs, P., Khan, M.S., Kalamas, R.L., et al. (1992). Process for the preparation of 9-beta-D-arabinofuranosyl-2-fluoroadenine 5′-phosphate. WO1992000312, filed 26 June 1991 and issued 9 January 1992.

15 Ikemoto, T., Haze, A., Hatano, H. et al. (1995). Phosphorylation of nucleosides with phosphorus oxychloride in trialkyl phosphate. *Chem. Pharm. Bull.* 43: 210–215.

16 Burgess, K. and Cook, D. (2000). Syntheses of nucleoside triphosphates. *Chem. Rev.* 100: 2047–2059.

17 Moreau, C., Ashamu, G.A., Bailey, V.C. et al. (2011). Synthesis of cyclic adenosine 5′-diphosphate ribose analogues: a C2′ endo/syn "southern" ribose conformation underlies activity at the sea urchin cADPR receptor. *Org. Biomol. Chem.* 9: 278–290.

18 Vaghefi, M. (2005). Chemical synthesis of nucleoside 5′-triphosphate. In: *Nucleoside Triphosphates and their Analogs. Chemistry, Biotechnology, and Biological Applications*, 1e (ed. M. Vaghefi), 1–22. Taylor & Francis.

19 Eriksson, S. (2005). Enzymatic synthesis of nucleoside triphosphates. In: *Nucleoside Triphosphates and Their Analogs. Chemistry, Biotechnology, and Biological Applications*, 1e (ed. M. Vaghefi), 23–37. Taylor & Francis.

20 Faber, K. (2011). Biocatalytic applications. In: *Biotransformations in Organic Chemistry. A Textbook*, 6e, 111–120. Springer.

21 Wever, R., Babich, L., and Hartog, A.F. (2015). Transphosphorylation. In: *Science of Synthesis*, Biocatalysis in Organic Synthesis, vol. 1 (ed. K. Faber, W.D. Fessner and N.J. Turner), 223–254. Thieme Verlagsgruppe.

22 Widlanski, T.S. and Taylor, W. (1999). Chemistry and enzymology of phosphatases. In: *Comprehensive Natural Products Chemistry* (ed. D.C. Poulter), 139–162. Elsevier.

23 Stukey, J. and Carman, G.M. (1997). Identification of a novel phosphatase sequence motif. *Protein Sci.* 6: 469–472.

24 Rossolini, G.M., Schippa, S., Riccio, M.L. et al. (1998). Bacterial nonspecific acid phosphohydrolases: physiology, evolution and use as tools in microbial biotechnology. *Cell. Mol. Life Sci.* 54: 833–850.

25 Medici, R., Garaycoechea, J.I., Valino, A.L. et al. (2014). A comparative study on phosphotransferase activity of acid phosphatases from *Raoultella planticola* and *Enterobacter aerogenes* on nucleosides, sugars, and related compounds. *Appl. Microbiol. Biotechnol.* 98: 3013–3022.

26 Mihara, Y., Utagawa, T., Yamada, H., and Asano, Y. (2001). Acid phosphatase/phosphotransferases from enteric bacteria. *J. Biosci. Bioeng.* 92: 50–54.

27 Ishikawa, K., Mihara, Y., Gondoh, K. et al. (2000). X-ray structures of a novel acid phosphatase from *Escherichia blattae* and its complex with the transition-state analog molybdate. *EMBO J.* 19: 2412–2423.

28 Brawerman, G. and Chargaff, E. (1954). On the synthesis of nucleotides by nucleoside phosphotransferases. *Biochim. Biophys. Acta* 15: 549–559.

29 Brawerman, G. and Chargaff, E. (1955). On the distribution and biological significance of the nucleoside phosphotransferases. *Biochim. Biophys. Acta* 16: 524–532.

30 Marutzky, R., Peterssen-Borstel, H., Flossdorf, J., and Kula, M.-R. (1974). Large scale enzymatic synthesis of nucleoside-5′-monophosphates using a phosphotransferase from carrots. *Biotechnol. Bioeng.* 16: 1449–1458.

31 Katagiri, H., Yamada, H., Mitsugi, K., and Tsunoda, T. (1964). Bacterial synthesis of nucleotides. Part I. Nucleoside phosphotransferase of *Escherichia coli*. *Agr. Biol. Chem.* 28: 577–585.

32 Zinchenko, A.I., Barai, V.N., Zalashko, L.M. et al. (1990). Enzymatic synthesis of nucleoside 5′-mono and -triphosphates. *FEBS Lett.* 260: 254–256.

33 Mitsugi, K., Komagata, K., Takahashi, M. et al. (1964). Bacterial synthesis of nucleotides. Part II. Distribution of nucleoside phosphotransferases in bacteria. *Agric. Biol. Chem.* 28: 586–600.

34 Asano, Y., Mihara, Y., and Yamada, H. (1999). A new enzymatic method of selective phosphorylation of nucleosides. *J. Mol. Catal. B: Enzym.* 6: 271–277.

35 Asano, Y., Mihara, Y., and Yamada, H. (1999). A novel selective nucleoside phosphorylating enzyme from *Morganella morganii*. *J. Biosci. Bioeng.* 87: 732–738.

36 Mihara, Y., Utagawa, T., Yamada, H., and Asano, Y. (2000). Phosphorylation of nucleosides by the mutated acid phosphatase from *Morganella morganii*. *Appl. Environ. Microbiol.* 66: 2811–2816.

37 Tasnádi, G., Lukesch, M., Zechner, M. et al. (2016). Exploiting acid phosphatases in the synthesis of phosphorylated monoalcohols and diols. *Eur. J. Org. Chem.* 45–50.

38 Tasnádi, G., Zechner, M., Hall, M. et al. (2017). Investigation of acid phosphatase variants for the synthesis of phosphate monoesters. *Biotechnol. Bioeng.* 114: 2187–2195.

39 Ishikawa, K., Mihara, Y., Shimba, N. et al. (2002). Enhancement of nucleoside phosphorylation activity in an acid phosphatase. *Protein Eng.* 15: 539–543.

40 Asano, Y. (2002). Overview of screening for new microbial catalysts and their uses in organic synthesis – selection and optimization of biocatalysts. *J. Biotechnol.* 94: 65–72.

41 Mihara, Y., Ishikawa, K., Suzuki, E., and Asano, Y. (2004). Improving the pyrophosphate-inosine phosphotransferase activity of *Escherichia blattae* acid phosphatase by sequential site-directed mutagenesis. *Biosci. Biotechnol., Biochem.* 68: 1046–1050.

42 Liu, Z.Q., Zhang, L., Sun, L.H. et al. (2012). Enzymatic production of 5′-inosinic acid by a newly synthesised acid phosphatase/phosphotransferase. *Food Chem.* 134: 948–956.

43 Babich, L., Hartog, A.F., van der Horst, M.A., and Wever, R. (2012). Continuous-flow reactor-based enzymatic synthesis of phosphorylated compounds on a large scale. *Chem. Eur. J.* 18: 6604–6609.

44 Tanaka, N., Hasan, Z., Hartog, A.F. et al. (2003). Phosphorylation and dephosphorylation of polyhydroxy compounds by class A bacterial acid phosphatases. *Org. Biomol. Chem.* 1: 2833–2839.

45 Tasnádi, G., Hall, M., Baldenius, K. et al. (2016). Biocatalytic functionalization of hydroxyalkyl acrylates and phenoxyethanol via phosphorylation. *J. Biotechnol.* 233: 219–227.

46 Slot Christiansen, L., Munch-Petersen, B., and Knecht, W. (2015). Non-viral deoxyribonucleoside kinases – diversity and practical use. *J. Genet. Genomics* 42: 235–248.

47 Jordheim, L.P. and Dumontet, C. (2007). Review of recent studies on resistance to cytotoxic deoxynucleoside analogues. *Biochim. Biophys. Acta* 1776: 138–159.

48 Hebrard, C., Dumontet, C., and Jordheim, L.P. (2009). Development of gene therapy in association with clinically used cytotoxic deoxynucleoside analogues. *Cancer Gene Ther.* 16: 541–550.

49 Fillat, C., Carrio, M., Cascante, A., and Sangro, B. (2003). Suicide gene therapy mediated by the *herpes simplex virus* thymidine kinase gene/ganciclovir system: fifteen years of application. *Curr. Gene Ther.* 3: 13–26.

50 Khan, Z., Knecht, W., Willer, M. et al. (2010). Plant thymidine kinase 1: a novel efficient suicide gene for malignant glioma therapy. *Neuro Oncol.* 12: 549–558.

51 Larsen, N.B., Munch-Petersen, B., and Piskur, J. (2014). Tomato thymidine kinase is subject to inefficient TTP feedback regulation. *Nucleosides Nucleotides Nucleic Acids* 33: 287–290.

52 Zheng, X., Johansson, M., and Karlsson, A. (2001). Nucleoside analog cytotoxicity and bystander cell killing of cancer cells expressing *Drosophila*

melanogaster deoxyribonucleoside kinase in the nucleus or cytosol. *Biochem. Biophys. Res. Commun.* 289: 229–233.

53 Kamiya, H., Ochiai, H., Harashima, H. et al. (2006). Transient expression of *Drosophila melanogaster* deoxynucleoside kinase gene enhances cytotoxycity of nucleoside analogs. *Nucleosides Nucleotides Nucleic Acids* 25: 553–560.

54 Ardiani, A., Johnson, A.J., Ruan, H. et al. (2012). Enzymes to die for: exploiting nucleotide metabolizing enzymes for cancer gene therapy. *Curr. Gene Ther.* 12: 77–91.

55 Eriksson, S., Munch-Petersen, B., Johansson, K., and Eklund, H. (2002). Structure and function of cellular deoxyribonucleoside kinases. *Cell. Mol. Life Sci.* 59: 1327–1346.

56 Eriksson, S. and Wang, L. (2008). Molecular mechanisms of mitochondrial DNA depletion diseases caused by deficiencies in enzymes in purine and pyrimidine metabolism. *Nucleosides Nucleotides Nucleic Acids* 27: 800–808.

57 Munch-Petersen, B., Piškur, J., and Søndergaard, L. (1998). Four deoxynucleoside kinase activities from *Drosophila melanogaster* are contained within a single monomeric enzyme, a new multifunctional deoxynucleoside kinase. *J. Biol. Chem.* 273: 3926–3931.

58 Munch-Petersen, B., Piškur, J., and Søndergaard, L. (1998). The single deoxynucleoside kinase in *Drosophila melanogaster*, *Dm*-dNK, is multifunctional and differs from the mammalian deoxynucleoside kinases. *Adv. Exp. Med. Biol.* 431: 465–469.

59 Munch-Petersen, B., Knecht, W., Lenz, C. et al. (2000). Functional expression of a multisubstrate deoxyribonucleoside kinase from *Drosophila melanogaster* and its C-terminal deletion mutants. *J. Biol. Chem.* 275: 6673–6679.

60 Knecht, W., Petersen, G.E., Munch-Petersen, B., and Piškur, J. (2002). Deoxyribonucleoside kinases belonging to the thymidine kinase 2 (TK2)-like group vary significantly in substrate specificity, kinetics and feed-back regulation. *J. Mol. Biol.* 315: 529–540.

61 Knecht, W., Petersen, G.E., Sandrini, M.P. et al. (2003). Mosquito has a single multisubstrate deoxyribonucleoside kinase characterized by unique substrate specificity. *Nucleic Acids Res.* 31: 1665–1672.

62 Ihlenfeldt, H.-G., Munch-Petersen, B., Piškur, J., et al. (2000). Drosophila melanogaster deoxynucleoside kinase, its recombinant production, enzymic activity and use in synthesis of nucleoside monophosphates. EP0999275 A2, filed 6 October 1999 and issued 10 May 2000.

63 Serra, I., Conti, S., Piškur, J. et al. (2014). Immobilized *Drosophila melanogaster* deoxyribonucleoside kinase (*Dm*dNK) as a high performing biocatalyst for the synthesis of purine arabinonucleotides. *Adv. Synth. Catal.* 356: 563–570.

64 Plunkett, W., Huang, P., and Gandhi, V. (1990). Metabolism and action of fludarabine phosphate. *Semin. Oncol.* 17: 3–17.

65 Sandrini, M.P.B., Soederbom, F., Mikkelsen, N.E., and Piškur, J. (2007). *Dictyostelium discoideum* salvages purine deoxyribonucleosides by highly specific bacterial-like deoxyribonucleoside kinases. *J. Mol. Biol.* 369: 653–664.

66 Serra, I., Ubiali, D., Piškur, J. et al. (2017). Immobilization of deoxyadenosine kinase from *Dictyostelium discoideum* (*Dd*dAK) and its application in the 5′-phosphorylation of arabinosyladenine and arabinosyl-2-fluoroadenine. *Chemistry Select* 2: 5403–5408.

67 Truppo, M.D. (2012). Cofactor recycling for enzyme catalyzed processes. In: *Comprehensive Chirality* (ed. H. Yamamoto and E.M. Carreira), 46–70. Elsevier.

3

Enzymatic Synthesis of Nucleic Acid Derivatives Using Whole Cells

Elizabeth S. Lewkowicz and Adolfo M. Iribarren

Universidad Nacional de Quilmes, Laboratorio de Biocatálisis y Biotransformaciones, Department of Science and Technology, Roque Saenz Peña 352, B1876BXD Bernal, Buenos Aires, Argentina

3.1 Introduction

Nucleosides and their analogues are key molecules in the generation of energy, in cell signaling, and in the storage and transmission of genetic information.

Besides other activities, they can act as antiviral and anticancer drugs by inhibiting replication and regulating transcription processes [1]. To take advantage of these characteristics, many efforts have been taken to generate more active and selective drugs, giving rise to a vast family of compounds: the nucleoside analogues (NAs) [2] (Scheme 3.1).

The first approved antiviral nucleoside was idoxuridine; (5-iodo-2′-deoxyuridine) [3], following which many other analogues have also been authorized for clinical practices by different drug agencies. Among the most representative examples are 2′,3′-dideoxyinosine (didanosine), 3′-azido-3′-deoxythymidine (AZT, zidovudine), L-2′-deoxy-3′-thiacytidine (lamivudine), 2′,3′-dideoxycytidine (zalcitabine), and 6-cyclopropylamino-2′,3′-didehydro-2′,3′-dideoxyguanosine (abacavir), which are employed for the treatment of infections caused by different viruses, such as human immunodeficiency virus (HIV), hepatitis-B virus (HBV), hepatitis-C virus (HCV), human cytomegalovirus (HCMV), herpes simplex virus (HSV), human papilloma virus (HPV), respiratory syncytial virus (RSV), and varicella zoster virus (VZV).

There are also several examples of well-known anticancer drugs based on NAs [4], such as: 2-fluoro-9-β-D-arabinofuranosyladenine (fludarabine), 2-chloro-2′-deoxy-2′-fluoro-9-β-D-arabinofuranosyladenine (clofarabine), 6-methoxy-9-β-D-arabinofuranosylguanine (nelarabine), 1-β-D-arabinofuranosylcytosine (cytarabine), 2′,2′-difluorodeoxycytidine (gemcitabine), and 2-chloro-2′-deoxy-adenosine (cladribine).

Besides, NAs find applications in other fields. Inosine-5′-monophosphate (IMP) and guanosine-5′-monophosphate (GMP) are excellent flavor enhancers and are also used as beef extract substitutes in the food industry [5]; tuberculosinyl adenosine, a diterpene nucleoside present in *Mycobacterium tuberculosis*, is explored as a new component for diagnostic test for tuberculosis [6],

Enzymatic and Chemical Synthesis of Nucleic Acid Derivatives, First Edition.
Edited by Jesús Fernández-Lucas and María-José Camarasa Rius.

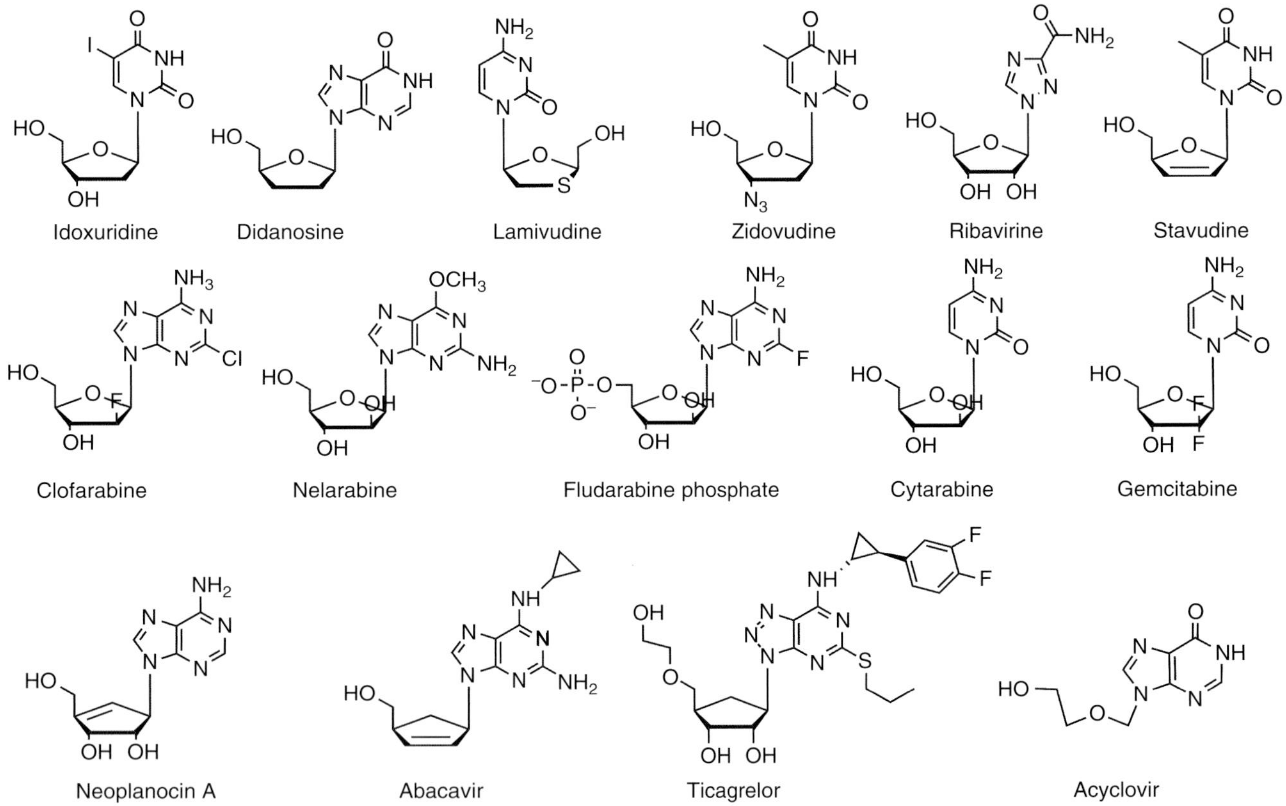

Scheme 3.1 Pharmacologically active nucleoside analogues.

and AZT-prolinamide analogues were described as new organocatalysts for enantioselective aldol condensations [7].

The traditional synthesis of NAs involves cumbersome processes that involve high costs [8]. Therefore, biocatalyzed strategies may provide alternative routes with unique benefits, such as mild reaction conditions, high stereo- and regioselectivities, and environmental-friendly procedures [9–11].

The choice of commercially available enzymes is limited, and therefore, the search for new alternatives through screening of microorganism collections provides an attractive proposition [12]. Moreover, the direct use of microbial whole cells as biocatalysts avoids enzyme isolation and stabilizes enzyme activity by keeping them in their natural environment, leading to easier and cheaper routes [13] that could be further improved by genetic manipulation [14].

This chapter intends to summarize the available information regarding the biocatalyzed whole cell synthesis of nucleosides and their analogues addressing relevant and recent researches.

3.2 Nucleoside Synthesis Mediated by Microbial Nucleoside Phosphorylases

The most widely spread enzymes used as biocatalysts for the synthesis of natural and non-natural nucleosides via a base exchange between combinations of purine and pyrimidine nucleosides are *N*-glycosyltransferases (GTs) [3]. Whole cells containing nucleoside 2′-deoxyribosyltransferases (NdRTs) or nucleoside phosphorylases (NPs) have been employed in one-pot, one-enzyme, or one-pot, two-enzymes, respectively, allowing access to antiviral and antitumor drugs [9, 10, 15–18]. Since whole cells with NP activities found greater applications than cells with NdRTs, NA synthesis using only microbial NPs as biocatalysts will be herein discussed.

NPs are essential enzymes for the salvage and catabolism pathways of nucleotides. They reversibly catalyze the cleavage of the glycosidic bond of ribo- or deoxyribonucleosides in the presence of inorganic phosphate to produce ribose- or deoxyribose-1-phosphate and the corresponding base [9]. Both pyrimidine nucleoside phosphorylase (PyNP) and purine nucleoside phosphorylase (PNP) exist. PNPs (EC 2.4.2.1) recognize 9-(β-D-pentofuranosyl) purines as substrates, being mammalian PNPs specific for 6-oxopurines while bacterial ones show broad specificity, accepting both 6-oxo- and 6-aminopurines. Thymidine phosphorylase (TP, EC 2.4.2.4) reversibly catalyzes the phosphorolysis of 1-(2′-deoxy-β-D-ribofuranosyl)-pyrimidines whereas uridine phosphorylase (UP, EC 2.4.2.3) accepts many 1-(β-D-pentofuranosyl) pyrimidines as substrates. Several organisms contain a single PyNP (EC 2.4.2.2) with broad specificity for the 2′-position of the ribose. Cytosine and its nucleosides are not substrates for any PyNP.

New sources of NPs are continually being discovered in order to expand the range of accepted substrates and improve the efficiency and productivity of these biocatalysts. In particular, microorganisms offer a large diversity of biocatalysts with different substrate specificities. Nowadays, whole bacterial

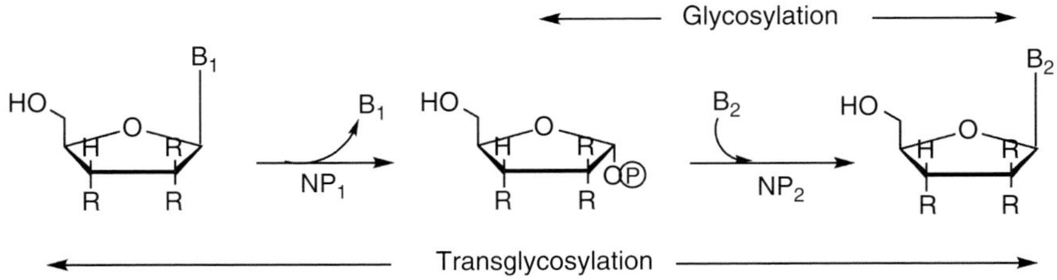

Scheme 3.2 Nucleoside synthesis mediated by nucleoside phosphorylases (NPs). R : H, OH; B_1,B_2: pyrimidine, or purine bases.

cells over-expressing recombinant NPs are replacing wild type strains, providing shorter reaction times and requiring less biomass [19].

The thermodynamic equilibrium for PNP, but not for PyNP, is shifted toward nucleoside synthesis, and therefore, the combination of both enzymes makes the generation of purine nucleosides from pyrimidine ones in two sequential steps possible. The so-called *transglycosylation,* is a one-pot cascade reaction that consists of a first conversion of a pyrimidine nucleoside to α-pentose-1-phosphate (P1P), catalyzed by a PyNP and a second step where a PNP catalyzes the transfer reaction between P1P and a purine base affording the corresponding purine nucleoside (Scheme 3.2).

Microbial transglycosylation proved to be an efficient methodology for the synthesis of several NAs with biological and pharmaceutical importance [20, 21], by using natural or modified pyrimidines nucleoside as pentofuranose donors as well as by using natural or artificial heterocyclic bases as acceptors.

Transglycosylation reaction temperature is an important parameter that needs to be optimized for particular substrates and biocatalysts. High temperatures favor the substrates solubility in aqueous solution and also increase the reaction rate [22]. In addition, when whole cells are used as biocatalysts, deleterious enzymes can be inhibited by increasing the reaction temperature. In most of the microorganisms, UP is usually active up to 80 °C, while TP is only active below 50 °C. Besides, PNP activity drops above 65 °C. Therefore, typically, transglycosylations are performed at 60 °C, since most of the NPs retain over 70% of their activity while the deleterious action of some enzymes such as deaminases (see Section 3.3.1) can be minimized. However, reactions involving TP are usually carried out at 45 °C. Then, to avoid deaminase interferences, 2′-deoxyuridine (dU) instead of thymidine can be used as an alternative source of the deoxyribosyl moiety at 60 °C.

Ribavirin, an NA that displays activity against a variety of DNA and RNA viruses [23, 24] as well as exhibits anticancer activity [25], was efficiently synthesized using *Aeromonas hydrophila* CECT4226 in 77% conversion starting from uridine and 1,2,4-triazole-3-carboxamide (TCA), at 60 °C for 26 hours [26]. Ribavirin preparation was also reported using whole cells of *Escherichia coli* ATCC 12407 at 30 °C in 86% conversion representing an increase of more than 40% of the conversion obtained at 60 °C with the same system [27].

An alternative to overcome temperature limitations is to employ thermophilic microorganisms such as *Geobacillus stearothermophilus* [28] and *Thermus thermophilus* [29]. Almendros et al. [30] identified several strains of *T. thermophilus* capable of synthesizing purine nucleoside employing adenine or hypoxanthine as bases. High yields in short times were obtained by carrying out the biotransformation at 65 °C. After cloning and isolation, three NPs from *T. thermophilus* with complementary substrate specificities were identified [31]: two purine phosphorylases, *Tt*PNPI (showing specificity for 6-oxopurines) and *Tt*PNPII (displaying 6-aminopurines specificity), and one pyrimidine phosphorylase *Tt*PyNP that uses thymine as well as uracil but not cytosine as substrates. Tomoike et al. [32] concluded that this enzymatic behavior is conserved in most thermophilic species belonging to the *Deinococcus-Thermus* phylum and in other thermophilic organisms such as *G. stearothermophilus.* Both ribo- and 2′-deoxyribosides of 6-halopurines were obtained with a conversion that reached 90% using *G. stearothermophilus* CECT43 [33]. This strain also catalyzed the synthesis of 2,6-diaminopurine (DAP) – riboside and – 2′ deoxyriboside at 30 °C in 90% and 83% conversion, respectively [28]. Similar conversions for the same products were previously informed by us [34] but by employing 60 °C, obtaining additionally 2,6-diaminopurine-2′,3′-dideoxyriboside (DAPddR) and 2,6-diaminopurine-arabinoside (DAPA). The synthesis of this range of DAP analogues was possible by using different mesophilic bacterial whole cells, which unlike thermophilic ones, are able to accept 2′,3′-dideoxyuridine and uracil arabinoside (AraU) as substrates.

Other examples of pharmaceutical drugs that can be prepared by whole cell biocatalysis are adenine arabinoside (9-β-D-arabinofuranosyladenine, vidarabine), an antiviral agent used in the treatment of human diseases caused by HSV, CMV, and HBV [35], and fludarabine, a fluoro analogue prescribed for chronic lymphocytic leukemia therapies [36]. Whole cells of *Enterobacter aerogenes* HXY2222 were reported as biocatalysts for vidarabine synthesis starting from AraU and adenine at 60 °C achieving 83% yield after 30 hours [37]. In this regard, we also reported the production, among other nucleosides, of fludarabine (58% yield in 14 hours), vidarabine (71% yield in 26 hours), and DAPA (77% yield in 24 hours), from AraU and the corresponding bases by *Citrobacter koseri* whole cells [38].

Genetically engineered bacteria that express high levels of recombinant NPs have been used to improve the production of NAs. These reactions require fewer amounts of biomass and shorter reaction times. NAs such as DAP-2′-deoxyriboside [39], 2-chloro-2′-deoxyadenosine [40], 2′-deoxyadenosine (dA) and 6-methylpurine-2′-deoxyriboside (MePdR) [41] have been successfully synthesized using *E. coli* recombinant strains.

Co-expression of different enzymes in a single strain provides an efficient method to simplify both biocatalyst preparation and biotransformation process, especially in large-scale industrial production. In order to obtain DAP-nucleosides, Ge et al. [42] designed two recombinant *E. coli* BL21(DE3) strains harboring double recombinant plasmids, using genes encoding PNP, TP, and UP from *E. coli* K12: PNP-UP (DUD) and PNP-TP (DAD) and two other strains containing tandem recombinant plasmids (TDU and TDA). After IPTG induction, DAD and TDA were employed to prepare DAP-2′-deoxyriboside from DAP and thymidine, reaching yields of 40.2% and 51.8%, respectively. TDU and

DUD showed higher activity, transforming uridine to DAP-riboside in 88.2% and 58.0% yield, respectively. When lactose was used instead of IPTG for inducing the expression of the same four recombinant strains, DAP-nucleosides production was improved. In addition, some other ribo- and 2′-deoxyribonucleosides were synthesized with TAD, TUD, DUD, or DAD, such as 5-methyluridine (5-MU), 5-fluorouridine, uridine, adenosine, azauridine, ribavirin, dA, dI, and 2′-deoxyribavirin in yields ranging from 50% to 85% [39].

3.3 Nucleoside Analogues Synthesis by the Combined Action of Microbial Nucleoside Phosphorylases and Other Enzymes

3.3.1 Nucleoside Phosphorylases Coupled to Deaminases

Transglycosylation using guanine as base donor usually fails due to its poor solubility. To overcome this drawback, some alternative strategies have been reported including the use of more soluble guanine derivatives like glyoxal-guanine or guanosine [43], the use of basic pHs, the increase of reaction temperature employing thermostable NPs, and the use of multienzymatic systems, where transglycosylation is coupled to other biocatalyzed steps either to use more suitable reagents for NPs or to shift the reaction equilibrium [44].

Deaminases are hydrolytic enzymes that naturally catalyze the hydrolysis of amino groups of nucleosides or their bases. The most used and widespread enzyme of this group is adenosine deaminase (ADA; EC 3.5.4.4), which is commercially available from mammalian sources, and catalyzes the irreversible hydrolysis of adenosine to inosine and ammonia. Other well-known deaminases are guanine deaminase (GD, EC 3.5.4.3); cytidine deaminase (CDA, EC 3.5.4.5); and adenine deaminase (AAH, EC 3.5.4.2), this being the last enzyme that has not been identified up to now in higher eukaryotic organisms [45].

As mentioned above, DAP-nucleosides are used as pharmaceutical drugs or prodrugs in anticancer and antiviral therapies [46, 47]. As prodrugs, DAP-nucleosides are converted *in vivo* into guanosine analogues by deamination performed by endogenous ADA [48]. In a similar approach, guanine arabinoside (AraG), a powerful antileukemic agent [49], was obtained *in vitro* by a one-pot strategy based on the action of two whole cells to perform the simultaneous transglycosylation and deamination reactions [44]. The deamination step was carried out by *Arthrobacter oxydans* whole cells [45], whose ADA has the peculiarity of remaining active at 60 °C (Scheme 3.3). The broad specificity of *A. oxydans* made it possible to obtain other pharmacologically active compounds such as didanosine. The presence of AAH and GD in *A. oxydans* was also demonstrated.

3.3.2 Nucleoside Phosphorylases Coupled to Phosphopentomutase

As discussed previously, biocatalyzed transglycosylation proceeds through the formation of the corresponding P1P and requires pyrimidine nucleosides and

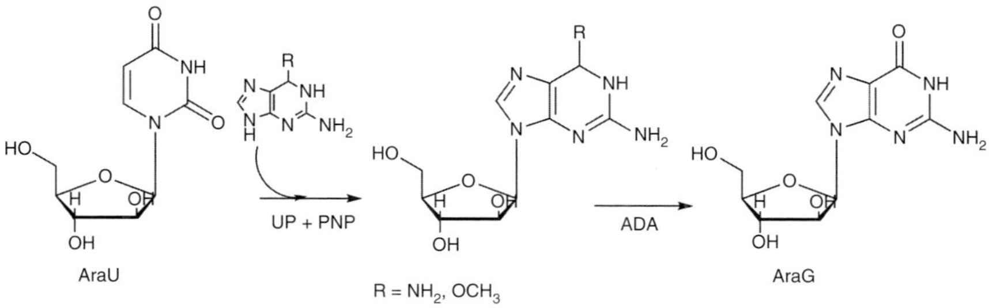

Scheme 3.3 One-pot AraG synthesis catalyzed by whole cells containing nucleosides phosphorylases (UP and PNP) and adenosine deaminase (ADA) activities.

inorganic phosphate as starting materials. An alternative approach (referred as *glycosylation*) is to employ only one NP and P1P intermediates as starting materials, which strategy is also suitable for the preparation of pyrimidine nucleoside. The synthesis of some interesting arabinonucleosides was described by the Mikhailopulo's and coworkers [50] using α-D-arabinofuranose 1-phosphate and recombinant *E. coli* PNP as the biocatalyst.

However, the availability of P1P is limited due to their intrinsic instability and their cumbersome synthesis [51, 52]. These compounds can be enzymatically obtained from pentose 5-phosphates (P5P) employing phosphopentomutase (PPM, EC 5.4.2.7). PPM takes part in the pentose pathway in bacteria and in mammalian tissues, catalyzing the transfer of a phosphate group between the hydroxyls of positions 5 and 1 of ribose and 2-deoxyribose [11]. After the preparation of P5P by a chemoenzymatic synthesis, which involved lipase-biocatalyzed regioselective protection and deprotection of the corresponding sugars, these compounds were employed for the synthesis of a set of natural and modified nucleosides using an over-expressed PPM from *E. coli* and commercial PNP or TP as biocatalysts [53].

3.3.3 Nucleoside Phosphorylases Coupled to Phosphopentomutase and Other Enzymes

Additionally, P5P can be biocatalytically prepared through different enzymatic approaches (Scheme 3.4). Chuvikovsky et al. [54] used *E. coli* ribokinase (RK, EC2.7.1.15) over-expressed in *E. coli* ER2566 as the biocatalyst for the first step of the cascade transformation of different pentoses into nucleosides using adenosine 5′-triphosphate (ATP) as phosphate donor. Notably, cladribine, a standard drug for the treatment of hairy cell leukemia was prepared from 2-deoxyribose and 2-chloroadenine in 90% yield [55].

Sugars can also be regioselectively phosphorylated using bacterial nonspecific acid phosphatases (NSAPs, EC 3.1.3.2) and disodium acid pyrophosphate (PPi) as the phosphate source [56]. NSAPs are a group of enzymes that are able to hydrolyze a broad range of organic phosphoesters, displaying optimal activity in acidic or neutral pH. In addition to their intrinsic phosphatase activity, some enzymes also exhibit phosphotransferase activity. In this sense, we employed recombinant *E. coli* BL21(DE3) strains carrying *En. aerogenes* and *Raoultella*

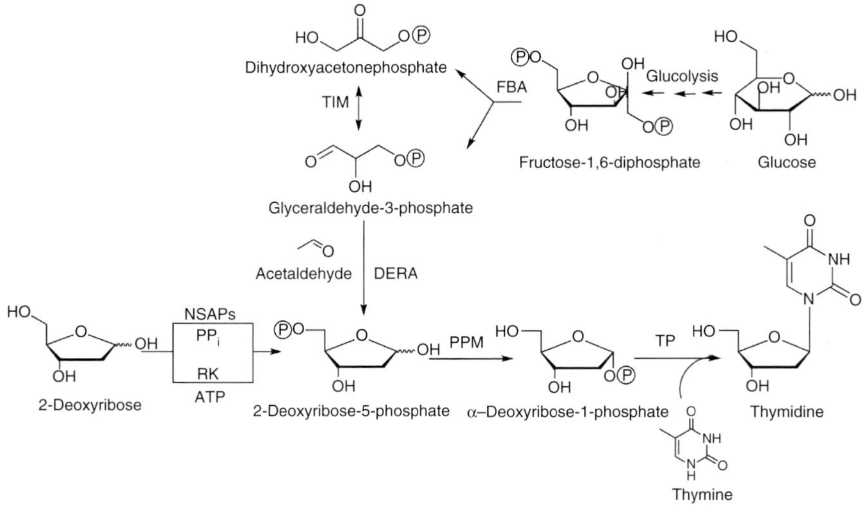

Scheme 3.4 Biocatalyzed strategies for the synthesis of 2-deoxyribose-5-phosphate and its subsequent conversion to pyrimidine 2′-deoxynucleosides. FBA: fructose-1,6-diphosphate aldolase. TIM: triose isomerase. DERA: 2-deoxy-5-phosphate aldolase. PPM: phosphopentomutase. NSAPs: nonspecific acid phosphatases. TP: thymidine phosphorylase. RK: ribokinase.

planticola acid phosphatases for the synthesis of ribose-, 2-deoxyribose-, and arabinose 5-phosphates[57].

A multienzymatic process to prepare 2-deoxyribose 5-phosphate (DR5P) that involved glycolytic enzymes was developed by Ogawa et al. [58]. In this route, D-glyceraldehyde 3-phosphate (G3P) was prepared from glucose and further converted to DR5P through condensation with acetaldehyde catalyzed by 2-deoxy-D-ribose-5-phosphate aldolase (DERA, EC 4.1.2.4). DERA is the only known member of the acetaldehyde-dependent aldolase family and is also the only known aldolase that catalyzes the aldol reaction between two aldehydes by a stereoselective and reversible reaction [59, 60]. The practical application of DERA is limited due to its low affinity toward acetaldehyde and low stability at high concentrations of this reagent [61]. Different DERA variants were isolated from microbial sources [62–69] and from environmental DNA libraries [70], and mutated via several random mutagenesis strategies to increase its tolerance to high aldehyde concentrations [71] and also to improve its affinity to non-phosphorylated substrates [72, 73]. For the purpose of industrial applications, a DERA over-expressing *E. coli* has been employed as a whole cell DERA biocatalyst [74].

For the first time, the Shimizu and coworkers reported the synthesis of dI [75] from glucose [76] or fructose-1,6-diphosphate (FBP), produced by yeast glucose fermentation [77], acetaldehyde, and adenine through three microbial steps without expensive energy sources such as ATP. In addition to yeast fermentation, this route comprised a recombinant phosphatase-negative *E. coli* 10B5 cells expressing DERA from *Klebsiella pneumoniae, E. coli* BL21 cells over-expressing PPM from *E. coli,* and a commercial PNP. It is noteworthy that dI was produced from adenine due to the ADA activity displayed by the *E. coli* transforming strains. Unfortunately, PPM from *E. coli* is markedly inhibited by phosphorylated compounds, such as FBP and G3P and also by inorganic phosphate. Therefore, after optimization of the reaction conditions in order to achieve low concentration of phosphate to prevent PPM inhibition, the authors were able to successfully produce dI in a one-pot process [78]. To improve this biotransformation, an *E. coli* BL21 co-expressing *K. pneumoniae* DERA and *E. coli* PPM was constructed, showing specific activities that were 9.8- and 7.0-fold higher than the host strain, respectively [79]. Later, the authors screened microorganisms isolated from soil and stock of bacterial and fungal cultures, selecting *Bacillus sphaericus* AKU229 as a source of both acetaldehyde and phosphorylated compounds tolerant PPM [80]. Recombinant *B. sphaericus* PPM resulted in being two times more active than the *E. coli* PPM under the same conditions.

In order to select new DERA sources with both high synthetic activity and tolerance to elevated acetaldehyde concentration, we identified *Erwinia carotovora* through a hierarchical screening procedure [81]. Starting from glucose and acetaldehyde, *Er. carotovora* whole cells were coupled to a recombinant PPM isolated from *E. coli* and commercial TP to synthesize thymidine. After DR5P formation, the final reaction medium was lyophilized to completely remove acetaldehyde, and the residue re-dissolved to fit PPM and TP requirements. After two hours of the addition of thymine, thymidine was obtained in 85% conversion relative to DR5P [82]. This strategy was also applied to the synthesis of 5-bromo-2′-deoxyuridine achieving 100% conversion after 45 minutes reaction [53].

3.3.4 Nucleoside Phosphorylases Coupled to Other Enzymes

Another interesting strategy for the multienzymatic synthesis of NAs consists of converting the base released during transglycosylation into a compound unable to be accepted by NPs in order to shift the reaction equilibrium toward the product. Horinouchi et al. [83] described the first detailed oxidative pyrimidine metabolism in *Rhodococcus erythropolis* showing that pyrimidine bases are initially oxidized to barbituric acid analogues by uracil/thymine dehydrogenase (EC 1.17.99.4), and then the barbituric acid analogues are further hydrolyzed to ureidomalonic acid by the barbiturase (EC 3.5.2.1) action. Finally, a ureidomalonase (EC 3.5.1.95) catalyzes the amidohydrolysis to urea and malonic acid. Therefore, *R. erythropolis* JCM 3132 whole cells can be used to carry out the removal of the pyrimidine base released during the transglycosylation reaction, increasing the productivity of NAs synthesis.

3.4 Chemoenzymatic Preparation of Nonconventional Nucleoside Analogues Involving Whole Cell Biocatalyzed Key Steps

The continuous appearance of emerging diseases as well as the resistance developed to already available drugs makes it mandatory for the constant

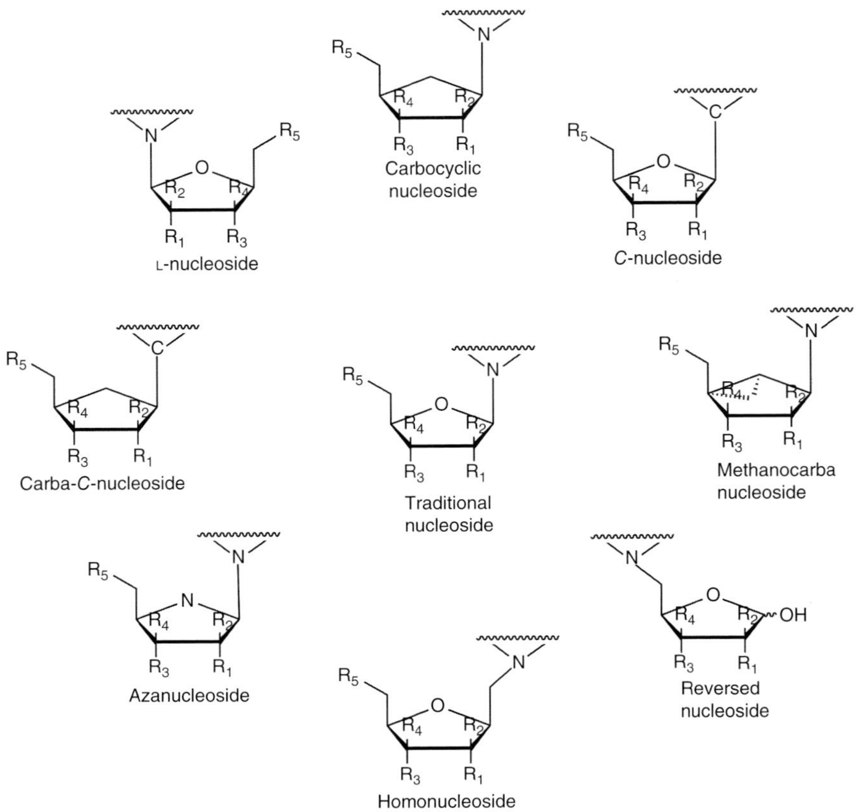

Scheme 3.5 Nonconventional nucleoside analogues.

generation of new active compounds [3, 84]. Although NPs show broad substrate specificity, there are many structural modifications of the heterocyclic bases and/or of the sugar moiety present in nonconventional NAs that are not recognized by these enzymes (Scheme 3.5). Whole cell biocatalyzed reactions involved in key steps of the synthetic routes of some nonconventional NAs are discussed in this section.

3.4.1 L-Nucleosides

Since the discovery of lamivudine, the first L-nucleoside approved against HIV and HBV, in the beginning of the 1990s, a large number of L-NAs have been synthesized, playing an effective role in the treatment of HBV, HCV, Epstein–Barr virus, and some other diseases [85].

Not only NAs but also their oligomers, the oligonucleotides, have received much attention as potential therapeutic agents [86]. These functional oligonucelotides like antisense, ribozymes, DNAzymes, siRNAs, and aptamers, among others, show different structures and ways of action. They find applications in the treatment of different diseases and also in fields such as molecular biology and diagnosis and in the development of new sensors. However, these molecules require chemical modifications in order to show proper stability in biological media. In this sense, L-oligonucleotides have been largely studied and in particular as proof of their wide range of application, we have recently applied this kind of oligonucleotides as catalysts in the oxidation of diverse compounds [87].

L-Ribose is a non-natural sugar that can be used as an intermediate for the synthesis of L–NAs. Because of its high price, some biocatalyzed methods from cheap raw materials under environmentally friendly conditions were developed [88–90]. Two multienzymatic strategies have been reported starting from L-arabinose or from glucose (Scheme 3.6). Both routes proceed via a common intermediate, L-ribulose, which can be converted into L-ribose by the catalytic action of L-ribose isomerase (L-RI, EC 5.3.1.B3) or mannose-6-phosphate isomerase (MPI, EC 5.3.1.8) [91].

From D-glucose, L-ribulose was obtained through a first fermentation step by *Trichosporonoides oedocephalis* or *Trichosporonoides megachillensis* followed by a second step where the formed ribitol was oxidized using washed cells of *Acetobacter aceti* [92] or *Gluconobacter oxydans* [93] but the overall yield was low.

On the other hand, L-ribulose is an intermediate in the pathway for L-arabinose metabolism in many bacteria, L-arabinose isomerase (L-AI, EC 5.3.1.5) being the enzyme responsible for this transformation. Helanto et al. [94] designed a ribulokinase-deficient *Lactobacillus plantarum* strain, to avoid further phosphorylation to L-ribulose-5-phosphate, and they applied resting cells of this mutant to the production of L-ribulose by the action of its endogenous L-AI. Borate was added to the reaction medium, which forms a complex with ketoses resulting in a shift of chemical equilibrium and an enhanced final yield. L-AI genes from *E. coli* [95] and from *Bacillus licheniformis* [96] were also cloned and expressed in *E. coli*. Compared to other L-AIs, *B. licheniformis* L-AI has a wider pH range, a higher substrate specificity, and better catalytic efficiency for L-arabinose conversion.

L-RI is an aldose–ketose isomerase that was first found in *Acinetobacter* sp. DL-28 by Kagawa and Okayama [97]. Enzymes from different organisms showed broad substrate specificity, an optimal temperature in the range of 30–40 °C and optimal pH between 7.0 and 9.0. New sources of L-RI were explored with the aim of improving thermostability and productivity. Among them, a thermostable *Paenibacillus* RI-39 L-RI was reported by Pyun et al. [98], an L-RI from *Raoultella ornithinolytica* MB426 was described by Izumori et al. [99], and a thermoactive and thermostable L-RI from *Cellulomonas parahominis* MB426, a strain isolated from soil, was further expressed in *E. coli* [100]. Helanto et al. [101] introduced an L-RI into the L-ribulokinase-deficient mutant of *L. plantarum*, and resting cells of this recombinant strain and L-arabinose were used to prepare L-ribose at 39 °C in 3-(*N*-morpholino)propanesulfonic acid (MOPS) buffer pH 8 in conversions higher than those obtained using isolated enzymes [102].

MPI catalyzes the reversible isomerization of β-D-mannose-6-phosphate to D-fructose-6-phosphate [103]. After Yeom et al. reported MPI ability to convert L-ribulose into L-ribose [104], different MPIs were discovered and genetically engineered to enhance isomerase activity, selectivity toward pentoses with respect to hexoses, and thermostability [105], including those from

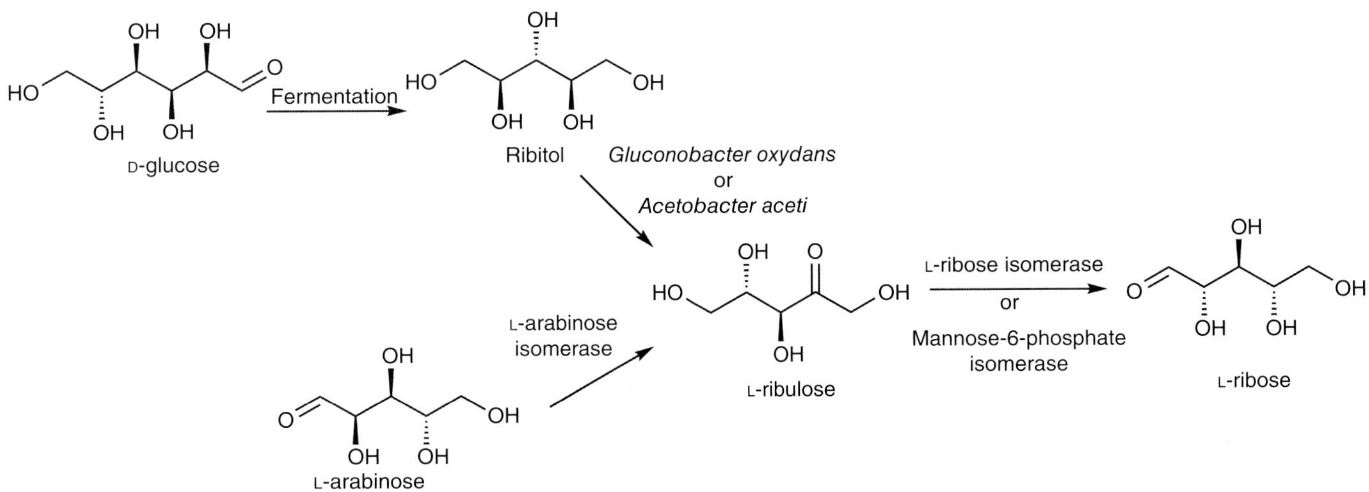

Scheme 3.6 Biocatalyzed approaches to synthesize L-ribose.

T. thermophilus [106], *Geobacillus thermodenitrificans* [107], and *B. subtilis* [104]. In particular, Kim et al. [108] developed a triple-site variant MPI from *G. thermodenitrificans* that showed a catalytic efficiency 7.1-fold higher than that of the wild-type [109], this mutant being further co-expressed together with L-AI, from the same origin, in *E. coli* ER2566 (NEB) cells. This system was applied to the conversion of L-arabinose to L-ribose, being the first report about MPI used as whole cell biocatalyst and the highest conversion reported among processes using L-arabinose as the substrate.

On the other hand, it is well known that aldolases have been employed to prepare rare sugars [110]. Recently, the above mentioned recombinant DERA from *K. pneumonia* [78] was subjected to multisite-directed mutagenesis strategy and the generated mutant exhibited a 3.15-fold improvement in enzyme activity and a 1.54-fold increase in substrate tolerance [72]. Using L-glyceraldehyde, not only the natural product 2-deoxy-L-xylose (3*R*, 4*S*) but also the non-natural product 2-deoxy-L-ribose (3*S*, 4*S*), were produced [111].

3.4.2 Carbocyclic Nucleosides

Carbocyclic nucleosides (CNs) are NAs in which a methylene group replaces the oxygen atom of the furanose moiety, having applications as antiviral, antitumor, or antibiotic drugs [112]. They can either be found in nature like neplanocinA [113], or be the result of synthetic routes such as abacavir [114]. Abacavir is a carbocyclic guanosine analogue that is used in the treatment of human HIV and HBV, acting as a prodrug of carbovir with improved toxicity profile and higher oral bioavailability [115]. The structural modification present in CNs results in an increase in the chemical stability of the *N*-glycosidic bond and confers metabolic resistance to the action of several enzymes such as NPs.

Chiral Vince lactam (γ-lactam, 2-azabicyclo [2.2.1] hept-5-en-3-one) has been used as one of the most successful intermediates for the synthesis of CNs for more than 30 years [116] since this intermediary has the distinctive characteristic of possessing the required cis orientation between the hydroxymethyl and the nucleobase functions [117]. In particular, (−)-γ-lactam is a common precursor for the synthesis of carbovir, abacavir, and also peramivir, a promising long-acting neuraminidase inhibitor for the treatment and prophylaxis of human influenza virus infection [118].

Enzymatic resolution of γ-lactam has been widely studied, and there are several enzymes that provide excellent enantioselectivity. Among them, γ-lactamase, is the preferred one for obtaining optically pure γ-lactam, and both (+)-γ- and (−)-γ-lactamases (EC 3.5.2.B2 and EC 3.5.2.B3 respectively) exist [119]. The resolution of γ-lactam to both optical forms has been performed by Taylor et al. [120] in very high optical purity using whole cell catalysts for the first time. Two microorganisms, ENZA-1 (*Rhodococcus equi* NCIB 40213) and ENZA-20 (*Pseudomonas solanacearum* NCIB 40249), were isolated from the environment and mutants of these strains with improved stability and activity have been constructed and used as whole cell biocatalysts to produce (+)-γ-lactam and (−)-γ-lactam, respectively after three hours reaction with >98% ee [121, 122]. There are few microorganisms reported to date with (+)-γ-lactamase activity and only *Microbacterium hydrocarbonoxydans, Sulfolobus solfataricus, Comamonas acidovorans, Pseudomonas cepacia, P. solanacearum, Pseudomonas fluorescens, Delftia* sp. CGMCC 5755, and *Pseudomonas granadensis* B6 have been used as whole cell biocatalysts [123–128]. Among them, the last one, a non-thermophilic bacterial strain that has been recently discovered, is expected to be suitable for large-scale (−)-γ-lactam production since it does not require high reaction temperatures, can tolerate high substrate concentration (up to 300 $g\,l^{-1}$), and the ee could reach 99.9%.

Wang et al. [129] and Zhu et al. [130] reported the existence of two totally different enantioselective γ-lactamases in *M. hydrocarbonoxydans* and *Bradyrhizobium japonicum* USDA6. Both enzymes were able to be easily over-expressed in a different host to prepare both optically pure lactam enantiomers.

Ticagrelor is a P_2Y_{12} receptor antagonist recently approved by FDA for the prevention of thrombotic events such as stroke or heart attack [131]. The synthesis of Ticagrelor reported by Singh et al. [132] involved the reduction of 1-(3,4-difluorophenyl)-3-nitropropan-1-one to the corresponding alcohol via a whole cell biocatalytic process. Sixteen different microorganisms were screened and tested both, in growing or resting, for bioreduction. *Candida* species demonstrated superior activity and selectivity in comparison to other strains and among them, *Candida parapsilosis* exhibited high conversion (>99.0%) and enantioselectivity (98.0%) for the *S*-enantiomer with the advantage of not requiring external cofactors or cofactor recycling systems.

3.4.3 *C*-Nucleosides

NAs in which the C—N bond between the sugar and base is replaced by a C—C bond are called *C*-nucleosides (Scheme 3.5). This structural characteristic confers increased hydrolytic and enzymatic stability; in addition, these compounds exhibit significant therapeutic properties such as antiviral, antitumoral, and antibiotic activities [133]. Since the discovery of biological activity of natural *C*-nucleosides, such as showdomycin, pseudouridine, and formycin, several synthetic analogues have been synthesized [134]. On the other hand, carba-*C*-nucleosides, which are *C*-NAs where the tetrahydrofuran (THF) ring is substituted by a cyclopentane, are only sporadically documented in the literature. It is expected that carba *C*-nucleosides become a class of potential bio-active compounds by combining the common features of classical *C*-nucleosides and a noncarbohydrate core [135].

Mihovilovic's and coworkers recently carried out the chemoenzymatic synthesis of some *C*-nucleosides such as showdomycin [136] and their corresponding carba analogues, involving a key stereoselective step of enantiodivergent bio-oxidation [137] mediated by two different Baeyer–Villiger monooxygenases (BVMOs): cyclopentanone monooxygenase (EC1.14.13.16) [138] from *Comamonas* sp. NCIMB9872 and cyclohexanone monooxygenase from *Xanthobacter* sp. ZL5 (EC1.14.13.22) [139]. Recombinant *E. coli* whole cells containing these enzymes were applied as biocatalysts to prepare both antipodals of the chiral lactone intermediates with high chiral purity [140–142], which were useful building blocks for the synthesis of *C*-nucleosides, homonucleosides – a methylene group is placed between the ring and the nucleobase and other analogues.

By analogy with *C*- and carba-*C*-nucleosides, aza-*C*-nucleosides are NAs in which the nucleobase is linked to a pyrrolidine through a C—C bond, and display anticancer and antiviral activities, and increased resistance toward nucleases [143]. A very promising class of aza-*C*-nucleosides are immucillins [144] and among them, adenosine analogue is currently being developed as a potential candidate in the treatment of Ebola virus infection [145].

To prepare aza-*C*-nucleosides, the synthetic strategy involves the addition of the heterocyclic base to a pyrrolidine ring containing an electrophilic functionality. Both, kinetic resolution and desymmetrization approaches mediated by amidase containing *R. erythropolis* AJ270 whole cells, were used to resolve racemic and meso pyrrolidine-2,5-dicarboxamide, a versatile synthetic intermediate to achieve enantiomerically pure disubstituted pyrrolidines [146, 147]. *R. Erythropolis* AJ270 amidase exhibited excellent 2*R*-enantioselectivity, giving products with >99.5% ee.

3.5 Nucleoside Prodrugs Preparation by Whole Cell Systems

Some important properties of nucleosides can be modulated by using prodrugs, which are conjugates that, *in vivo,* release the active drug. There are two main groups of nucleoside prodrugs: hydrophilic (e.g. nucleoside phosphates) that increase their water solubility, and lipophilic (e.g. acylnucleosides) that improve their passive uptake.

3.5.1 Acylnucleosides

Acylated nucleosides show enhanced uptake, increased absorption by oral delivery, improved circulation times, and better chemical or *in vivo* stability. Well-known members of this family of compounds are acylated cytarabine analogues since the parent drug, cytarabine [148], is rapidly inactivated *in vivo* by the action of cytosine deaminase and is also ineffective against solid tumors due to its high hydrophilicity [149].

The biocatalytic acetylation of cytarabine using whole cell of *Aspergillus oryzae* was reported by Li et al. [150], observing regioselectivity toward the 3′-hydroxyl group. The authors tested 11 different lipids as carbon sources, which acted also as inducers, affecting microorganism growth and the production of cell-bond lipases. When Spans were used, higher 3′-regioselectivity was observed, indicating a favored production of enzymes with 3′-OH preference.

Acylation of nucleosides in nonaqueous media has been hampered due to their poor solubility inorganic solvents with low cytotoxicity [151]. The first report of a regioselective whole cell biocatalyzed acetylation of cytarabine in nonaqueous media used freeze-dried *P. fluorescens* [152], but in contrast to the results mentioned above, this biocatalyst exhibited 5′-regioselectivity. This distinct regioselective behavior might be attributed to the structural difference between extracellular and intracellular lipases [153, 154].

Li et al. [155] further analyzed the influence of organic solvents in the regioselective synthesis of monoacylated cytarabine catalyzed by lyophilized *P. fluorescens* cells. This system showed an enhancement of the catalytic activity with the increase of the organic solvent polarity, with the exception of the acetonitrile–pyridine mixture. Among all the tested solvents, isopropyl ether (IPE)–pyridine exhibited the best results, showing the cells also good operational and thermal stabilities. Under optimal conditions, 5′-*O*-propionyl cytarabine was obtained in 77% yield with 97.3% 5′-*O*-regioselectivity. Performance comparison between bacterial cells and immobilized *Candida antarctica* lipase B (Novozyme 435), showed similar regioselectivity, suggesting environmental and cost advantages of the whole cell biocatalyst.

Another study regarding the synthesis of 3′-*O*-propionyl cytarabine using dehydrated whole cells of *A. Oryzae* and nonaqueous media was reported by Yang et al. [156]. Six different IPE/pyridine ratios were tested and other variables were optimized. Using the optimal conditions found by these authors, 3′-*O*-propionyl cytarabine was prepared in 88% yield and 70% regioselectivity. The whole cell biocatalyst also exhibited good thermal stability in both IPE–pyridine and hexane–pyridine mixtures.

Nucleoside esters carrying longer acyl chains are expected to exhibit higher bioactivity [157]. Having this in mind a whole cell based method for the 5′-regioselective preparation of arabinocytosine laurate was developed [158]. In order to explore substrate recognition of microbial enzymes [159, 160], a screening of different microorganisms was carried out. Among the tested strains, *P. fluorescens* gave the highest productivity and regioselectivity. The influence of the solvent was also assessed, and the best results were found in IPE/pyridine followed by those in isopentanol/pyridine. However, the whole cells showed much lower thermostability in IPE/pyridine than in THF-pyridine. To understand more deeply the deleterious effects of the organic solvents, growing cells and whole cells were further examined. This analysis established the presence of changes in the surface morphology of whole and growing cells cultured in media containing different organic solvents, revealing differences in surface smoothness, bulges, and modified cell sizes.

Owing to toxicity to the environment, volatileness, and lower biocatalyst of organic solvents, ionic liquids (ILs) were evaluated as alternative solvents for nonaqueous biocatalysis [161]. ILs are liquids comprising ions with a melting point below 100 °C and, preferentially, liquids at room temperature. They show unique characteristics such as non-volatility, nonflammability, and excellent chemical and thermal stability. There are several examples of biocatalyzed reactions using ILs [162] and some useful generalizations are available. Hydrophilic ILs with strong hydrogen-bonding capacity are water-mimicking liquids that can dissolve enzymes while maintaining their activity [163, 164]. Regarding catalytic processes using whole cells in ILs, a review involving this subject has been recently published [165].

Particularly, the synthesis of long-chain nucleoside esters using ILs as solvents has been reported [166]. The synthesis of laurate ester of cytarabine was chosen as the model reaction using *P. fluorescens* GIM1.209 whole cells as biocatalyst. The influence of different IL types on biocatalyzed reactions and their effects on the cell morphology were evaluated. The obtained results showed that their effects were closely related with both the anions and cations of the ILs, 10% butylmethylimidazolium hexafluorophosphate ([BMI] [PF6])/THF being the solvent that afforded the highest reaction efficiency (81% yield and 5′-regioselectivity was higher than 99%). In addition, ILs modified the cell surface morphology and thus facilitated substrate mass transfer. This result highlights the potential of ILs as a promising reaction medium for efficient and regioselective whole cell catalysis particularly in the field of nucleoside chemistry.

3.5.2 Nucleoside Phosphates

Nucleoside monophosphates (NMPs) show enhanced solubility with respect to the parent nucleosides and are rapidly hydrolyzed by endogenous phosphatases, and for these reasons, they are well-established nucleoside prodrugs [167].

The most representative example is fludarabine monophosphate (FaraAMP); after dephosphorylation its parent drug is transported into cells where it is re-phosphorylated by deoxycytidine kinase to regenerate FaraAMP and subsequently accumulated as its biologically active 5′-triphosphate analogue [168].

5′-Monophosphorylation is the rate-limiting step in the transformation to the corresponding 5′-triphosphates of most nucleosides; therefore, the involved kinases constitute a powerful toolbox for the enzymatic synthesis of non-natural NMPs. In addition, other enzymes with phosphorylating activity such as phosphotransferases, phosphoribosyltransferases, and acid phosphatases have been explored as biocatalysts for NMP synthesis. A review dealing with nucleoside prodrugs covers the biocatalyzed synthesis of NMPs and can be also consulted for this purpose [169]. Regarding whole cells methodologies, different strategies have been proposed for the synthesis of NMPs, which are summarized in Scheme 3.7.

One of the pioneer papers that dealt with the whole cells preparation of NMPs was published by Zinchenko et al. [170]. They used the nucleoside phosphotransferase (NPT, EC 2.7.1.77) activity present in *Pantoea agglomerans* 47/3 (previously *Erwinia herbicola* 47/3) whole cells and *p*-nitrophenylphosphate as the phosphate source for the synthesis of 5′-monophosphates of ribonucleosides, 2′-deoxyribonucleosides, 3′-fluoro-3′-deoxyribonucleosides and 3′-fluoro-2′,3′-dideoxyribonucleosides.

The same methodology was applied to the phosphorylation of 2′,3′-dideoxyuridine and 2′,3′-dideocytidine, observing also the selective formation of 5′-monophosphorylated analogues, although in lower yields. The pH range of activity, cellular localization, and phosphohydrolase activity of this enzyme probably suggest a wrong classification and it may actually belong to a broader group of enzymes with phosphatase/phosphotransferase activity such as acid phosphatases (Class B) family.

The use of whole cells carrying NSAPs was first assessed by Asano et al. [171]. They studied the production of IMP using several microorganisms from their

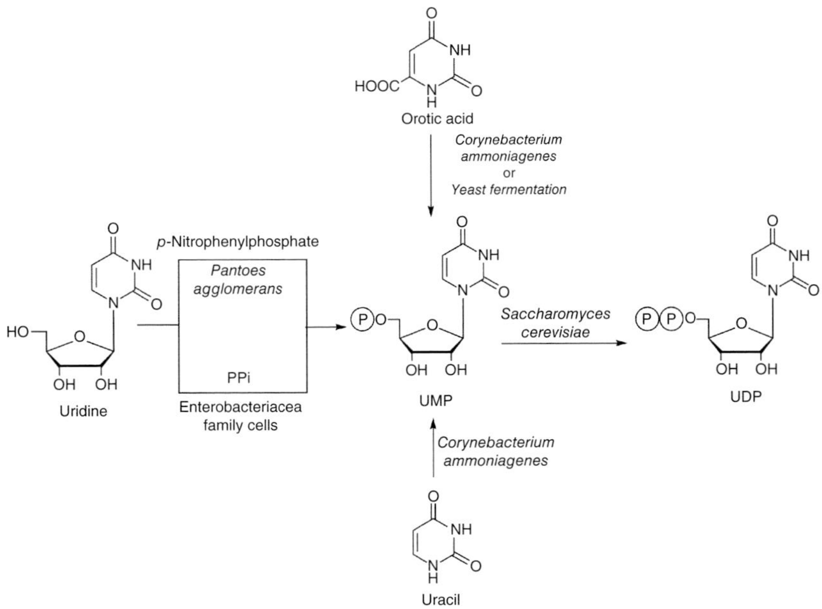

Scheme 3.7 Microbial strategies for UMP and UDP synthesis.

culture collection and PPi as the phosphate donor. Phosphotransferase activity specific to the 5′-position was detected among the bacteria belonging to the Enterobacteriacea family, *Morganella morganii* NCIMB10466 being selected for further studies that showed that this strain was able to produce 6.02 mg ml^{-1} (11.4 mM) of IMP under chosen conditions.

Wang et al. successfully developed a two-step biocatalytic process for uridine 5′-monophosphate (UMP) production from orotic acid using *Corynebacterium ammoniagenes* ATCC 6872 [172]. In this synthetic strategy, 5-phosphoribosyl-1-pyrophosphate (PRPP) formation was the rate-limiting step. Accumulation of the final product afforded 38.5 mM of UMP in 24 hours.

Since this approach employed large amounts of cells (40 g dry cell weight l^{-1}), an alternative route using *Saccharomyces cerevisiae* as biocatalyst was developed by Chen et al. [173]. UMP was synthesized by means of yeast-catalyzed fermentation or biotransformation, but this last methodology was chosen for large-scale UMP production in view of its accessibility. The enzymes involved in this process are orotatephosphoribosyltransferase (OPRTase, EC 2.4.2.10) and orotidine 5′-monophosphate decarboxylase (ODCase, EC 4.1.1.23).

We have also informed the phosphotransferase activity of NSAPs in the phosphorylation of nucleosides, using wild type and recombinant strains [87]. A screening of 26 bacteria belonging to the genera *Citrobacter*, *Escherichia*, *Erwinia*, *Proteus*, *Klebsiella*, and *Serratia* was carried out using PPi as the phosphate source for the 5′-regioselective phosphorylation of different nucleosides. *R. planticola* and *En. aerogenes* strains exhibited the best performances and were further studied for the synthesis of natural and modified NMPs. In general terms, *En. aerogenes* was more effective in the phosphorylation of purine nucleosides, whereas *R. planticola* cells showed higher activity on 2′-deoxynucleosides and arabinosides. NSAPs from *En. aerogenes* and *R. planticola* were later heterologously expressed in acid phosphatase deficient *E. coli* BL21 cells and further employed as whole cell biocatalysts. In most of the cases, the use of these recombinant strains afforded increased yields and significant reductions in the reaction times (from days to no more than 2.5 hours). The observed higher yields could be the result of the circumvention of competitive side reactions that occurred inside the cells. In particular, 2.1 mM of FaraAMP was obtained in 50 minutes.

Within the group of NMPs, cyclic adenosine-3′,5′-monophosphate (cAMP) is of high importance due to its biological relevance. It acts as second messenger in cellular signal transduction and takes part in many biological activities in both prokaryotes and eukaryotes [174] and has several applications in human clinics. The first report that addressed the production of cAMP [228] made use of the adenylate cyclase activity (EC 4.6.1.1). This enzyme catalyzes the synthesis of cAMP from ATP. Adenylate cyclase from *Arthrobacte*r was over-expressed in *E. coli* Rosetta and the recombinant *E. coli* was further permeabilized using Triton-X 100 and directly employed as whole cell catalyst for the mentioned biotransformation, showing a productivity of 2.49 g h^{-1}.

3.6 Other Nucleoside Derivatives

3.6.1 NDP

Classical strategies employed for the preparation of nucleoside 5′-diphosphates (NDPs) are cumbersome; therefore, we [82] explored multistep enzymatic systems starting from readily available reagents. We assessed different routes to prepare uridine- and thymidine-5′-diphosphates as ribo- and deoxyribonucleoside models, respectively. Using a simple methodology that made use of commercial yeast and biocatalytically *in situ* prepared UMP, uridine diphosphate (UDP) was synthesized with conversions ranging from 38% to 66%. *C. ammoniagenes* ATCC 19350 was employed for the first time as the biocatalyst to synthesize UMP using uracil and orotic acid as starting materials, while *R. planticola* was selected as the biocatalyst for UMP preparation from uridine. Although the overall performances of all the assessed approaches were similar, the use of uracil led to a more appropriate and cheaper process. For thymidine monophosphate (TDP) synthesis, two consecutive one-pot multistep enzyme systems were assayed. In one, DR5P was formed by the action of *Er. carotovora* whole cells starting from glucose, as mentioned above; subsequent action of PPM and TP afforded thymidine. Finally, in the second one-pot reaction, thymidine was converted to the corresponding 5′-diphosphate by the combined action of *E. coli* BL21 pET22b-phoRp and *S. cerevisiae*.

3.6.2 NDP-sugar

NDP analogues of glucose, galactose, and mannose, among other sugars, are the naturally activated substrates of nucleotide sugar-dependent (“Leloir”)

GTs in cellular biosynthetic processes [175]. This bioconversion is involved in the synthesis of many glycocompounds, such as oligo- and polysaccharides, glycosylated natural products, glycolipids, and glycoproteins [176]. Since GTs efficiently glycosylate unprotected substrates, they are important tools for the production of fine chemical glycosides, which possess relevant characteristics with respect to the corresponding aglycones.

Uridine diphosphate glucose (UDP-glucose) is the most employed glucosyltransferases substrate and also the precursor of other UDP-sugars, like UDP-galactose, UDP-xylose, and UDP-glucuronic acid (UDP-GA). Therefore, the synthesis of UDP-glucose has been the subject of several researches, most of them dealing with whole cells approaches since the regeneration of several cofactors is mandatory. For extending information related to the synthesis of UDP-sugar analogues, and also to the biotechnological glycosylation of small molecules, the review reported by De Bruyn et al. should be consulted [177]. Several methods that were developed so far for the preparation of UDP-sugar analogues made use of synthases, sucrose and trehalose synthases being the most studied biocatalysts. This is an attractive strategy since the reversible reactions catalyzed by these enzymes generate UDP-sugars starting from a disaccharide.

With the aim of producing NDP-sugars, Ryu and Lee [178] explored the use of trehalose synthase (EC 2.4.1.245) by constructing recombinant *E. coli* MC1061 cells harboring this enzyme. Using trehalose as a glucosyl donor and NDPs such as UDP, adenosine diphosphate (ADP), and GDP, this whole cell biocatalyst was capable of producing the corresponding NDP-glucoses in the extracellular medium. This system produced ADP-glucose from ADP in 60% yield after 12 hours. The recombinant cells also produced GDP and UDP in yields of 82% and 27%, respectively. In addition, the authors assessed the performance of a whole cell system having a UDP-galactose-4-epimerase (pGALE) for the synthesis of UDP-galactose, reporting a yield of 26% based on the employed amount of UDP-glucose. Sucrose synthase (SuSy, EC 2.4.1.13), has received much more attention by different research groups (Scheme 3.8). Schmölzer et al. addressed the study of UDP-glucose production catalyzed by recombinant SuSy from *Acidithiobacillus caldu*, using an integrated whole cell process [179], providing excellent results: UDP-glucose could be isolated in a single batch with 90% purity and in 73% isolated yield. Gutmann and Nidetzky [180] also proposed the use of SuSy obtaining UDP-glucose in 85% yield from UDP. The strategy proposed by De Bruyn et al. [181] using *E. coli* W involved the introduction of a different sucrose metabolism using sucrose phosphorylase (EC 2.4.1.7) for the direct and effective synthesis of glucose 1-phosphate (G1P) as the precursor for UDP-glucose generation and fructose, which can be further used as the carbon source for growth. By targeted gene deletions, a split metabolism is created, resulting in the reduced or eliminated use of G1P for growth, saving it for UDP-glucose formation. A multistep synthesis of UDP-glucose was also reported [182] that makes use of recombinant *E. coli* cells containing over-expressed sucrose phosphorylase of *Leuconostoc mesenteroides*. Further treatment of the cells with surfactants, generated "enzyme bags" that can be used for multienzyme synthesis under defined conditions [183]. Using this methodology, the direct synthesis of UDP-glucose using UMP as the starting

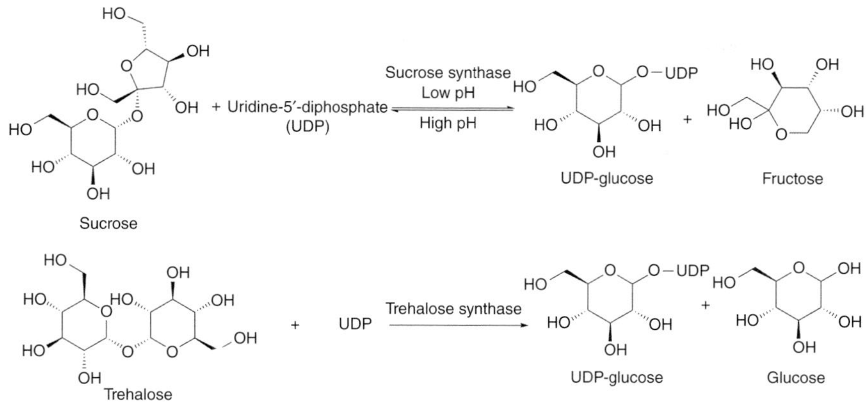

Scheme 3.8 Synthesis of UDP-glucose catalyzed by sugar synthases.

material with simultaneous regeneration of ATP and G1P was carried out. Employing optimized conditions, UDP-glucose yield with respect to UMP rose to 60%.

UDP-GA is a donor of glucuronic acid in detoxification processes and is also an intermediate in polysaccharide biosynthesis [184]. In addition, the formation of glucuronide metabolites, with increased drug activity, is a potential toxicological problem for patients [185]. The synthesis of glucuronated compounds is particularly attractive because it requires highly regiospecific oxidation reactions. In this line, a pioneer study [186], developed a successful whole cell biotransformation using recombinant human glucuronosyltransferase (UGTs) expressed in the fission yeast *Schizosaccharomyces pombe* for the synthesis of glucuronosides without the need of the cofactor UDP-GA. Using also permeabilized cells of *S. pombe* that expressed human uridine diphosphate-glucose 6-dehydrogenase (UGDH), Weyler et al. explored the selective oxidation of UDP-glucose to UDP-GA [187]. UDP-glucose and NAD^+ were used as substrates, obtaining quantitative conversion to UDP-GA within three hours without observable formation of by-products.

GDP-L-fucose is a key starting material for the fucosylation of relevant metabolites like 2′-*O*-fucosyllactose (2-FL), present in human milk, which provides protection against enteric pathogen infections. Lee et al. [188] carried out the combinatorial over-expression of the genes involved in the biosynthesis of GDP-D-mannose, a precursor of GDP-L-fucose, observing a 4.4-fold enhancement in the final GDP-L-fucose concentration when using a glucose-limited fed-batch fermentation of a recombinant *E. coli* BL21star(DE3) strain, over-expressing *manB, manC, gmd,* and *wcaG* genes, as compared to the control strain that over-expressed only *gmd* and *wcaG* genes. Since the biosynthesis of GDP-L-fucose requires NADPH as a reducing cofactor, Lee et al. [189] also studied the influence of the over-expression in recombinant *E. coli* of endogenous NADPH regenerating enzymes such as glucose-6-phosphate dehydrogenase (G6PDH), isocitrate dehydrogenase (Icd), and NADP(+)-dependent malate dehydrogenase (MaeB) on GDP-L-fucose production. In batch fermentations the over-expression of G6PDH afforded the highest GDP-L-fucose production. Finally, taking into account that guanosine 5′-triphosphate (GTP) is an essential substrate for the biosynthesis of GDP-L-fucose, Lee et al. [190] explored the effect of the over-expression of the enzymes involved in the biosynthetic pathway for guanosine nucleotides. IMP-dehydrogenase, GMP-synthetase (GuaB and GuaA), GMP-reductase (GuaC), and guanosine–inosine kinase (Gsk) were over-expressed in a recombinant *E. coli* producing GDP-L-fucose. Among the evaluated-enzymes, over-expression of Gsk led to a 58% enhancement in GDP-L-fucose production compared with the control strain.

Another NMP analogue that has been synthesized via whole cells biocatalysis is cytidine 5-monophosphate *N*-acetylneuraminic acid (CMP-NeuAc). Unlike the other sugar nucleotides, CMP-NeuAc can be synthesized without the intervention of a sugar-1-P intermediate by *N*-acylneuraminatecytidylyltransferase (CMP-NeuAc synthetase, EC 2.7.7.43) [191]. CMP-NeuAc is the essential, and very expensive, substrate of sialyltransferases, enzymes involved in the sialylation of many oligosaccharides and glycoconjugates. Sialic acids are a complex group of nine-carbon monosaccharides that are present at the end of many glycans and secreted proteins. They are involved in the modulation of several biological processes, being a determinant aspect in the design of glycoprotein therapeutics.

For the preparation of CMP-NeuAc, Lee et al. [192] developed a whole cell biocatalyst based on an *E. coli* strain that expressed three recombinant enzymes: CMP-kinase, NeuAc-aldolase, and CMP-NeuAc synthetase (Scheme 3.9). This system catalyzed the production of CMP-NeuAc starting from CMP, *N*-acetylmannosamine, pyruvate, ATP, and acetylphosphate, in 90% conversion based on the initial CMP concentration. Endo et al. [193] designed a whole cell reaction combining the activity of *E. coli* and *C. ammoniagenes* recombinant strains. They used two recombinant *E. coli* that over-expressed the genes of cytidine triphosphate (CTP)-synthetase and CMP-Neu5Ac synthetase. *C. ammoniagenes* catalyzed the transformation of orotic acid to uridine triphosphate (UTP). Applying this procedure, CMP-NeuAc was accumulated at 27 mM

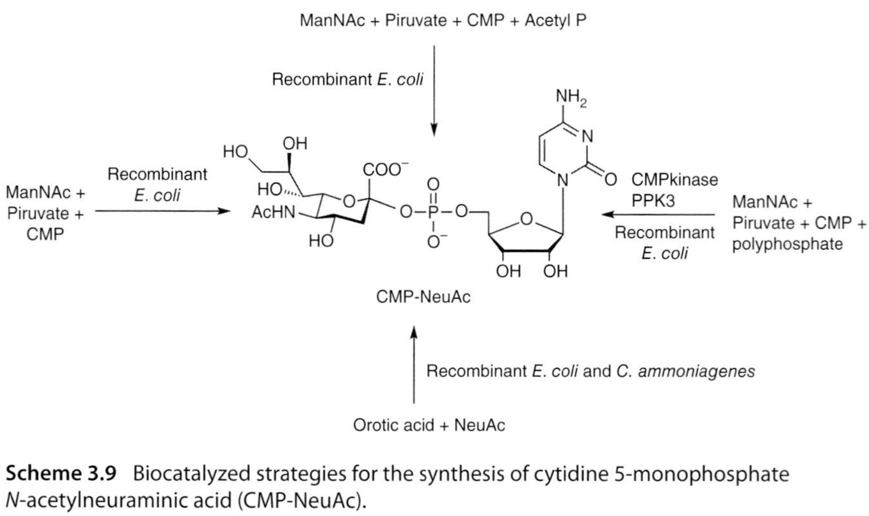

Scheme 3.9 Biocatalyzed strategies for the synthesis of cytidine 5-monophosphate *N*-acetylneuraminic acid (CMP-NeuAc).

($17\,g\,l^{-1}$) after 27 hours, the reaction using orotic acid and *N*-acetylneuraminic acid as starting materials. Song et al. [194] also reported a large-scale production system of CMP-NeuAc but by employing a single bacterial strain. They assembled the genes of two enzymes: NeuAc-aldolase and CMP-NeuAc synthetase to generate an artificial gene cluster in a single plasmid. To the same whole cell biocatalyst developed by Song et al., Nahálka and Pätoprstý coupled the activity of a novel polyphosphate kinase (PPK3), allowing the synthesis of CMP-NeuAc and 3′-sialyllactose using cheap polyphosphates [195]. Antoine et al. generated CMP-NeuAc from exogenous sialic acid, which was transported into the cells by the permease NanT [196]. Considering that sialic acid is expensive the same research group [197] used a genetically engineered *E. coli* K12 capable of generating CMP-NeuAc employing its own internal metabolism.

3.7 Perspectives

The collected work displayed above shows that whole cell biocatalysts are very useful and efficient tools for the synthesis of a varied range of NAs.

Although the results listed herein show a great advance in the biocatalyzed synthesis of nucleoside, it is expected that future progresses in the fields of protein engineering and nucleoside metabolism, among other disciplines, will enhance the already vast number of contributions provided by whole cell biocatalysis to nucleoside chemistry.

References

1 De Clercq, E. (2009). The history of antiretrovirals: key discoveries over the past 25 years. *Rev. Med. Virol.* 19 (5): 287–299.
2 Ichikawa, E. and Kato, K. (2001). Sugar-modified nucleosides in past 10 years, a review. *Curr. Med. Chem.* 8 (4): 385–423.
3 De Clercq, E. and Li, G. (2016). Approved antiviral drugs over the past 50 years. *Clin. Microbiol. Rev.* 29 (3): 695–747.
4 Galmarini, C.M., Mackey, J.R., and Dumontet, C. (2002). Nucleoside analogues and nucleobases in cancer treatment. *Lancet Oncol.* 3 (7): 415–424.
5 Tikk, M., Tikk, K., Tørngren, M.A. et al. (2006). Development of inosine monophosphate and its degradation products during aging of pork of different qualities in relation to basic taste and retronasal flavor perception of the meat. *J. Agric. Food. Chem.* 54 (20): 7769–7777.
6 Young, D.C., Layre, E., Pan, S.J. et al. (2015). *In vivo* biosynthesis of terpene nucleosides provides unique chemical markers of *Mycobacterium tuberculosis* infection. *Chem. Biol.* 22 (4): 516–526.
7 Naresh, T., Kumar, T.P., Haribabu, K., and Chandrasekhar, S. (2014). AZT-prolinamide: the nucleoside derived pyrrolidine catalysts for asymmetric aldol reactions using water as solvent. *Tetrahedron: Asymmetry* 25 (18–19): 1340–1345.
8 Mandala, D., Thompson, W.A., and Watts, P. (2016). Synthesis routes to anti-HIV drugs. *Tetrahedron* 72 (24): 3389–3420.

9 Lewkowicz, E.S. and Iribarren, A.M. (2006). Nucleoside phosphorylases. *Curr. Org. Chem.* 10 (11): 1197–1215.
10 Li, N., Smith, T.J., and Zong, M.H. (2010). Biocatalytic transformation of nucleoside analogues. *Biotechnol. Adv.* 28 (3): 348–366.
11 Mikhailopulo, I.A. (2007). Biotechnology of nucleic acid constituents – state of the art and perspectives. *Curr. Org. Chem.* 11 (4): 317–335.
12 Perkins, C., Siddique, S., Puri, M., and Demain, A.L. (2016). Biotechnological applications of microbial bioconversions. *Crit. Rev. Biotechnol.* 36 (6): 1050–1065.
13 Boryski, J. (2008). Reactions of transglycosylation in the nucleoside chemistry. *Curr. Org. Chem.* 12 (4): 309–325.
14 Demain, A.L. (2000). Microbial biotechnology. *Trends Biotechnol.* 18 (1): 26–31.
15 Mikhailopulo, I.A. and Miroshnikov, A.I. (2011). Biologically important nucleosides: modern trends in biotechnology and application. *Mendeleev Commun.* 21 (2): 57–68.
16 Fernández-Lucas, J. (2015). Multienzymatic synthesis of nucleic acid analogues: a general perspective. *Appl. Microbiol. Biotechnol.* 99 (11): 4615–4627.
17 Fresco-Taboada, A., de la Mata, I., Arroyo, M., and Fernández-Lucas, J. (2013). New insights on nucleoside 2′-deoxyribosyltransferases: a versatile biocatalyst for one-pot one-step synthesis of nucleoside analogs. *Appl. Microbiol. Biotechnol.* 97 (9): 3773–3785.
18 Lewkowicz, E.S. and Iribarren, A.M. (2017). Whole cell biocatalysts for the preparation of nucleosides and their derivatives. *Curr. Pharm. Des.* doi: https://doi.org/10.2174/1381612823666171011101133.
19 Spoldi, E., Ghisotti, D., Cali, S. et al. (2001). Recombinant bacterial cells as efficient biocatalysts for the production of nucleosides. *Nucleosides Nucleotides Nucleic Acids* 20 (4–7): 977–979.
20 Lapponi, M.J., Rivero, C.W., Zinni, M.A. et al. (2016). New developments in nucleoside analogues biosynthesis: a review. *J. Mol. Catal. B: Enzym.* 133: 218–233.
21 Mikhailopulo, I.A. and Miroshnikov, A.I. (2017). New trends in nucleoside biotechnology. *Acta Naturae* 2 (5): 36–58.
22 Gordon, G.E.R., Visser, D.F., Brady, D. et al. (2011). Defining a process operating window for the synthesis of 5-methyluridine by transglycosylation of guanosine and thymine. *J. Biotechnol.* 151 (1): 108–113.
23 Wu, J.Z., Larson, G., Walker, H. et al. (2005). Phosphorylation of ribavirin and viramidine by adenosine kinase and cytosolic 5′-nucleotidase II: implications for ribavirin metabolism in erythrocytes. *Antimicrob. Agents Chemother.* 49 (6): 2164–2171.
24 Kamiyama, N., Soma, R., Hidano, S. et al. (2017). Ribavirin inhibits Zika virus (ZIKV) replication *in vitro* and suppresses viremia in ZIKV-infected STAT1-deficient mice. *Antiviral Res.* 146: 1–11.

25 Pettersson, F., Yau, C., Dobocan, M.C. et al. (2011). Ribavirin treatment effects on breast cancers overexpressing eIF4E, a biomarker with prognostic specificity for luminal B-type breast cancer. *Clin. Cancer Res.* 17 (9): 2874–2884.
26 Nóbile, M., Terreni, M., Iribarren, A.M., and Lewkowicz, E.S. (2010). *Aeromonas hydrophila* strains as biocatalysts in the transglycosylation reaction. *Biocatal. Biotransform.* 28 (05–06): 395–402.
27 Rivero, C.W., De Benedetti, E.C., Lozano, M.E., and Trelles, J.A. (2015). Bioproduction of ribavirin by green microbial biotransformation. *Process Biochem.* 50 (6): 935–940.
28 De Benedetti, E.C., Rivero, C.W., Britos, C.N. et al. (2012). Biotransformation of 2,6-diaminopurine nucleosides by immobilized *Geobacillus stearothermophilus*. *Biotechnol. Progr.* 28 (5): 1251–1256.
29 Del Arco, J., Martinez, M., Donday, M. et al. (2017). Cloning, expression and biochemical characterization of xanthine and adenine phosphoribosyltransferases from *Thermus thermophilus* HB8. *Biocatal. Biotransform.* doi: https://doi.org/10.1080/10242422.2017.1313837.
30 Almendros, M., Sinisterra, J.V., and Berenguer, J. (2009). *Thermus thermophilus* strains active in purine nucleoside synthesis. *Molecules* 14 (3): 1279–1287.
31 Almendros, M., Berenguer, J., and Sinisterra, J.V. (2012). *Thermus thermophilus* nucleoside phosphorylases active in the synthesis of nucleoside analogues. *Appl. Environ. Microbiol.* 78 (9): 3128–3135.
32 Tomoike, F., Kuramitsu, S., and Masui, R. (2013). Unique substrate specificity of purine nucleoside phosphorylases from *Thermus thermophilus*. *Extremophiles* 17 (3): 505–514.
33 Rivero, C.W., De Benedetti, E.C., Sambeth, J.E. et al. (2012). Biosynthesis of anti-HCV compounds using thermophilic microorganisms. *Bioorg. Med. Chem. Lett.* 22 (19): 6059–6062.
34 Médici, R., Lewkowicz, E.S., and Iribarren, A.M. (2006). Microbial synthesis of 2,6-diaminopurine nucleosides. *J. Mol. Catal. B: Enzym.* 39 (1–4): 40–44.
35 Miwa, N., Kurosaki, K., Yoshida, Y. et al. (2005). Comparative efficacy of acyclovir and vidarabine on the replication of varicella-zoster virus. *Antiviral Res.* 65 (1): 49–55.
36 Robak, P. and Robak, T. (2013). Older and new purine nucleoside analogs for patients with acute leukemias. *Cancer Treat. Rev.* 39 (8): 851–861.
37 Feng, K. and Cui, L. (2012). Biosynthesis of adenine arabinoside by whole cells of *Enterobacter aerogenes* HXY2222. *Adv. Mater. Res.* 396, 1417–398, 1421.
38 Nóbile, M., Médici, R., Terreni, M. et al. (2012). Use of *Citrobacter koseri* whole cells for the production of arabinonucleosides, a larger scale approach. *Process Biochem.* 47 (12): 2182–2188.
39 Ding, Q.B., Ou, L., Wei, D.Z. et al. (2010). Enzymatic synthesis of nucleosides by nucleoside phosphorylase co-expressed in *Escherichia coli*. *J. Zhejiang Univ. Sci. B* 11 (11): 880–888.
40 Taran, S.A., Verevkina, K.N., Esikova, T.Z. et al. (2008). Synthesis of 2-chloro-2′-deoxyadenosine by microbiological transglycosylation using

a recombinant *Escherichia coli* strain. *Appl. Biochem. Microbiol.* 44 (2): 162–166.

41 Liang, S., Li, W., Gao, T. et al. (2010). Enzymatic synthesis of 2′-deoxyadenosine and 6-methylpurine-2′-deoxyriboside by *Escherichia coli* DH5α overexpressing nucleoside phosphorylases from *Escherichia coli* BL21. *J. Biosci. Bioeng.* 110 (2): 165–168.

42 Ge, C.T., OuYang, L.M., Ding, Q.B., and Ou, L. (2009). Co-expression of recombinant nucleoside phosphorylase from *Escherichia coli* and its application. *Appl. Biochem. Biotechnol.* 159 (1): 168–177.

43 Yokozeki, K. and Tsuji, T.J. (2000). A novel enzymatic method for the production of purine-2′-deoxyribonucleosides. *J. Mol. Catal. B: Enzym.* 10 (1–3): 207–213.

44 Médici, R., Iribarren, A.M., and Lewkowicz, E.S. (2009). Synthesis of 9-β-D-arabinofuranosylguanine by combined use of two whole cell biocatalysts. *Bioorg. Med. Chem. Lett.* 19 (15): 4210–4212.

45 Médici, R., Lewkowicz, E.S., and Iribarren, A.M. (2008). *Arthrobacter oxydans* as a biocatalyst for purine deamination. *FEMS Microbiol. Lett.* 289 (1): 20–26.

46 Rosenbohm, C., Pedersen, D.S., Frieden, M. et al. (2004). LNA guanine and 2,6-diaminopurine. Synthesis, characterization and hybridization properties of LNA 2,6-diaminopurine containing oligonucleotides. *Bioorg. Med. Chem.* 12 (9): 2385–2396.

47 Herdewijn, P. (2004). Heterocyclic modifications of oligonucleotides and antisense technology. *Antisense Nucleic Acid Drug Dev.* 10 (4): 297–310.

48 Mahmoudian, M., Eaddy, J., and Dawson, M. (1999). Enzymatic acylation of 506U78 (2-amino-9-β-D-arabinofuranosyl-6-methoxy-9*H*-purine), a powerful new anti-leukaemic agent. *Biotechnol. Appl. Biochem.* 29 (3): 229–233.

49 Gandhi, V., Keating, M.J., Bate, G., and Kirkpatrick, P. (2006). Nelarabine. *Nat. Rev. Drug Discovery* 5 (1): 17–18.

50 Konstantinova, I.D., Antonov, K.V., Fateev, I.V. et al. (2011). A chemo-enzymatic synthesis of β-D-arabinofuranosyl purine nucleosides. *Synthesis* 2011 (10): 1555–1560.

51 Hamon, N., Quintiliani, M., Balzarini, J., and McGuigan, C. (2013). Synthesis and biological evaluation of prodrugs of 2-fluoro-2-deoxyribose-1-phosphate and 2,2-difluoro-2-deoxyribose-1-phosphate. *Bioorg. Med. Chem. Lett.* 23 (9): 2555–2559.

52 MacDonald, D.L. and Fletcher, H.G. (1962). 2-Deoxy-D-ribose. VIII. Synthesis of the anomeric 2-deoxy-D-ribofuranose 1-phosphates. *J. Am. Chem. Soc.* 84 (7): 1262–1265.

53 Taverna-Porro, M., Bouvier, L.A., Pereira, C.A. et al. (2008). Chemoenzymatic preparation of nucleosides from furanoses. *Tetrahedron Lett.* 49 (16): 2642–2645.

54 Chuvikovsky, D.V., Esipov, R.S., Skoblov, Y.S. et al. (2006). Ribokinase from *E. coli*: expression, purification, and substrate specificity. *Bioorg. Med. Chem.* 14 (18): 6327–6332.

55 Miroshnikov, A.I., Esipov, R.S., Muravyova, T.I. et al. (2010). A new strategy for the synthesis of nucleosides: one-pot enzymatic transformation of D-pentoses into nucleosides. *Open Conf. Proc. J.* 1: 98–102.

56 van Herk, T., Hartog, A.F., van der Burg, A.M., and Wever, R. (2005). Regioselective phosphorylation of carbohydrates and various alcohols by bacterial acid phosphatases, probing the substrate specificity of the enzyme from *Shigella flexneri*. *Adv. Synth. Catal.* 347 (7–8): 1155–1162.

57 Médici, R., Garaycoechea, J.I., Valino, A.L. et al. (2014). A comparative study on phosphotransferase activity of acid phosphatases from *Raoultella planticola* and *Enterobacter aerogenes* on nucleosides, sugars, and related compounds. *Appl. Microbiol. Biotechnol.* 98 (7): 3013–3022.

58 Ogawa, J., Saito, K., Sakai, T. et al. (2003). Microbial production of 2-deoxyribose 5-phosphate from acetaldehyde and triosephosphate for the synthesis of 2′-deoxyribonucleosides. *Biosci. Biotechnol., Biochem.* 67 (4): 933–936.

59 Brovetto, M., Gamenara, D., Saenz Méndez, P., and Seoane, G.A. (2011). C—C bond-forming lyases in organic synthesis. *Chem. Rev.* 111 (7): 4346–4403.

60 Barbas, C.F., Wang, Y.F., and Wong, C.H. (1990). Deoxyribose-5-phosphate aldolase as a synthetic catalyst. *J. Am. Chem. Soc.* 112 (5): 2013–2014.

61 Jennewein, S., Schürmann, M., Wolberg, M. et al. (2006). Directed evolution of an industrial biocatalyst: 2-deoxy-D-ribose 5-phosphate aldolase. *Biotechnol. J.* 1 (5): 537–548.

62 Kim, Y.M., Choi, N.S., Kim, Y.O. et al. (2010). Expression and characterization of a novel deoxyribose 5-phosphate aldolase from *Paenibacillus* sp. EA001. *J. Microbiol. Biotechnol.* 20 (6): 995–1000.

63 Kim, Y.M., Chang, Y.H., Choi, N.S. et al. (2009). Cloning, expression, and characterization of new deoxyribose 5-phosphate aldolase from *Yersinia* sp. EA015. *Protein Expression Purif.* 68 (2): 196–200.

64 Sakuraba, H., Yoneda, K., Yoshihara, K. et al. (2007). Sequential aldol condensation catalyzed by hyperthermophilic 2-deoxy-D-ribose-5-phosphate aldolase. *Appl. Environ. Microbiol.* 73 (22): 7427–7434.

65 Lokanath, N.K., Shiromizu, I., Ohshima, N. et al. (2004). Structure of aldolase from *Thermus thermophilus* HB8 showing the contribution of oligomeric state to thermostability. *Acta Crystallogr., Sect. D: Struct. Biol.* 60: 1816–1823.

66 Han, T.K., Zhu, Z., and Dao, M.L. (2004). Identification, molecular cloning, and sequence analysis of a deoxyribose aldolase in *Streptococcus mutans* GS-5. *Curr. Microbiol.* 48 (3): 230–236.

67 Woo, M.H., Kim, M.S., Chung, N., and Kim, J.S. (2014). Expression and characterization of a novel 2-deoxyribose-5-phosphate aldolase from *Haemophilus influenzae* Rd KW20. *J. Korean Soc. Appl. Biol. Chem.* 57 (5): 655–660.

68 Cao, T.P., Kim, J.S., Woo, M.H. et al. (2016). Structural insight for substrate tolerance to 2-deoxyribose-5-phosphate aldolase from the pathogen *Streptococcus suis*. *J. Microbiol.* 54 (4): 311–321.

69 Dick, M., Weiergräber, O.H., Classen, T. et al. (2016). Trading off stability against activity in extremophilic aldolases. *Sci. Rep.* 6: 17908.

70 Franke, D., Hsu, C.C., and Wong, C.H. (2004). Directed evolution of aldolases. *Methods Enzymol.* 388: 224–238.

71 Dick, M., Hartmann, R., Weiergraber, O.H. et al. (2016). Mechanism-based inhibition of an aldolase at high concentrations of its natural substrate acetaldehyde: structural insights and protective strategies. *Chem. Sci.* 7: 4492–4502.

72 Li, J., Yang, J., Men, Y. et al. (2015). Biosynthesis of 2-deoxysugars using whole-cell catalyst expressing 2-deoxy-D-ribose 5-phosphate aldolase. *Appl. Microbiol. Biotechnol.* 99 (19): 7963–7972.

73 De Santis, G., Liu, J., Clark, D.P. et al. (2003). Structure-based mutagenesis approaches toward expanding the substrate specificity of D-2-deoxyribose-5-phosphate aldolase. *Bioorg. Med. Chem.* 11 (1): 43–52.

74 Oslaj, M., Cluzeau, J., Orkic, D. et al. (2013). A highly productive, whole-cell DERA chemoenzymatic process for production of key lactonized side-chain intermediates in statin synthesis. *PLoS One* 8 (5): 62250.

75 Horinouchi, N., Ogawa, J., Kawano, T. et al. (2006). Biochemical retrosynthesis of 2′-deoxyribonucleosides from glucose, acetaldehyde, and a nucleobase. *Appl. Microbiol. Biotechnol.* 71: 615–662.

76 Horinouchi, N., Ogawa, J., Sakai, T. et al. (2003). Construction of deoxyriboaldolase-overexpressing *Escherichia coli* and its application to 2-deoxyribose 5-phosphate synthesis from glucose and acetaldehyde for 2′-deoxyribonucleoside production. *Appl. Environ. Microbiol.* 69 (7): 3791–3797.

77 Horinouchi, N., Ogawa, J., Kawano, T. et al. (2006). Efficient production of 2-deoxyribose 5-phosphate from glucose and acetaldehyde by coupling of the alcoholic fermentation system of Baker's yeast and deoxyriboaldolase-expressing *Escherichia coli*. *Biosci. Biotechnol., Biochem.* 70 (6): 1371–1378.

78 Horinouchi, N., Ogawa, J., Kawano, T. et al. (2006). One-pot microbial synthesis of 2′-deoxyribonucleoside from glucose, acetaldehyde, and a nucleobase. *Biotechnol. Lett.* 28 (12): 877–881.

79 Horinouchi, N., Sakai, T., Kawano, T. et al. (2012). Construction of microbial platform for an energy-requiring bioprocess: practical 2′-deoxyribonucleoside production involving a C–C coupling reaction with high energy substrates. *Microb. Cell Fact.* 11: 82.

80 Horinouchi, N., Kawano, T., Sakai, T. et al. (2009). Screening and characterization of a phosphopentomutase useful for enzymatic production of 2-deoxyribonucleoside. *J. Biotechnol.* 26 (1–2): 75–82.

81 Valino, A.L., Palazzolo, M.A., Iribarren, A.M., and Lewkowicz, E.S. (2012). Selection of a new whole cell biocatalyst for the synthesis of 2-deoxyribose 5-phosphate. *Appl. Biochem. Biotechnol.* 166 (2): 300–308.

82 Valino, A.L., Iribarren, A.M., and Lewkowicz, E.S. (2015). New biocatalysts for one pot multistep enzymatic synthesis of pyrimidine nucleoside diphosphates from readily available reagents. *J. Mol. Catal. B: Enzym.* 114: 58–64.

83 Horinouchi, N., Soon, C.L., Shimizu, S., and Ogawa, J. (2012). Oxidative pyrimidine metabolism in *Rhodococcus erythropolis* useful for valuable nucleoside synthesis, discovery of a novel amidohydrolase, ureidomalonase. *Biocatal. Agric. Biotechnol.* 1 (3): 264–266.

84 Lamberth, C. (2002). Nucleosides with a carbon bridge between sugar and nucleobase, the chemistry of 1′-homonucleosides and reversed nucleosides. A review. *Org. Prep. Proced. Int.* 34 (2): 149–167.

85 Mathé, C. and Gosselin, G. (2006). L-Nucleoside enantiomers as antivirals drugs: a mini-review. *Antiviral Res.* 71 (2–3): 276–281.

86 Bennett, C.F. and Swayze, E.E. (2010). RNA targeting therapeutics, molecular mechanisms of antisense oligonucleotides as a therapeutic platform. *Annu. Rev. Pharmacol. Toxicol.* 50: 259–293.

87 Dellafiore, M., Montserrat, J., and Iribarren, A. (2015). Exploring the use of a guanine-rich catalytic DNA for sulfoxide preparation. *PLoS One* 10 (6): e0129695.

88 Yun, M.K., Moon, H.R., Kim, H.O. et al. (2005). A highly efficient synthesis of unnatural L-sugars from D-ribose. *Tetrahedron Lett.* 46 (35): 5903–5905.

89 Dumon, C., Song, L., Bozonnet, S. et al. (2012). Progress and future prospects for pentose-specific biocatalysts in biorefining. *Process Biochem.* 47 (3): 346–357.

90 Hu, C., Li, L., Zheng, Y. et al. (2011). Perspectives of biotechnological production of L-ribose and its purification. *Appl. Microbiol. Biotechnol.* 92 (3): 449–455.

91 Yeom, S.J., Kim, N.H., Park, C.S., and Oh, D.K. (2009). L-Ribose production from L-arabinose by using purified L-arabinose isomerase and mannose-6-phosphate isomerase from *Geobacillus thermodenitrificans*. *Appl. Environ. Microbiol.* 75 (21): 6941–6943.

92 Ahmed, Z., Shimonishi, T., Bhuiyan, S.H. et al. (1999). Biochemical preparation of L-ribose and L-arabinose from ribitol, a new approach. *J. Biosci. Bioeng.* 88 (4): 444–448.

93 De Muynck, C., Pereira, C., Soetaert, W., and Vandamme, E. (2006). Dehydrogenation of ribitol with *Gluconobacter oxydans*, production and stability of L-ribulose. *J. Biotechnol.* 125 (3): 408–415.

94 Helanto, M., Kiviharju, K., Leisola, M., and Nyyssölä, A. (2007). Metabolic engineering of *Lactobacillus plantarum* for production of L-ribulose. *Appl. Environ. Microbiol.* 73 (21): 7083–7091.

95 Zhang, Y.W., Jeya, M., and Lee, J.K. (2010). L-Ribulose production by an *Escherichia coli* harboring L-arabinose isomerase from *Bacillus licheniformis*. *Appl. Microbiol. Biotechnol.* 87 (6): 1993–1999.

96 Prabhu, P., Jeya, M., and Lee, J.K. (2010). Probing the molecular determinant for the catalytic efficiency of L-arabinose isomerase from *Bacillus licheniformis*. *Appl. Environ. Microbiol.* 76 (5): 1653–1660.

97 Kagawa, K.I. and Okayama, K.T. (2001). L-Ribose isomerase, its preparation and uses. US Patent 6, 294,369 B.

98 Pyun, Y.R., Cho, E.A., Seo, M.J. et al. (2006). Thermophilic L-ribose isomerase and use thereof KR Patent 2006110647.

99 Izumori, K., Morimoto, K., Takata, G. et al. (2009). Thermostable L-ribose isomerase and methods for producing same and use of same. US Patent 2009325231.

100 Morimoto, K., Terami, Y., Maeda, Y. et al. (2013). Cloning and characterization of the L-ribose isomerase gene from *Cellulomonas parahominis* MB426. *J. Biosci. Bioeng.* 115 (4): 377–381.

101 Helanto, M., Kiviharju, K., Granström, T. et al. (2009). Biotechnological production of L-ribose from L-arabinose. *Appl. Microbiol. Biotechnol.* 83 (1): 77–83.

102 Hung, X., Yu, M., Chen, Y., and Fang, T. (2015). Characterization of a recombinant L-ribose isomerase from *Geodermatophilus Obscurus* DSM 43160 and application of this enzyme to the production of L-ribose from L-arabinose. *J. Mar. Sci. Tech. Taiwan* 23 (4): 558–566.

103 Cleasby, A., Wonacott, A., Skarzynski, T. et al. (1996). The X-ray crystal structure of phosphomannose isomerase from *Candida albicans* at 1.7 Å resolution. *Nat. Struct. Mol. Biol.* 3 (5): 470–479.

104 Yeom, S., Ji, J., Kim, N. et al. (2009). Substrate specificity of a mannose-6-phosphate isomerase from *Bacillus subtilis* and its application in the production of L-ribose. *Appl. Environ. Microbiol.* 75 (14): 4705–4710.

105 Xu, Z., Sha, Y., Liu, C. et al. (2016). L-Ribose isomerase and mannose-6-phosphate isomerase, properties and applications for L-ribose production. *Appl. Microbiol. Biotechnol.* 100 (21): 9003–9011.

106 Yeom, S., Seo, E., Kim, B. et al. (2011). Characterization of a mannose-6-phosphate isomerase from *Thermus thermophilus* and increased L-ribose production by its R142N mutant. *Appl. Environ. Microbiol.* 77 (3): 762–767.

107 Yeom, S., Kim, N., Yoon, R. et al. (2009). Characterization of a mannose-6-phosphate isomerase from *Geobacillus thermodenitrificans* that converts monosaccharides. *Biotechnol. Lett.* 31 (8): 1273–1278.

108 Kim, K., Seo, E., and Oh, D. (2014). L-Ribose production from L-arabinose by immobilized recombinant *Escherichia coli* co-expressing the L-arabinose isomerase and mannose-6-phosphate isomerase genes from *Geobacillus thermodenitrificans*. *Appl. Biochem. Biotechnol.* 172 (1): 275–288.

109 Lim, Y., Yeom, S., and Oh, D.K. (2012). Production of L-ribose from L-ribulose by a triple-site variant of mannose-6-phosphate isomerase from *Geobacillus thermodenitrificans*. *Appl. Environ. Microbiol.* 78 (11): 3880–3884.

110 Li, Z., Cai, L., Qi, Q. et al. (2011). Synthesis of rare sugars with L-fuculose-1-phosphate aldolase (FucA) from *Thermus thermophilus* HB8. *Bioorg. Med. Chem. Lett.* 21 (17): 5084–5087.

111 Li, Z., Cai, L., Qi, Q., and Wang, P.G. (2011). Enzymatic synthesis of D-sorbose and D-psicose with aldolase RhaD, effect of acceptor configuration on enzyme stereoselectivity. *Bioorg. Med. Chem. Lett.* 21 (23): 7081–7084.

112 Boutureira, O., Matheu, M.I., Diaz, Y., and Castillon, S. (2013). Advances in the enantioselective synthesis of carbocyclic nucleosides. *Chem. Soc. Rev.* 42 (12): 5056–5072.

113 Arai, Y., Hayashi, Y., Yamamoto, M. et al. (1987). A new route to (−)-aristeromycin and (−)-neplanocin A via the asymmetric Diels–Alder cycloaddition. *Chem. Lett.* 16 (1): 185–186.

114 Hervey, P.S. and Perry, C.M. (2000). Abacavir, a review of its clinical potential in patients with HIV infection. *Drugs* 60 (2): 447–479.

115 Staszewski, S., Keiser, P., Montaner, J. et al. (2001). Abacavir-lamivudine-zidovudine vs indinavir-lamivudine-zidovudine in antiretroviral-naive HIV-infected adults, a randomized equivalence trial. *JAMA* 285 (9): 1155–1163.

116 Taylor, S.J.C., McCague, R., Wisdom, R. et al. (1993). Development of the biocatalytic resolution of 2-azabicyclo[2.2.1]hept-5-en-3-one as an entry to single-enantiomer carbocyclic nucleosides. *Tetrahedron: Asymmetry* 4 (6): 1117–1128.

117 Singh, R. and Vince, R. (2012). 2-Azabicyclo [2.2.1] hept-5-en-3-one, chemical profile of a versatile synthetic building block and its impact on the development of therapeutics. *Chem. Rev.* 112 (8): 4642–4686.

118 Laborda, P., Wang, S.Y., and Voglmeir, J. (2016). Influenza neuraminidase inhibitors, synthetic approaches, derivatives and biological activity. *Molecules* 21 (11): 1513.

119 Connelly, S., Line, K., Isupov, M.N., and Littlechild, J.A. (2005). Synthesis and characterisation of a ligand that forms a stable tetrahedral intermediate in the active site of the *Aureobacterium species* (−) γ-lactamase. *Org. Biomol. Chem.* 3 (18): 3260–3262.

120 Taylor, S.J.C., Sutherland, A.G., Lee, C. et al. (1990). Chemoenzymatic synthesis of (−)-carbovir utilizing a whole cell catalysed resolution of 2-azabicyclo[2.2.1] hept-5-en-3-one. *J. Chem. Soc., Chem. Commun.* 16: 1120–1121.

121 Wang, J.J., Zhu, J.G., Min, C., and Wu, S. (2014). CBD binding domain fused γ-lactamase from *Sulfolobus solfataricus is* an efficient catalyst for (−) γ-lactam production. *BMC Biotech.* 14: 40–48.

122 Xue, T.Y., Xu, G.C., Han, R.Z., and Ni, Y. (2015). Soluble expression of (+)-γ-lactamase in *Bacillus subtilis* for the enantioselective preparation of abacavir precursor. *Appl. Biochem. Biotechnol.* 176 (6): 1687–1699.

123 Chen, H.G., Ni, Y., and Sun, Z.H. (2012). Isolation and fermentation optimization of a (+)-γ-lactamase-producing strain. *China Biotechnol.* 32: 41–47.

124 Qin, X.X., Wang, J.J., and Zheng, G.J. (2010). Enantioselective resolution of γ-lactam by a whole cell of *Microbacterium hydrocarbonoxydans* (L29-9) immobilized in polymer of PVA–alginate–boric acid. *Appl. Biochem. Biotechnol.* 162 (8): 2345–2354.

125 Wang, J., Zhang, X., Min, C. et al. (2011). Single-step purification and immobilization of γ-lactamase and on-column transformation of 2-azabicyclo [2.2.1] hept-5-en-3-one. *Process Biochem.* 46 (1): 81–87.

126 Line, K., Isupov, M.N., and Littlechild, J.A. (2004). The crystal structure of a (−) γ-lactamase from an *Aureobacterium* species reveals a tetrahedral intermediate in the active site. *J. Mol. Biol.* 338 (3): 519–532.

127 Holt-Tiffin, K.E. (2009). (+)- and (−)-2-Azabicyclo [2.2.1]hept-5-en-3-one, extremely useful synthons. *Chim. Oggi* 27 (4): 23–25.

128 Wang, J., Cheng, D., Wang, J., and Jin, L. (2016). Isolation of *Pseudomonas granadensis* B6 with (+)-γ-lactamase and whole cell resolution of racemic γ-lactam. *J. Mol. Catal. B: Enzym.* doi: http://doi.org/10.1016/j.molcatb.2016.12.004.

129 Wang, J.J., Zhu, Y.X., Zhao, G.G. et al. (2015). Characterization of a recombinant (+)-γ-lactamase from *Microbacterium hydrocarbonoxydans* which provides evidence that two enantiocomplementary γ-lactamases are in the strain. *Appl. Microbiol. Biotechnol.* 99 (7): 3069–3080.

130 Zhu, S., Gong, C., Song, D. et al. (2012). Discovery of a novel (+)-γ-lactamase from *Bradyrhizobium japonicum* USDA 6 by rational genome mining. *Appl. Environ. Microbiol.* 78 (20): 7492–7495.

131 Włodarczyk, J., Wolan, A., Rakowiecki, M. et al. (2015). Synthesis of an all-cis intermediate of ticagrelor. *Tetrahedron Lett.* 56 (44): 6093–6096.

132 Singh, M., Krishnen, H., Neelam, U.K. et al. (2016). Highly enantioselective bioreduction of 1-(3,4-difluorophenyl)-3-nitropropan-1-one, key intermediate of ticagrelor. *RSC Adv.* 6 (41): 35086–35090.

133 Stambasky, J., Hocek, M., and Kocovsky, P. (2009). *C*-nucleosides, synthetic strategies and biological applications. *Chem. Rev.* 109 (12): 6729–6764.

134 Wellington, K.W. and Benner, S.A. (2006). A review, synthesis of aryl *C*-glycosides via the heck coupling reaction. *Nucleosides Nucleotides Nucleic Acids* 25 (12): 1309–1333.

135 Maier, L., Khirsariya, P., Hylse, O. et al. (2017). Diastereoselective flexible synthesis of carbocyclic *C*-nucleosides. *J. Organomet. Chem.* 82 (7): 3382–33402.

136 Rudroff, F., Bianchi, D.A., Moran-Ramallal, R. et al. (2016). Synthesis of tetrahydrofuran-based natural products and their carba analogs via stereoselective enzyme mediated Baeyer–Villiger oxidation. *Tetrahedron* 72 (46): 7212–7221.

137 Mihovilovic, M.D., Rudroff, F., Grotzl, B. et al. (2005). Family clustering of Baeyer–Villiger monooxygenases based on protein sequence and stereopreference. *Angew. Chem. Int. Ed.* 117 (23): 3675–3679.

138 Iwaki, H., Hasegawa, Y., Wang, S. et al. (2002). Cloning and characterization of a gene cluster involved in cyclopentanol metabolism in *Comamonas* sp. strain NCIMB 9872 and biotransformations effected by *Escherichia coli*-expressed cyclopentanone 1,2-monooxygenase. *Appl. Environ. Microbiol.* 68 (11): 5671–5684.

139 Van Beilen, J.B., Mourlane, F., Seeger, M.A. et al. (2003). Cloning of Baeyer–Villiger monooxygenases from *Comamonas, Xanthobacter* and *Rhodococcus* using polymerase chain reaction with highly degenerate primers. *Environ. Microbiol.* 5 (3): 174–182.

140 Chen, G., Kayser, M.M., Mihovilovic, M.D. et al. (1999). Asymmetric oxidations at sulfur catalyzed by engineered strains that overexpress cyclohexanone monooxygenase. *New J. Chem.* 23 (8): 827–832.

141 Bučko, M., Gemeiner, P., Schenkmayerová, A. et al. (2016). Baeyer–Villiger oxidations, biotechnological approach. *Appl. Microbiol. Biotechnol.* 100 (15): 6585–6599.

142 Bianchi, D.A., Moran-Ramallal, R., Iqbal, N. et al. (2013). Enantiocomplementary access to carba-analogs of *C*-nucleoside derivatives by recombinant Baeyer–Villiger monooxygenases. *Bioorg. Med. Chem. Lett.* 23 (9): 2718–2720.

143 Tejero, T., Delso, I., and Merino, P. Recent advances in the chemical synthesis of aza-nucleosides. *Chem. Synth. Nucleoside Analogues* 15: 699–725.

144 Schramm, V.L. (2003). Enzymatic transition state poise and transition state analogues. *Acc. Chem. Res.* 36 (8): 588–596.

145 Warren, T.K., Wells, J., Panchal, R.G. et al. (2013). Protection against filovirus diseases by a novel broad-spectrum nucleoside analogue BCX4430. *Nature* 2014, 508 (7496): 402–405.

146 Chen, P., Gao, M., Wang, D.X. et al. (2012). Enantioselective biotransformations of racemic and meso pyrrolidine-2,5-dicarboxamides and their application in organic synthesis. *J. Organomet. Chem.* 77 (8): 4063–4072.

147 Chen, P., Gao, M., Wang, D.X. et al. (2012). Practical biocatalytic desymmetrization of *meso-N*-heterocyclic dicarboxamides and their application in the construction of aza-sugar containing nucleoside analogs. *Chem. Commun.* 48 (29): 3482–3484.

148 Wagner, C.R., Iyer, V.V., and McIntee, E.J. (2000). Pronucleotides, toward the *in vivo* delivery of antiviral and anticancer nucleotides. *Med. Res. Rev.* 20 (6): 417–451.

149 Bergman, A.M., Kuiper, C.M., Voorn, D.A. et al. (2004). Antiproliferative activity and mechanism of action of fatty acid analogues of arabinofuranosylcytosine in leukemia and solid tumor cell lines. *Biochem. Pharmacol.* 67 (3): 503–511.

150 Li, X.F., Zhu, Z., Zhao, G.L. et al. (2012). A novel biocatalytic approach to acetylation of 1-β-D-arabinofuranosylcytosine by *Aspergillus oryzae* whole cell in organic solvents. *Appl. Microbiol. Biotechnol.* 93 (1): 143–150.

151 Schmid, A., Dordick, J.S., Hauer, B. et al. (2001). Industrial biocatalysis today and tomorrow. *Nature* 409: 258–268.

152 Li, X.F., Lu, Z.H., Liu, H.W. et al. (2012). Dependence of catalytic performance of a freeze-dried whole-cell biocatalyst of *Pseudomonas fluorescens* in regioselective acetylation of 1-β-D-arabinofuranosylcytosine on growth conditions. *Process Biochem.* 47 (12): 1826–1831.

153 Molinari, F., Gandolfi, R., Converti, A., and Zilli, M. (2000). Mycelium-bound carboxylesterase from *Aspergillus oryzae*, an efficient catalyst for acetylation in organic solvent. *Enzyme Microb. Technol.* 27 (8): 626–630.

154 Lavandera, I., Fernández, S., Magdalena, J. et al. (2006). Remote interactions explain the unusual regioselectivity of lipase from *Pseudomonas cepacia* toward the secondary hydroxyl of 2′-deoxynucleosides. *ChemBioChem* 7 (4): 693–698.

155 Li, X.F., Lu, Z.H., Zhao, G.L. et al. (2012). A facile whole-cell biocatalytic approach to regioselective synthesis of monoacylated 1-β-D-arabinofuranosylcytosine, influence of organic solvents. *Bioresour. Technol.* 114: 6–11.

156 Yang, M.Y., Wu, H., Lu, Z.H. et al. (2014). Regioselective synthesis of cytarabine monopropionate by using a fungal whole-cell biocatalyst in nonaqueous medium. *Bioorg. Med. Chem. Lett.* 24 (15): 3377–3380.

157 Ferrero, M. and Gotor, V. (2000). Chemoenzymatic transformations in nucleoside chemistry. *Monatsh. Chem.* 131: 585–616.

158 Yang, M., Wu, H., Lian, Y. et al. (2014). Influence of organic solvents on catalytic behaviors and cell morphology of whole-cell biocatalysts for synthesis of 5′-arabinocytosine laurate. *PLoS One* 9 (8): 104847.

159 Bornscheuer, U.T. (2002). Microbial carboxyl esterases, classification, properties and application in biocatalysis. *FEMS Microbiol. Rev.* 26 (1): 73–81.

160 Bornscheuer, U.T., Bessler, C., Srinivas, R., and Hari Krishna, S. (2002). Optimizing lipases and related enzymes for efficient application. *Trends Biotechnol.* 20 (10): 433–437.

161 Sheldon, R.A. (2005). Green solvents for sustainable organic synthesis, state of the art. *Green Chem.* 7 (5): 267–278.

162 Potdar, M.K., Kelso, G.F., Schwarz, L. et al. (2015). Recent developments in chemical synthesis with biocatalysts in ionic liquids. *Molecules* 20 (9): 16788–16816.

163 De Gonzalo, G., Lavandera, I., Durchschein, K. et al. (2007). Asymmetric biocatalytic reduction of ketones using hydroxy-functionalised water-miscible ionic liquids as solvents. *Tetrahedron: Asymmetry* 18 (21): 2541–2546.

164 Lau, R.M., Sorgedrager, M.J., Carrea, G. et al. (2004). Dissolution of *Candida antarctica* lipase B in ionic liquids, effects on structure and activity. *Green Chem.* 6 (9): 483–487.

165 de Carvalho, C.C.C.R. (2017). Whole cell biocatalysts, essential workers from Nature to the industry. *Microb. Biotechnol.* 10 (2): 250–263.

166 Yang, M., Wu, H., Lian, Y. et al. (2014). Using ionic liquids in whole-cell biocatalysis for the nucleoside acylation. *Microb. Cell Fact.* 13: 143–154.

167 Heimbach, T., Fleischer, D., and Kaddoumi, A. (2007). Overcoming poor aqueous solubility of drugs for oral delivery. In: *Prodrugs, Challenges and Rewards*, vol. 5 (ed. V. Stella, R. Borchardt, M. Hageman, et al.), 157–215. Springer.

168 Ricci, F., Tedeschi, A., Morra, E., and Montillo, M. (2009). Fludarabine in the treatment of chronic lymphocytic leukemia, a review. *Ther. Clin. Risk. Manage.* 5: 187–207.

169 Iglesias, L.E., Lewkowicz, E.S., Medici, R. et al. (2015). Biocatalytic approaches applied to the synthesis of nucleoside prodrugs. *Biotechnol. Adv.* 33 (5): 412–434.

170 Zinchenko, A.I., Barai, V.N., Zaiashko, L. et al. (1990). Enzymatic synthesis of nucleoside 5′-mono and -triphosphates. *FEBS Lett.* 260 (2): 254–256.

171 Asano, Y., Mihara, Y., and Yamada, H. (1999). A new enzymatic method of selective phosphorylation of nucleosides. *J. Mol. Catal. B: Enzym.* 6 (3): 271–277.

172 Wang, X., Wang, X.W., Yin, M.X. et al. (2007). Production of uridine 5′-monophosphate by *Corynebacterium ammoniagenes* ATCC 6872 using a statistically improved biocatalytic process. *Appl. Microbiol. Biotechnol.* 76: 321–328.

173 Chen, Y., Li, S., Xiong, J. et al. (2010). The mechanisms of citrate on regulating the distribution of carbon flux in the biosynthesis of uridine 5′-monophosphate by *Saccharomyces cerevisiae*. *Appl. Microbiol. Biotechnol.* 86 (1): 75–81.

174 Sands, W.A. and Palmer, T.M. (2008). Regulating gene transcription in response to cyclic AMP elevation. *Cell. Signalling* 20 (3): 460–466.

175 Thibodeaux, C.J., MelanÅon, C.E. III,, and Liu, H.W. (2008). Natural-product sugar biosynthesis and enzymatic glycodiversification. *Angew. Chem. Int. Ed.* 47 (51): 9814–9859.

176 Krasnova, L. and Wong, C.H. Understanding the chemistry and biology of glycosylation with glycan synthesis. *Annu. Rev. Biochem.* 85: 599–630.

177 De Bruyn, F., Maertens, J., Beauprez, J. et al. (2015). Biotechnological advances in UDP-sugar based glycosylation of small molecules. *Biotechnol. Adv.* 33 (2): 288–302.

178 Ryu, S.I. and Lee, S.B. (2013). Synthesis of nucleotide sugars and α-galacto-oligosaccharides by recombinant *Escherichia coli* cells with trehalose substrate. *Enzyme Microb. Technol.* 53 (5): 359–363.

179 Schmölzer, K., Lemmerer, M., Gutmann, A., and Nidetzky, B. (2017). Integrated process design for biocatalytic synthesis by a Leloir glycosyltransferase, UDP-glucose production with sucrose synthase. *Biotechnol. Bioeng.* 114 (4): 924–928.

180 Gutmann, A. and Nidetzky, B. (2016). Unlocking the potential of Leloir glycosyltransferases for applied biocatalysis, efficient synthesis of uridine 5′-diphosphate-glucose by sucrose synthase. *Adv. Synth. Catal.* 358 (22): 3600–3609.

181 De Bruyn, F., De Paepe, B., Maertens, J. et al. (2015). Development of an *in vivo* glucosylation platform by coupling production to growth, production of phenolic glucosides by a glycosyltransferase of *Vitis vinifera*. *Biotechnol. Bioeng.* 112 (8): 1594–1603.

182 Weyler, C. and Heinzle, E. (2015). Multistep synthesis of UDP-glucose using tailored, permeabilized cells of *E. coli*. *Appl. Biochem. Biotechnol.* 175 (8): 3729–3736.

183 Krauser, S., Weyler, C., Blaß, L.K., and Heinzle, E. (2013). Directed multistep biocatalysis using tailored permeabilized cells. *Adv. Biochem. Eng./Biotechnol.* 137: 185–234.

184 Egger, S., Chaikuad, A., Kavanagh, K.L. et al. (2010). UDP-glucose dehydrogenase, structure and function of a potential drug target. *Biochem. Soc. Trans.* 38 (5): 1378–1385.

185 Coller, J.K., Christrup, L.L., and Somogyi, A.A. (2009). Role of active metabolites in the use of opioids. *Eur. J. Clin. Pharmacol.* 65 (2): 121–139.

186 Dragan, C.A., Buchheit, D., Bischoff, D. et al. (2010). Glucuronide production by whole-cell biotransformation using genetically engineered fission yeast *Schizosaccharomyces pombe*. *Drug Metab. Dispos.* 38 (3): 509–515.

187 Weyler, C., Bureik, M., and Heinzle, E. (2016). Selective oxidation of UDP-glucose to UDP-glucuronic acid using permeabilized *Schizosaccharomyces pombe* expressing human UDP-glucose 6-dehydrogenase. *Biotechnol. Lett.* 38 (3): 477–481.

188 Lee, W.H., Han, N.S., Park, Y.C., and Seo, J.H. (2009). Modulation of guanosine 5′-diphosphate-D-mannose metabolism in recombinant *Escherichia coli* for production of guanosine 5′-diphosphate-L-fucose. *Bioresour. Technol.* 100 (24): 6143–6148.

189 Lee, W.H., Chin, Y.W., Han, N.S. et al. (2011). Enhanced production of GDP-L-fucose by overexpression of NADPH regenerator in recombinant *Escherichia coli. Appl. Microbiol. Biotechnol.* 91 (4): 967–976.

190 Lee, W.H., Shin, S.Y., Kim, M.D. et al. (2012). Modulation of guanosine nucleotides biosynthetic pathways enhanced GDP-L-fucose production in recombinant *Escherichia coli. Appl. Microbiol. Biotechnol.* 93 (6): 2327–2334.

191 Munster, A.K., Eckhardt, M., Potvin, B. et al. (1998). Mammalian cytidine 5′-monophosphate *N*-acetylneuraminic acid synthetase, a nuclear protein with evolutionarily conserved structural motifs. *Proc. Natl. Acad. Sci. U.S.A.* 95 (16): 9140–9145.

192 Lee, S.G., Lee, J.O., Yi, J.K., and Kim, B.G. (2002). Production of cytidine 5-monophosphate *N*-acetylneuraminic acid using recombinant *Escherichia coli* as a biocatalyst. *Biotechnol. Bioeng.* 80 (5): 516–524.

193 Endo, T., Koizumi, S., Tabata, K., and Ozaki, A. (2000). Large-scale production of CMP-NeuAc and sialylated oligosaccharides through bacterial coupling. *Microb. Biotechnol.* 53 (3): 257–261.

194 Song, J., Zhang, H., Wu, B. et al. (2003). Reassembled biosynthetic pathway for a large-scale synthesis of CMP-Neu5Ac. *Mar. Drugs* 1 (4): 34–45.

195 Nahálka, J. and Pätoprstý, V. (2009). Enzymatic synthesis of sialylation substrates powered by a novel polyphosphate kinase (PPK3). *Org. Biomol. Chem.* 7 (9): 1778–1780.

196 Antoine, T., Heyraud, A., Bosso, C., and Samain, E. (2005). Highly efficient biosynthesis of the oligosaccharide moiety of the GD3 ganglioside by using metabolically engineered *Escherichia coli. Angew. Chem. Int. Ed.* 44 (9): 1350–1352.

197 Fierfort, N. and Samain, E. (2008). Genetic engineering of *Escherichia coli* for the economical production of sialylated oligosaccharides. *J. Biotechnol.* 134 (3–4): 261–265.

4

Enzymatic Synthesis of Nucleic Acid Derivatives by Immobilized Cells

Jorge A. Trelles[1,2], *Cintia W. Rivero*[1,2], *Claudia N. Britos*[1], *and María J. Lapponi*[1,2]

[1] *National University of Quilmes, Laboratory of Sustainable Biotechnology (LIBioS), Department of Science and Technology, Roque Saenz Peña 352, Bernal B1868BXD, Argentina*
[2] *National Scientific and Technical Research Council (CONICET), Godoy Cruz 2290, CABA C1425FQB, Argentina*

4.1 Introduction

Nucleic acids, deoxyribonucleic and ribonucleic acids (DNA and RNA), are one of the most important biomolecules in living systems. Nucleotides represent the basic unit of the nucleic acids and both nucleosides and nucleotides represent the derivatives of the nucleic acids. As components of these macromolecules, nucleosides are involved in all cellular processes related to the synthesis of RNA and DNA, cell signaling, enzymatic regulation, and metabolism [1, 2]. Natural nucleosides consist of a purine (adenine and guanine) or a pyrimidine (cytosine, thymine, and uracil) base joined with a pentose residue, β-D-deoxyribofuranose or β-D-ribofuranose for DNA or RNA, respectively (Figure 4.1). Analogues of these nucleosides are synthetic compounds specially designed to mimic the physiological functions of their natural counterparts in order to interfere with the cellular metabolism [1, 3]. Nucleoside analogues (NAs) enter the nascent DNA and RNA chains during cellular division or viral replication interfering with these processes. Once included in the DNA, a proofreading machinery detects these errors and suppression of chain elongation and apoptosis induction occur, and lethal mutations accumulate in the viral progeny [4]. These actions have potential therapeutic benefits such as inhibition of cancer cell growth and viral replication [4]. Apart from these inhibitory mechanisms, nucleosides and their analogues can interact and inhibit enzymes such as essential and human viral polymerases, kinases, ribonucleotide reductase, DNA methyltransferases, purine and pyrimidine nucleoside phosphorylases (NPs), and thymidylate synthase [2].

NAs are mainly synthesized by chemical processes that involve numerous stages of protection and deprotection of functional groups, involving the use of many toxic organic solvents, harsh reaction conditions, and the appearance of species that are very difficult to purificate (racemic mixtures) [5]. In view of what has been mentioned previously, biocatalytic synthesis of nucleosides (employing enzymes either isolated or in whole cells) appeared as a different approach for the bioproduction of NAs due to the nature of enzymatic reactivity: performing

Enzymatic and Chemical Synthesis of Nucleic Acid Derivatives, First Edition.
Edited by Jesús Fernández-Lucas and María-José Camarasa Rius.

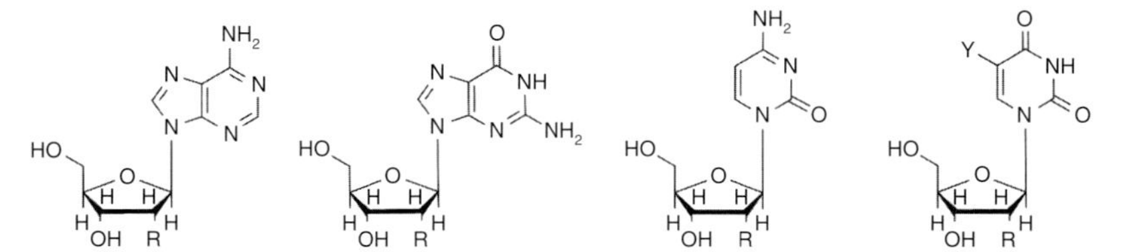

Figure 4.1 Chemical structure of natural nucleosides that compose DNA (R = H: β-D-deoxyribofuranose) and RNA (R = OH: β-D-ribofuranose).

reactions with high selectivity, efficiency and easy recovery, and reuse [1, 6]. The advantages of using these biocatalysts are high stereo- and regioselectivity, reactions that take place in a mild environment, offering a greener bioprocess [6].

There are different enzyme classes that can be used to synthesize nucleoside analogues derivatives: hidrolytic enzymes that act on the base like adenosine deaminase (ADA); nucleoside clorination or fluorination by halogenases, hydrolytic enzymes such as lipases, proteases, and esterases; also oxygenases that can oxidize the CH_2OH group of the pentose ring to CO_2H and enzymes that catalyze nucleoside synthesis via transglycosylation [7]. The last mentioned enzymes are NPs and *N*-deoxyribosyltransferases (NDTs), which transfer glycosyl residues from nucleosides to acceptor bases, and are the ones that have been mostly employed for the synthesis of nucleic acid derivatives [2]. They can be obtained either by their isolation or by the use of whole cell microorganisms in wild type form or genetically engineered to over-express the enzyme of interest [8–11]. NPs catalyze the reversible phosphorolysis of both ribo- and deoxyribonucleosides by the cleavage of *N*-glycosidic bonds of nucleosides without addition of adenosine triphosphate (ATP), to form a free base and its respective activated pentose moiety (pentose-1-P), which is then coupled to the desired modified base either by the same or a different NP to give an NA [12]. They have substrate specificity and are, therefore, classified as purine nucleoside phosphorylases (PNPs; EC 2.4.2.1) or pyrimidine nucleoside phosphorylases (PyNPs; EC 2.4.2.2) [12, 13]. Alternatively, NDTs (EC 2.4.2.6) catalyze the exchange between the purine or pyrimidine base of 2′-deoxyribonucleosides and free purine or pyrimidine bases. In the reaction, an intermediate of a covalently bound 2-deoxy-α-D-ribofuranosyl moiety is formed, where the glycosidic hydroxyl of 2′-deoxyribose is esterified by a glutamic acid of the active site of NDTs [2, 14]. Besides, some studies have shown that NDTs accept other related sugar donors as substrates and have been used to produce nucleoside analogues [15]. Again, these enzymes are classified by their substrate specificity in NDT type I (also named purine deoxyribosyltransferase, PDT) that is specific for purines (Pur↔Pur) and NDT type II (NDT) that catalyzes the transfer between purines and/or pyrimidines (Pur↔Pur, Pur↔Pyr, Pyr↔Pyr).

ADA is an enzyme in the purine salvage pathway that catalyzes the deamination of adenosine and deoxyadenosine to inosine and deoxyinosine, respectively. This enzyme activity catalyzes the rapid and irreversible deamination of adenine nucleosides as well as other C6 purine analogues to the corresponding hypoxanthine ones. Besides deamination, this enzyme can perform dehalogenation, demethoxylation reactions, and diastereoisomeric resolution [16].

Regarding lipases and glycosidases, these enzymes are used for the modification of existing nucleoside analogues to enhance their activity and pharmacokinetic and pharmacodynamic properties. Lipases are one of the most important biocatalysts employed in the synthesis of various pharmaceutical drugs [17]. Lipases catalyze the hydrolysis of esters to alcohols in aqueous solution, whereas in organic environments they perform transesterification of the previously mentioned substrates [18]. Regarding glycosylation, the attractive feature is that it allows conversion of lipophilic compounds into hydrophilic ones, changing their pharmacological properties, obtaining novel and more effective drug delivery systems (prodrugs). Among the advantages of using enzymatic glycosylation instead of chemical modification is that the use of harsh reaction conditions or the use of toxic (heavy metals) catalysts are avoided [19]. Lastly, biocatalytic halogenation of nucleosides can be directed to the sugar moiety as well as to the base moiety, conferring the nucleosides important biological activities to the parent compounds [20].

4.2 Nucleic Acid Derivatives

Nucleic acid derivatives, as the name indicates, are compounds that are derived from nucleic acids and are basically natural nucleosides and nucleotides, which are the subunits that constitute the backbone of DNA and RNA. Other compounds that belong to this group include nucleoside analogues (as derivatives of nucleosides). As mentioned above, these compounds imitate their natural versions, confuse the cell machinery and are included within the DNA or RNA, thus interfering with cell division or viral replication. Currently, nucleic acid derivatives and particularly their analogues represent a major group of antimetabolite cytotoxic drugs in active clinical use [3]. These compounds are therapeutically employed for treating diverse human cancers such as lymphomas, leukemias, and pancreatic, bladder, colon, breast and kidney cancer, among others [4, 21]. They are also used for treating viral infections caused by DNA viruses such as herpes simplex virus (HSV), cytomegalovirus (CMV) [21]. Besides, another class of nucleoside derivatives is represented by nucleoside analogues derivatives; that are, in essence, nucleoside analogues that underwent some class of subsequent modification such as an enzymatic acylation or glycosylation [22]. In this case, the nucleoside analogue turns into a prodrug, which is defined as a drug derivative that undergoes biotransformation either enzymatically or nonenzymatically inside the body before exhibiting its therapeutic effect. In an ideal case, the prodrug is converted to the original drug as soon as the derivative reaches the site of action. The main purpose for the synthesis of prodrugs is bioavailability enhancement or a specific targeting of the drug. Prodrugs are usually more soluble and more stable in the living organism; moreover, they often have a specific affinity to some biological structures allowing targeting to specific tissues [23].

Concerning the mechanism of action of nucleic analogues, it is essential for NAs to exert their activity to get inside the cell, and this is done through specialized nucleoside transporter proteins because of their hydrophilic nature. There is growing evidence that the abundance and tissue distribution of nucleoside transport proteins contribute to cellular specificity and sensitivity to nucleoside analogues [24]. The clinically administered compounds are generally prodrugs that enter the cell and are phosphorylated by deoxynucleoside salvage pathways, and the triphosphate metabolite (nucleotide) is responsible for producing their activity by being incorporated into the DNA during replication or repair, acting as chain terminators and stopping viral DNA polymerase. However, each NA has unique drug–target interactions that help to explain the differences in their activity in various diseases. The most important modifications that can lead from a natural nucleoside to analogues involve modifications in the sugar ring such as an anomeric or a configuration inversion, ring opening, substitution in one or more sugar atoms and phosphorylation, or modifications in the nucleobase such as halogenations, N-conjugation, etc. [2, 25].

Nucleosides are molecules with multiple hydroxyl and amino groups, features that make them polyvalent molecules, very polar and hydrophilic, and also causes them to exhibit low oral bioavailability. They have a rather short plasma half-life because they suffer hydrolysis by NPs or deamination by nucleoside deaminases present in the body [26]. So, in order to maintain the plasma concentration that is required for effective treatment, large doses and frequent administration are necessary, which in many cases augment undesirable side effects. Therefore, there is continuous search for new strategies for NA modification. Chemical modification of parent drugs is one of the most successful strategies to overcome the drawbacks mentioned above. Consequently, nucleosides have been modified by different techniques such as glycosylation, halogenation, deamination, and acylation [6], of which the last strategy is one of the most studied to create derivatives with improved activity [23, 27, 28]. Different enzymes other than NPs and PNPs have been used for this purpose, for instance, lipases, glycosidases, deaminases (Figure 4.2) in addition to other strategies such as the development of lipid-nucleoside conjugates, prodrugs, and liposome preparations [3].

Acylation is not considered to be an effective approach because lipophilic nucleoside prodrugs are able to cross the cell membrane and liberate parent

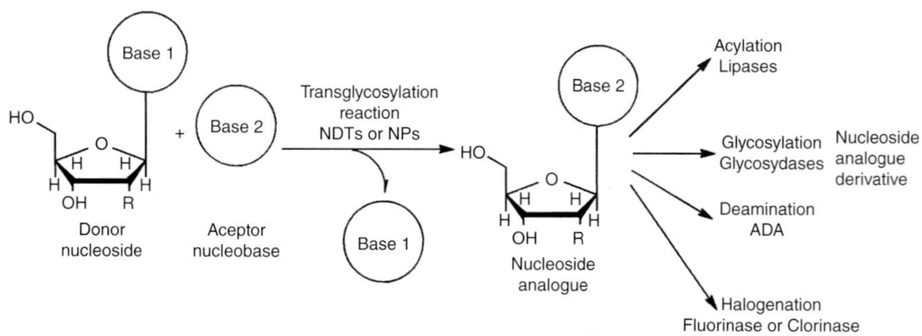

Figure 4.2 Strategies for the biocatalytic synthesis of nucleoside analogues and their derivatives.

agents [29]. The most commons examples are the clinically employed amino acid ester prodrugs of nucleosides valganciclovir and valacyclovir, which have been used as alternatives to the antiviral agents, ganciclovir (GCV) and acyclovir (ACV), due to their higher therapeutic efficacy [27]. However, selective modification of a single functional group using classic organic chemistry, when there are repetitive functional groups with the same chemical reactivity in a molecule, represents a problematic task. An example of such a difficulty is the regioselective acylation of polyhydroxyl compounds such as nucleosides [30]. The first reports in this field are from the 1950s and involved nonselective acylations with subsequent tedious separation processes, or time-consuming protection–deprotection steps in selective reactions on ribonucleosides and mainly in the 3′-O-acylation of deoxyribonucleosides. By employing traditional chemical methods, one-step acylations with high selectivity was only achieved by means of using bulky acylating reagents, which leave secondary hydroxyls free due to steric hindrances [31]. Nowadays, the use of enzymes such as lipases appears to be an excellent alternative for the modification of these polyfunctional compounds due to their high regioselectivity, mild reaction conditions, and environmental friendliness of the enzymatic processes [6, 32]. Also, glycosylated derivatives of many nucleoside drugs [33, 34] have been described as compounds that significantly mask the toxicity of the parent drugs and/or improve their pharmacokinetic properties. Besides, the cellular uptake of many glycosylated products can be enhanced due to the active absorption mediated by glucose transport systems [35]. Again, the regioselective glycosylation of nucleosides remains a tedious task for organic chemists, owing to the presence of multiple repeated functional groups (hydroxyl and amino groups) within the molecules. Nucleoside O-glycosylation represents a typical problem in chemical selectivity, considering that the nucleobase (an undesired site of reaction) is usually more nucleophilic than the hydroxyl (the desired site of reaction). This is why the use of enzymes for the reaction catalysis appears so attractive [36]. Glycosylation is an established method for the synthesis of prodrugs. A typical example is the targeting of glycosylated drugs on the colon, where they are specifically hydrolyzed by the local intestinal anerobic microflora [37].

As mentioned previously, the main medical use of these compounds is as anticancer and antiviral pharmacological drugs, but this is not the only use, as the potential use of these nucleosides as antibiotics [38–40], radiopharmaceuticals [41–43], or molecular probes [44] has also been reported. In cancer, the purine nucleoside analogues (PNAs) cladribine, fludarabine, clofarabine, and nelarabine are employed in the treatment of malignant disorders of the blood (Figure 4.3). On the other hand, pyrimidine nucleoside analogues (PyNAs), cytarabine is extensively used in the treatment of acute leukemia; gemcitabine is indicated for the treatment of several solid tumors; and decitabine is used for the treatment of myelodysplastic syndromes (MDS). The fluoropyrimidine floxuridine has activity against various types of cancer [2]. Also, antiviral NAs constitute the first-line therapy against specific viral infections, and the discovery of NAs effective in HIV therapy marked a breakthrough in acquired immune deficiency syndrome (AIDS) treatment. The cytidine analogues lamivudine and emtricitabine are used in combination with other drugs in HIV therapy [5]. As they can be used for long

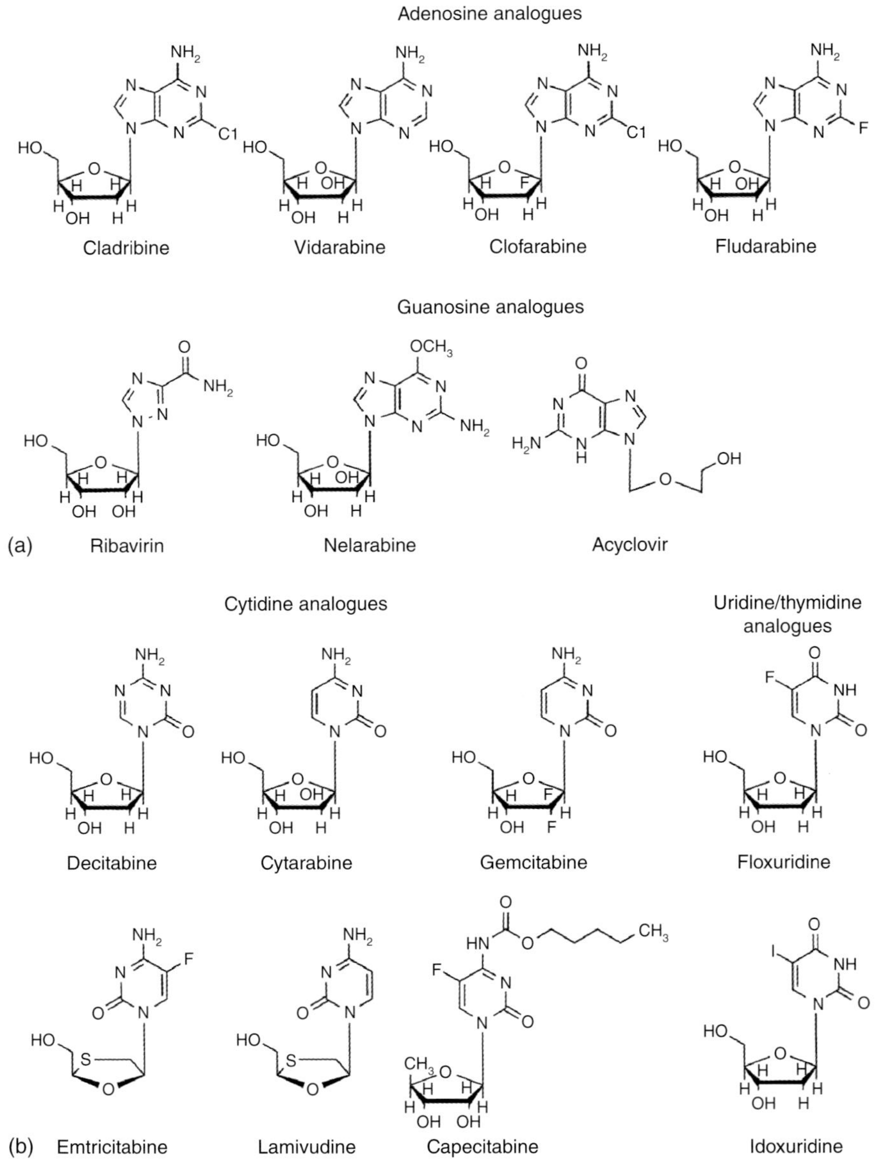

Figure 4.3 Nucleic acid derivatives in current clinical use: (a) PNAs and (b)PyNAs.

periods of time in chronic viral infections, NAs with improved bioavailability and reduced side effects have been extensively explored. NAs used in antiviral therapies include ribavirin, iduviran, vidarabine, ACV, among others [2].

Bearing in mind that the research and clinical trials required for optimal pharmaceutical use of a new NA involve very long and costly processes, biocatalytic modification of existing nucleosides with proven therapeutical activity present an attractive approach. So, in this chapter we will discuss the advances in the current available as well as alternative methods for the production of nucleoside analogues and their derivatives by whole cell immobilized biocatalysts.

4.3 Whole Cell Immobilization: Generalities

Immobilization describes diverse ways of enzyme or cell confinement to different types of matrices or supports. In terms of advantages, biocatalyst immobilization stabilizes, allows easy separation from the reaction medium, favors reuse, and also makes the purification of the final product easier. Different techniques have been employed for whole cell and enzyme immobilization [45]. Broadly, they can be classified into entrapment, adsorption, covalent, and cross-linking stabilization, also, the combination of any of these give rise to a combined methodology (Figure 4.4).

Microbial whole cells are efficient, ecological, and low-cost catalysts that have been successfully applied in the pharmaceutical, environmental, and food industries, among others [45–48]. There are a great variety of methods for immobilizing whole bacterial cells, but generally whole cell immobilization can be performed by entrapment in porous polymers or microcapsules or by interacting with inorganic or organic supports [49, 50]. The interaction of cells is caused by van der Waals forces and ionic or covalent bonding, especially when microbial exopolymers participate in the procedure [50]. Cell entrapment is the most widely used technique for whole cell immobilization. Basically, cells are included within a rigid network porous enough to allow the diffusion of substrates and products and protect the selected microorganism from the reaction medium [45]. Cells are entrapped within a gel in the free spaces of a polymeric net composed of natural or synthetic polymer. Gelation is then reached by the supplementation of one or more cross-linking agents. The mixture of the soluble polymer and the cells is passed through a needle or micropipette tip into a gelling solution to create the beads [51]. There is gelation induced by temperature and by ions (ionotropic). There are three types of matrices used for microorganism stabilization and can be divided into hydrogels (alginate, κ-carrageenan, chitosan), thermogels (agar, agarose, cellulose), and synthetic polymers (polyacrylamide, polyvinyl alcohol, polyurethane) [45, 52, 53]. Of these matrices agarose, alginate, and polyacrylamide have been widely used for biocatalyst immobilization for NA synthesis. Additionally, it has been reported that the addition of nanocomposites to immobilization process improves the mechanical properties of biocatalysts obtained [54] facilitating the design of bioreactors for further scale bioprocess [55]. Apart from being one of the most widely used, this method is considered to be a specially easy, simple, and safe technique for stabilization. It is important to

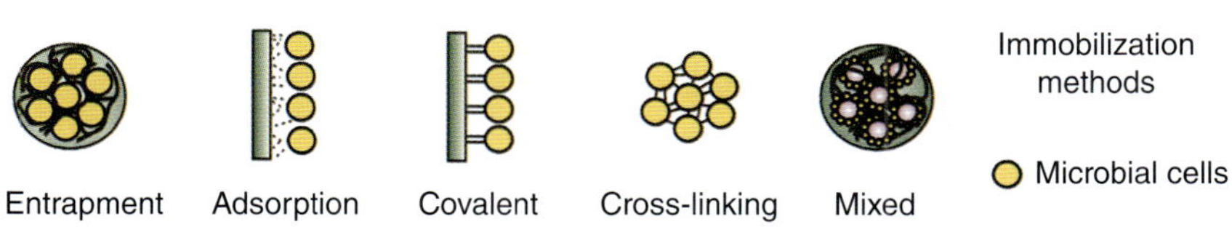

Figure 4.4 Different stabilization methodologies employed for microbial cell immobilization.

mention that there are a number of cell immobilization methods that are different from entrapment and are less used for biotransformations of NAs. These include adsorption or adhesion on polymeric surfaces, adhesion onto covalent supports, and cell encapsulation [47, 56, 57]. The bond established between the surface and the biocatalyst can be either reversible or irreversible, and the interactions can be grouped as adsorption, covalent, entrapment, and cross-linking immobilization [58, 59]. Entrapment is a versatile method for enzyme or whole cell stabilization. Whole cells are the preferred biocatalysts for entrapment immobilization method. Also, these methods can be combined, such as a biocatalyst being first immobilized by adsorption and then entrapped in a porous polymer [58].

As previously mentioned, two or three different immobilization methods can be combined to create a new one, which in many cases synergizes the advantages of the separate techniques. The classic example is immobilizing an enzyme first by adsorption, and then by entrapment in a porous polymer. Different approaches have been used to merge matrices, such as chitosan combined with alginate [60, 61] and polyacrylamide supplemented with natural chitosan nanofibers (CNFs) resulting in a mixed matrix with enhanced mechanical and biocatalytical properties (compressive strength, storage modulus, and swelling ratio) [62]. Also, polymer matrix based nanocomposites have gained plenty of attention in the nanotechnology area [63]. Nanocomposites have been used over the past decades for enzyme immobilization, and these compounds are composed of a variety of clay minerals [64]. Clay minerals are a naturally occurring group of hydrous aluminum or magnesium phyllosilicates with a two-dimensional layered structure on a nanometer scale [64]. This particular structure gave the clay minerals distinctive physicochemical features including a large surface area, swelling, and ion exchangeability [64]. Also, this particular structure permits these clay minerals to be altered by organic, polymeric, or biological molecules. Clay minerals include bentonite, montmorillonite, smectites, halloysite, sepiolite, laponite, palygorskite, among others. The integration of inorganic nanoparticles into a polymer matrix allows both properties from inorganic nanoparticles and polymer to be combined/enhanced matrix and thus advanced new functions are developed [63, 65].

At the moment of choosing a support or a technique to stabilize a biocatalyst, different parameters will define the appropriate selection, such as ability to retain the biocatalyst in the matrix, mass transfer (diffusion of substrates and products), and stability maintenance or increase of enzymatic, operational, and storage stability. Regarding the type of immobilization procedure, the advantages and disadvantages of each technology must be considered in order to fulfill the purpose sought (Table 4.1).

4.4 Synthesis of Nucleosides by Immobilized Cells

As mentioned, immobilized cells represent economical and versatile biocatalysts with the advantage of carrying their cofactors and accessorial machinery inside the cells. Also, the cost of enzyme isolation or purification is not an issue when

Table 4.1 Advantages and disadvantages of the different immobilization techniques.

	Advantages	Disadvantages
Entrapment	High operational stability	Diffusion issues between substrates and products
	Easy upstream separation	
	Bioprocess scale-up feasibility	
Adsorption	Mild immobilization methodology	Desorption may occur
	Reusability of the matrix	Nonspecific nature of the interaction with the matrix (van der Waals, ionic or hydrophobic bonding)
Covalent	Permanent nature of the bonds between the biocatalyst and the functional groups of the support	The matrix cannot be reused
	No biocatalyst leakage	Citotoxicity of the reagents employed, and cell damage after immobilization
Cross-linking	No matrix or support is required, only the organic solvents, and the bonds are established between the biocatalysts	Citotoxicity of the reagents employed, and cell damage after immobilization
		Bifunctional reagents may react with the enzymes present in the cells reducing their biocatalytic activities

using whole cells and that is why many times they are the first choice or the first step in a bioprocess design. When using an immobilized whole cell biocatalyst some considerations must be taken into account, such as assuring that no secondary reactions such as rapid product hydrolysis or deamination happen in the desired products. Also, in view of future bioprocesses, scale-up biocatalyst reusability remains an important factor. As mentioned previously, cell entrapment is the most widely used technique for cell immobilization. When one or more activities are desired to produce a nucleic acid derivative of clinical interest, two microorganisms can be co-immobilized or immobilized separately and perform a reaction in tandem in two steps. Combinations of these biocatalysts have been employed for the production of natural nucleosides, nucleosides analogues, and nucleosides analogues derivatives (with a lipid, sugar, or different motive further incorporated in their structure). Section 4.4.1 we will describe the methodologies, microorganisms, and immobilization techniques employed to bioproduce nucleic acid derivatives.

4.4.1 Natural Nucleoside Synthesis

Diverse methodologies for the biocatalytic synthesis of nucleoside have been employed using immobilized cells. Nucleoside synthesis by immobilized bacterial cells have been developed since the late 1980s [66, 67], but these have been revised previously [68], so in this chapter we will focus in the most recent and relevant.

One of them was achieved after a taxonomic screening using 176 microorganisms, *Escherichia coli* ATCC 47092, *Enterobacter gergoviae* CECT 857, and *Citrobacter amalonaticus* CECT 863 were selected after assaying for adenosine synthesis from uridine as a reaction model as they were the ones that afforded the highest yields. Whole cell entrapment immobilization was performed using different matrix such as alginate, agar, agarose, and polyacrylamide. Different parameters were optimized to obtain immobilized cells that were able to synthesize adenosine for longer times than did free cells, and were reused for more than 30 times without significant activity loss [69]. Each strain exhibited different behavior in the matrices tested, while *E. coli* gave better yields when stabilized in agarose or polyacrylamide, *En. gergoviae* performed better in polyacrylamide, while *C. amalonaticus* catalytic activity was not affected by the matrix. This different behavior has already been described and it may be due to differences in the cell wall and/or membrane composition [70]. Apart from adenosine different nucleosides of adenine (Ara-A, dA, ddA) and other purinic nucleosides were assayed.

Another report identified the strains *Bacillus coagulans* CECT 12, *Lactobacillus* sp. CECT 4219, *Bacillus psychrosaccharolyticus* CECT 4074, and *Psychrobacter immobilis* CECT 4492 as positive to deoxyribosyltransferase (type II) activity, wherein the last two psychrotrophic microorganisms activity was described for the first time. The selected strains were immobilized in calcium alginate or calcium pectate and were able to bioproduce 2′-deoxyadenosine at 57 °C with high yields and without secondary reactions, also selective for 6-aminopurine compared to 6-oxopurines and for 2′-deoxyuridine compared to thymidine. The synthesis of 2-deoxyribosylnucleosides was carried out in Tris/HCl at $T = 57\,°C$. Also, 2′-deoxyribosylnucleosides of 5-chloro- or 5-fluoruracil were obtained in a one-pot reaction [71].

4.4.2 Nucleoside Analogues Synthesis

As mentioned previously in this chapter, nucleoside analogues are synthesized by transglycosylation using two types of intracellular enzymes, namely NPs and NDTs. NPs catalyze the reversible phosphorolysis of nucleosides and the transferase reaction involving purine or pyrimidine bases. Both types of enzymes display a high regio- (N-1 glycosylation in pyrimidine and N-9 in purine) and stereoselectivity (β-anomers are exclusively formed).

Immobilized whole cells have been vastly employed in the synthesis of many of these nucleoside derivatives that will be discussed further. For example, the antitumoral compound fludarabine (9-β-D-arabinosyl-2-fluoroadenine, Ara-U, Fludara) was used clinically for the treatment of chronic lymphocytic leukemia (CLL) in patients who were refractory to prior treatment with alkylating agents [69, 72] and it is administered as the soluble prodrug fludarabine 5′-monophosphate, which is then triphosphorylated and activated inside the cell. This compound was obtained using *Citrobacter koseri* cells stabilized by entrapment in various matrices, with agarose being the one with the better performance. However, the best results were obtained with free cells yielding 58%

of fludarabine after 14 hours at 60 °C. Also, these cells were used for the production of other related arabinofuranosyl compounds such as vidarabine. Vidarabine (9-β-D-arabinofuranosyladenine, Ara-A, vira-A) was the first clinically approved antiviral drug, and it is currently only used topically for herpes virus keratitis. Regarding ara-A biosynthesis, the above-mentioned biocatalyst immobilized *C. koseri* and was able to produce vidarabine, obtaining 71% yield in 26 hours, and was successfully scaled up to 150 ml, obtaining 62.5% yield after 26 hours [73].

The guanosine (ribonucleic) analogue ribavirin (1-β-ribofuranosyl-1,2,4-triazole-3-carboxamide, Virazole) was originally employed as an antiviral agent in combination with (PEG)ylated interferon β in the treatment of hepatitis C [74, 75] exhibiting proven activity for the treatment of different types of cancer, including leukemia, lymphoma [76], and also breast cancer [77], making this compound relevant not only as an antiviral, but also as an antitumor agent. This compound was obtained with *E. coli* immobilized on agarose cells obtaining 72 mM ribavirin after four hours of reaction time at pH 7.0, 60 °C (Figure 4.5) [10]. Our group reported the biosynthesis of this compound using *E. coli* ATCC 12407 stabilized in agarose. In this work, a significant increase of over 40% in bioconversion was achieved when the reactions were carried out at 30 °C with respect to higher temperatures. It is noteworthy that this was the first time that ribavirin biotransformation was performed at 30 °C, favored by the use of the soluble substrate uridine, also allowing easier bioprocess scale-up in a packed-bed reactor, achieving 95 mg of ribavirin [72]. The developed immobilized biocatalyst was operationally active for more than 270 hours and could be stored for four months without activity loss. This microorganism was also stabilized in polyacrylamide; but this matrix was more labile than the agarose matrix, considering that integrity loss was apparent after 80 hours of continuous reaction. Regarding storage stability, the agarose immobilized biocatalyst remained stable for over 120 days while the polyacrylamide stabilized biocatalyst remained active for 80 days [72]. Also, our group reported the use of the extremophilic microorganism *Geobacillus kaustophilus* ATCC 8005 entrapped in an agarose matrix supplemented with bentonite for the bioproduction of ribavirin. The addition of the nanoclay bentonite to the conventional agarose matrix resulted in a biocatalyst with improved mechanical properties and operational stability (Ag-Bent, 580 hours) in comparison with the non-supplemented agarose matrix

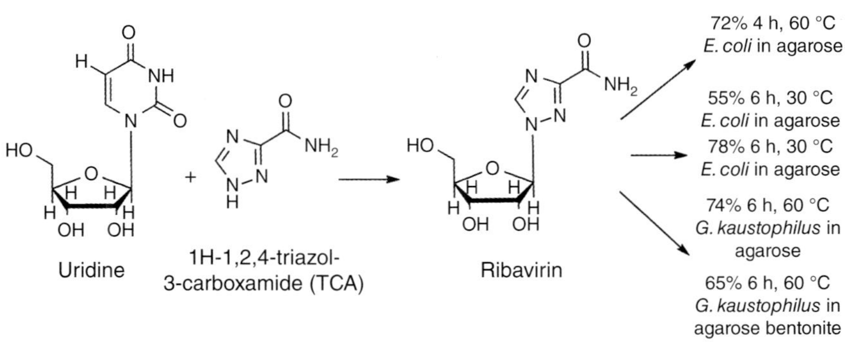

Figure 4.5 Ribavirin bioconversion using bacterial biocatalysts in different matrices.

(Ag-Ctrl, 150 hours). Additionally, a packed-bed bioreactor for bioprocess scale-up was designed, which was able to produce 370 mg l^{-1} of ribavirin [12].

Floxuridine (5-fluoro-2′-deoxuridine, FdUrD, FUDF) is a fluorinated pyrimidine antitumoral agent employed in the treatment of a wide variety of cancers such as colorectal, pancreatic, breast, head and neck [4, 78]. The microorganism *Aeromonas salmonicida* ATCC 27013 was employed for the synthesis of this compound and also other 5-halogenated pyrimidine nucleosides from thymidine. The immobilized biocatalyst achieved a 65% bioconversion in one hour, was stable for more than four months in storage conditions (4 °C) and could be reused at least 30 times without loss of its activity. This microorganism was able to biosynthesize of 5-chloro-2′-deoxyuridine, 60% in three hours. These halogenated pyrimidine 2′-deoxynucleosides are used as antitumoral agents [79]. Recently, our group reported several advances in the biosynthesis of this compound, using immobilized *Lactobacillus animalis* as the biocatalyst. In the first approach, the best results were obtained using alginate hydrogels, specifically calcium alginate 4% (w/v) (Figure 4.6). This biocatalyst showed an operational stability of 44 hours in batch process and 144 hours at a prepilot scale, obtaining 0.8 mg l^{-1} of floxuridine [80]. Different parameters must be set for each immobilization. Regarding ionotropic gelation of alginate, a different behavior of the same biocatalyst immobilized in this matrix using different divalent ions (Ca^{2+}, Ba^{2+}, and Sr^{2+}) as cross-linking solution has been reported. Various parameters such as exposure time and concentration of the solution in the gelation process were optimized, with Sr-alginate exhibiting better performance, obtaining a biocatalyst with better mechanical properties (compression strength, swelling ratio, and fracture frequency), which is favorable for future scale-up. Also, the obtained biocatalyst had higher operational stability in comparison with that obtained with Ca-alginate [81]. In the latter case, the natural matrix previously developed was improved by using strontium as the cross-linking agent, obtaining a novel biocatalytic system consisting of *L. animalis* immobilized in Sr-alginate. This biocatalyst showed bioconversion parameters higher than 80% and was also able to obtain a related nucleoside (5-bromouracil-2′-deoxyriboside) in nonconventional media at shorter reaction times. The developed Sr-alginate matrix was an efficient alternative to stabilize microorganisms and could be used to produce a broad spectrum of nucleoside analogues from substrates with low solubility in water [81].

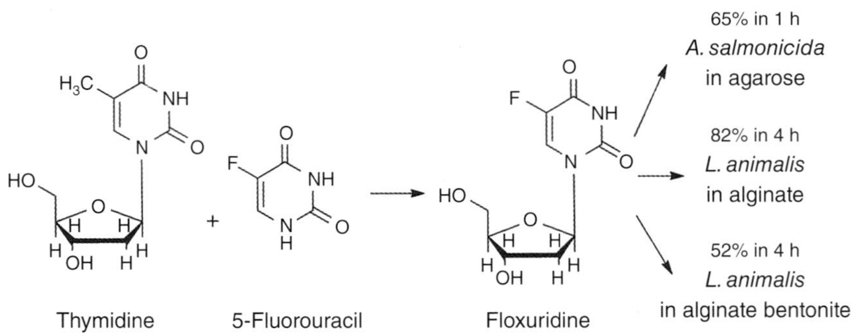

Figure 4.6 Floxuridine bioconversion using different immobilized biocatalysts.

Also, as previously mentioned, the use of bionanocomposites employing natural polysaccharides such as alginate and nanoclays such as bentonite are a promising alternative to developing stabilized biocatalysts. Different mechanical parameters such as swelling ratio, compressive strength, and fracture frequency were optimized, favoring scale-up. The developed and improved immobilized biocatalyst was efficiently used for bioprocess scale-up obtaining floxuridine, showing a productivity of 596 mg g^{-1} of biocatalyst. Additionally, the supplementation of the matrix with the nanoclay bentonite resulted in an improvement of more than 90% of storage stability, and the reusability of the biocatalysts was greater compared with control conditions, allowing bioprocess scale-up to obtain floxuridine with high productivity per gram of biocatalyst [55]. *L. animalis* was also successfully employed for the synthesis of 5-iodouracil-2′-deoxyriboside, known as idoxuridine, a nucleoside analogue with antiviral activity used in the treatment of viral infections in the absence of efficient vaccines. A bionanocomposite (alginate–nanoclay) was employed as matrix to stabilize *L. animalis* offering better stability in the presence of organic cosolvents. The obtained conversion was close to 90% in different nonconventional media using free cells. The bioprocess was scaled up affording a product yield close to 0.4 g l^{-1} in the presence of 20% of DMSO [82].

6-Substituted purine nucleosides are of interest as pharmaceutical targets and in particular, 6-chloropurine derivatives are useful intermediates in the synthetic routes of several biological active analogues. The 6-modified purine nucleosides, such as 6-chloro-, 6-methoxy-, 6-iodo-, and 6-mercaptopurine ribonucleoside were synthesized using *Bacillus stearothermophilus* ATCC 12980 entrapped in agar, agarose, and polyacrylamide, agar or agarose being the matrices that did not affect the biocatalyst activity. Interestingly, 6-chloropurine riboside was obtained in 75% yield when *B. stearothermophilus* was immobilized in agarose in comparison to free cells (72% after 48 hours) [83]. Another thermophilic microorganism, *Geobacillus stearothermophilus* CECT 43, was successfully used for the synthesis of 2,6-diaminopurine-2′-deoxyriboside (DAPurd-Rib), which is a precursor of 2′-deoxyguanosine (dGuo) analogues by deamination due to low solubility of guanine as the starting substrate. This compound is also used in cancer therapy as an antisense oligonucleotide and also to synthesize mRNA cap analogues for studies of *in vitro* genetic expression [84]. The selected biocatalyst was successfully stabilized in an agarose matrix obtaining 97% after 8 hours of reaction, and was used to produce up to 23.4 g of DAPurd-Rib in 240 hours of process [84].

The deoxyadenosine analogue Cladribine (2-chlorodeoxyadenosine, 2-CdA, Leustatin) has a chlorine atom instead of the hydrogen atom at the 2-position, substitution that makes this compound resistant to deamination by the enzyme ADA. There is much interest in this NA since it shows remarkable activity and produces long-lasting complete remissions, and has been the standard treatment in hairy cell leukemia (HCL), a chronic B-cell lymphoproliferative disorder [85]. Also, it has been shown to be effective in treating CLL [86], indolent non-Hodgkin lymphoma and acute myeloid leukemia [87, 88]. Enzymatic synthesis of this compound was achieved by direct transfer of 2-deoxyribofuranose moiety from dGuo to 2-chloroadenine (Cl-Ade) by *E. coli* BMT-1D/lA strain treated with glutaraldehyde (GA) [89]. For the reaction conditions, GA-treated cells were incubated with 10 mM K-phosphate buffer (pH 7.23), 10% DMSO (v/v), 5 mM Cl-Ade and 15 mM dGuo was incubated at 60 °C for 21 hours and formation of 2-CdA was monitored by thin layer chromatography (TLC). Similarly to previously reported, it was observed that the treatment of *E. coli* cells with GA enhances the stability of their phosphorylases and significantly improves the yields of desired nucleosides. The compound was obtained with a good yield (65%) after 21 hours of reaction. The obtained results are summarized in Table 4.2.

4.4.3 Nucleoside Analogues Derivatives Synthesis

These compounds are also called prodrugs, which by definition are drug derivatives that undergo a transformation once inside the body. Ideally, the prodrug is turned back into its original form as soon as the derivative arrives at the site of action. These compounds were developed to improve the well-known features of existing nucleosides analogues, such as bioavailability and tissue specificity, toward designing a specific target directed drug. Generally, these modified compounds turn out to be more soluble and stable in the living organism. As previously mentioned, the preferred reactions to generate a derivative or prodrug are acylation, glycosylation, deaminations, and halogenations [6].

Lipases are one of the most important biocatalysts employed in the synthesis of various pharmaceutical drugs [17]. Lipases are hydrolases, which act under aqueous conditions on the carboxyl ester bonds present in triacylglycerols to liberate fatty acids and glycerol. These enzymes catalyze the hydrolysis of esters to alcohols in aqueous solution, whereas in organic environments they perform transesterification of the previously mentioned substrates [18]. Regarding its structure, lipases are characterized by a common α/β hydrolase fold and a conserved catalytic triad (Ser-Asp/Glu-His) and also many of them possess the consensus motif G-X1-S- X 2-G [91]. Bacterial lipases are produced by submerged fermentation since they are mostly extracellular, and that is why they are considered to be commercially appealing, since their bulk production is much simpler [91]. These enzymes are generally purified by hydrophobic interaction chromatography, and by other modern approaches such as reverse micellar and aqueous two-phase systems. Of the most widely used lipases, it is accepted that *Pseudomonas cepacia* lipase (PSL) acylates in 3′-position, while *Candida antarctica* lipase B (CAL-B) leads to the acylation exclusively in the 5′-position [92]. It appears that the regioselectivity is only associated with the enzyme characteristics rather than with the acyl-donor structure and, therefore, by using these two lipases, CAL-B and PSL, it is possible to prepare the acylated or the carbonate compounds in 5′-position with CAL-B, while PSL is selective toward the secondary hydroxyl group [92]. The hyperactivation of lipases can be accomplished by adsorption of these enzymes on hydrophobic supports, which allows enhanced activity due to the correct immobilization of the open conformation of the lipases [51, 93].

There are a number of works that describe the use of microbial lipase in combination with different acyl donors (oxime esters and oxime carbonates) for nucleoside acylation with promising yields and selectivity. The great number

Table 4.2 Summary of different nucleic acid derivatives obtained by the use of immobilized whole cells.

Type of interaction	Matrix/support	Biocatalyst	Nucleic acid derivative	Reaction conditions	Reaction time (h)	Yield (%)	Operational stability (h)	References
Adsorption	DEAE-sepharose	*Lactobacillus animalis*		20 mM Tris–HCl buffer pH 7, 30 °C	—	0.7 mg l^{-1}	248	[80]
Entrapment	Alginate Agar Agarose Polyacrylamide	*Enterobacter gergoviae* CECT 857	*Adenosine*	Uridine 0.15 mM; adenine = 0.05 mM 60 °C	3	95 78.5 89 87	>30	[69]
	Agar Agarose Polyacrylamide	*Escherichia coli*				87 86 99		
	Agar Agarose Polyacrylamide	*Citrobacter amalonaticus*				93 92 91.5		
	Agar Agarose Calcium alginate Calcium pectate κ-carrageenan	*Bacillus coagulans* CECT 12	*Deoxyadenosine*	dAdo in Tris buffer. [dU] = [A] = 5 mM. T_a = 57 °C	0.50 0.50 0.66 0.25 0.25	3.5 5 23 25 18		[71]
	Calcium alginate Calcium pectate κ-carrageenan	*Bacillus psychrosaccharolyticus* CECT 4074			0.66 0.50 0.75	17 36 6	—	
	Agarose Calcium alginate Calcium pectate	*Lactobacillus* sp. CECT 4219			0.50 1 0.5	4.32 25 32	—	
	Calcium alginate Calcium pectate	*Psychrobacter immobilis* CECT 4492			0.66 0.50	22 45	—	
Entrapment	Agarose	*Aeromonas salmonicida ATCC 27013*	*Floxuridine and 5-chloro-2′-deoxyuridine*		1 3	65 60	30	[79]

(Continued)

Table 4.2 (Continued)

Type of interaction	Matrix/support	Biocatalyst	Nucleic acid derivative	Reaction conditions	Reaction time (h)	Yield (%)	Operational stability (h)	References
	Ca-Alginate	*L. animalis*	*Floxuridine*	20 mM Tris–HCl buffer pH 7, 30 °C	4	82	16	[81]
	Sr-Alginate	*L. animalis* In Sr-alginate				82	40	
	Alginate-Bentonite	*L. animalis*				38 ($mg\,l\,h^{-1}$)	80	[55]
	Alginate-Bentonite		Idoxuridine		2	90	–	[82]
		L. animalis		25 mM Tris-HCl buffer pH 6 DMSO 20% (v/v) 30 °C				
	Agarose	*Escherichia coli* ATCC 12407	Ribavirin	Phosphate buffer 30 mM pH 7.0, 30 °C	6	55	270	[72]
	Polyacrylamide					78	80	
	Agarose	*Geobacillus kaustophilus* ATCC 8005		Phosphate buffer 30 mM pH 7.0, 60 °C	6	74	150	[90]
	Agarose bentonite					65	580	
	Agarose	*E. coli*		Phosphate buffer 50 mM pH 7.0, 60 °C	4	72	480	[10]
	Agarose	*Geobacillus stearothermophilus* CECT 43	2,6-diaminopurine-2′-deoxyriboside		8		240	[84]
	Agarose	*Citrobacter koseri*	Fludarabine	Phosphate buffer 50 mM, 60 °C	14	58	—	[73]
			Vidarabine		26	62.5	—	
Cross-linking	—	*E. coli* BMT-1D/lA treated with gluraraldehyde	cladribine		21	65	—	[89]

of these reports uses commercial extracellular lipases that have been isolated by supernant concentration and purification, which is why many of these publications involve the use of immobilized isolated enzymes, a topic that will be addressed in further detail in Chapter 6. As mentioned, enzymatic acylation is a great alternative for the preparation of nucleosides analogues derivatives with increased activity and this task is generally performed by enzymatic preparations although there are a few reports using whole cells for their bioproduction. Whole cells are not often the preferred choice to perform these specific tasks because the number of enzyme activities carried inside the bacterial cell may interfere with the desired selectivity. And if whole cells cannot be used for the desired biocatalysis, the next step, which is immobilization, cannot be performed. However, there are some reports using whole cells; for example, lyophilized *Pseudomonas fluorescens* was employed for acylation of the antitumoral compound cytarabine (ara-C) using vinyl propionate as acyl donor in a mixture of organic solvents, obtaining good yields (77.1%) and excellent 5′-OH regioselectivity (97.3%) using isopropyl ether:pyridine (30 : 70) [94]. In a later work, the fungus *Aspergillus oryzae* was used as the whole cell catalyst for the acylation of ara-C with vinyl acetate in hexane–pyridine solvents. The activity of cell-bound lipase drastically increased in the early stages of cell growth and declined in the later stages, irrespective of the culture media used. These fungus cells showed more specificity toward the 3′-OH group, which was potentiated by the addition of the sorbitan fatty acid esters Span 80 (81% 3′-OH regioselectivity) [95]. The authors continued to optimize the reaction conditions using *A. oryzae* whole cell systems. In addition, the desired 3′-O-propional derivative of ara-C was synthesized with 88.3% yield and >70% regioselectivity [96]. These results showed that the use of whole cells presents a promising tool for the green and economic synthesis of nucleoside esters in the organic solvent mixture, considering the high cost of enzyme preparation and immobilization. Conversely, enzymatic hydrolysis of acetylated nucleosides was attempted using microbial whole cells but none of the tested microorganisms displayed a common deacetylation profile unlike that observed in CAL-B-catalyzed alcoholysis. Depending on the substrate and the biocatalyst used, 5′-selective deprotection or mixtures of mono O-acetylated products were obtained. Only *Cellulomonas cellulans*-catalyzed hydrolysis of 2′,3′,5′-tri-*O*-acetyladenosine and *Klebsiella* sp.-catalyzed hydrolysis of 2′,3′,5′-tri-*O*-acetyluridine were regioselective, furnishing the corresponding 5′-hydroxyl free derivatives in 79% and 71% yields, respectively [97]. Whole cell immobilized biocatalysts have been used for biodiesel production [98–100] that represents one of many of lipase's applications; some cell bound lipases have been used as whole cell biocatalysts for nucleoside selective regioacylation [101, 102], but so far immobilized whole cell biocatalysts have not been employed for fine pharmaceuticals productions, probably because of the unique selectivity required for this task and also because most lipases are extracellular [91, 92].

Glycosylation represents a very stabilized method for the synthesis of prodrugs, a typical example being the targeting of glycosylated drugs on the colon, where they are specifically hydrolyzed by the local intestinal anerobic microflora [103]. Nucleoside O-glycosylation represents an archetypal problem in chemical selectivity, in as much as the nucleobase (an undesired site of reaction) is usually more nucleophilic than the hydroxyl (the desired site of reaction) [36]. Different approaches have been used to glycosylate nucleosides using enzymes from animal (commercial β-galactosidase from bovine liver) for the regioselective galactosylation (92–100%) of pyrimidine 2′-deoxynucleosides with satisfactory yields (41–68%) [104], also acyclic nucleoside analogues derivatives were obtained using this enzyme. The important galactosylated antiviral agents ACV, GCV and penciclovir (PCV) were substrates of the enzyme with moderate yields of 29%, 26%, and 71% respectively [105]. Further, by purified bovine liver β-galactosidase with *p*-nitrophenyl and β-D-galactopyranoside as the substrate, the enzymatic synthesis of a group of β-fucosyl-containing disaccharide pyrimidine nucleosides, among which floxuridine was obtained with 44–60% yields, the enzyme could accept acyclic nucleoside analogues as the substrate and catalyze the enzymatic fucosylation of these nucleosides, achieving yields of 32–36% [106]. Also, glucosylation of a series of 2′-deoxynucleosides was described by using the crude β-glycosidase from bovine liver, which is less expensive and can be simply prepared in a standard organic laboratory. Glucosylation of 2′-deoxyuridine was used as a model reaction, different parameters were assayed obtaining the desired 5′-*O*-glucosylated derivatives of 2′-deoxynucleosides that were synthesized with yields of 22–72% and exclusive 5′-regioselectivities (>99%) [107]. Besides, β-galactosidases from bacterial sources have been successfully used for this purpose also to glycosylate acyclic nucleosides, using *E. coli* as a catalyst and lactose as a sugar donor [37], and also by using β-galactosidase from solvent-tolerant *Bacillus megaterium YZ08* β-galactosyl nucleoside analogues were successfully achieved in a 10% DMSO solvent system [108]. Glycosylation performed by whole living cells combines the advantages of the multienzyme approaches, e.g. high selectivity and cofactor regeneration and helps to avoid expensive and laborious isolation of the respective enzymes. Mostly, glycosyltransferases are responsible for the glycosyl transfer in these systems. There exist few examples, in which living microbial cells were used for glycosylations. There are also whole microbial cell biosystems possessing glycosidases with transglycosylation activity. These enzymes are usually bound to cells quite often in the periplasmatic space [19]. A special case of glycosylation by microorganisms is a preparation of *N*-2′-deoxy-β-ribosides of heterocyclic bases with NDTs, which will be thoroughly described in later chapters.

Regarding deamination, the enzymes ADA and adenylate deaminase (AMPDA) are the most important enzymes that perform deamination, dehalogenation, and demethoxylation of nucleosides [16, 20]. Although sometimes this activity is desired for the modification of specific NA, other times, it is avoided, then, adenosine deaminase activity should be considered during the screening stage of the candidates microorganisms. Intracellular ADA is widely used in the preparation of antiviral structures by specific deamination of 6-aminopurines, via and hydrolytic reaction that leaves the amino group of the 2′-amino sugar motif unaltered. Additionally it has been used with different substrates such as carbocyclonucleosides or acyclonucleosides [20]. This versatile enzyme has been employed in various preparations, and is traditionally used for the synthesis of guanosine or inosine nucleoside analogs [109] and also in the resolution of racemic nucleoside analogues [7, 110]. Regarding AMPDA, it has broader substrate specificity as it was successfully used for deamination of adenosine derivatives including phosphorylated cyclic, carbocyclic, and acyclic analogues. In addition, the enzyme can catalyze demethylation and dechlorination of purine ribonucleosides [20]. Whole cells microorganisms were used for the deamination of purine nucleosides, such as *Arthrobacter oxydans*, obtaining moderate yields [111], and coupled systems were also used first to transglycosylate and then to deaminate purine nucleosides since the use of guanine base for transglycosylation is difficult due to the low solubility of this compound [112]. Since long hours are needed to achieve the process, immobilization was not investigated further, as it may add hours to the final reaction time. To our knowledge, immobilized whole cells have not been used with this enzyme so far to modify nucleosides.

Finally, regarding halogenations, there are several reports in which biocatalytic halogenation of nucleosides is directed to the sugar moiety or toward the base moiety. Examples of biocatalytic halogenation to different substrates, such as elaboration of organofluorines in the presence of fluoride ion F^- have been described in 1986 [113], and in spite of such early acknowledgment less achievements have been made in this area. The enzyme that catalyzes this reaction is fluorinase and it was discovered in 2002 by O'Hagan et al. [114], and basically catalyzes the formation of C—F bonds through reversible transformation of *S*-adenosylmethionine into 5′-fluoro-5′-deoxyadenosine in the presence of F^-. This enzyme can also act as a chlorinase via the formation of the C—Cl bond, but does not accept Br^- and I^- to produce other compounds [114]. It has also been successfully employed in the synthesis of fluorinated analogues and coupled synthesis [115, 116]. The synthesis of halogenated nucleosides and nucleobases is of interest due to their chemical and pharmacological applications. The enzymatic halogenation of nucleobases and analogues catalyzed by using whole cell microorganisms and by chloroperoxidase from *Caldariomyces fumago* has been studied. The latter enzyme catalyzed the chlorination and bromination of indoline and uracil. *Pseudomonas*, *Citrobacter*, *Aeromonas*, *Streptomyces*, *Xanthomonas*, and *Bacillus* genera catalyzed the chlorination and/or bromination of indole and indoline. Different products were obtained depending on the substrate, the biocatalyst, and the halide used. In particular, 85% conversion from indole to 5-bromoindole was achieved using *Streptomyces cetonii* [117]. Recently, different marine microorganisms were identified as organisms capable of biocatalyzing halogenation reactions. In this study, two marine sediment bacteria were isolated from Mokpo, Jeollanam-do, South Korea, and showed 99% identity to parts of the 16S rDNA of bacterial species *Shewanella marisflavi* and *Pseudoalteromonas prydzensis*. These strains were capable of biohalogenation activity considered to have produced indigo and dibromoindigo from indole. This is the first report on marine sediment bacteria capable of biocatalyzing bromination of indole as resting whole cells [118]. The discovery of these new bacterial activities may lead to a broad range of future applications in the field of nucleoside halogenation. Similar to the lipase enzyme group, the use of enzymes or combined enzyme systems (clorinase-PNP, fluorinase- AMPDA) represents an interesting alternative for performing the reactions involving chlorinases.

4.5 Conclusion

Biocatalytic approaches represent an interesting plan of action for nucleic acid derivative synthesis. Much progress has been made since the early techniques for the preparation of these compounds, such as new immobilization protocols, the employment of friendly reaction conditions, and the shortening of reaction time, thereby reducing production time. Also, the use of biocatalysts allows for replacing multistep chemical reactions, and great progress in nucleoside analogue synthesis has been achieved by combining chemical methods and biochemical transformations, allowing the synthesis of a large variety of structurally diverse nucleosides. This combination represents a very promising technology and also shows high efficiency for biotechnological process development. Enzyme application in multistep reactions employing recombinant whole cells expressing multiple enzymes or sequential bio/chemocatalyzed transformations provides efficient pathways for the synthesis of complex compounds in a one-pot process.

Considering how important nucleoside analogues are in medicine and biotechnology, there is considerable interest in the development of simple and efficient synthesis of these compounds. The low cost and high productivity of biocatalysis technology allows obtaining modified nucleosides with potential use in cancer and antiviral therapies. The use of whole cells immobilized biocatalysts appears to be an excellent choice for manufacturing pharmaceuticals derivatives, owing to its ease of production, substrate specificity, green chemistry, low cost, and great and fast availability. They offer many advantages such as fast reaction conditions, and the enzymes are protected within the cell in their natural environment. The choice of support is of vital importance for a successful biocatalyst development; a number of matters must be taken into consideration, the most relevant being cost, nature and intended use of the biocatalyst, reaction conditions (temperature, pH, media), and scale of production, among others. The vast majority of reports using whole cells biocatalysts employs entrapment immobilization, being thermogels and hydrogels the preffered matrices.

In spite of the laborious and costly processes that must be carried on in order to obtain purified enzymes, these biocatalysts constitute a very important part bioprocess technology. Due to their high chemo-, stereo-, and regioselectivity, they are an excellent choice in the synthesis of fine pharmaceuticals, and in cases where only an enzymatic activity is desired they are a better alternative to whole cells, as by products can interfere in the reactions. So, these considerations must be taken before realizing a biocatalytical design. Nevertheless, the use of whole cell immobilized catalysts jointly with immobilized enzymes appears to be an interesting approach to overcome these potential problems.

References

1 Mikhailopulo, I.A. and Miroshnikov, A.I. (2010). New trends in nucleoside biotechnology. *Acta Naturae* 2 (2): 36–59.

2 Lapponi, M.J., Rivero, C.W., Zinni, M.A. et al. (2016). New developments in nucleoside analogues biosynthesis: a review. *J. Mol. Catal. B: Enzym.* 133: 218–233.

3 Jordheim, L.P., Durantel, D., Zoulim, F., and Dumontet, C. (2013). Advances in the development of nucleoside and nucleotide analogues for cancer and viral diseases. *Nat. Rev. Drug Discovery* 12 (6): 447–464.

4 Galmarini, C.M., Mackey, J.R., and Dumontet, C. (2002). Nucleoside analogues and nucleobases in cancer treatment. *Lancet Oncol.* 3 (7): 415–424.

5 Ichikawa, E. and Kato, K. (2001). Sugar-modified nucleosides in past 10 years, a review. *Curr. Med. Chem.* 8 (4): 385–423.

6 Li, N., Smith, T.J., and Zong, M.H. (2010). Biocatalytic transformation of nucleoside derivatives. *Biotechnol. Adv.* 28 (3): 348–366.

7 Fernández-Lucas, J. (2006). Enzymatic synthesis of modified nucleosides. In: *Biocatalysis in the Pharmaceutical and Biotechnology Industries* (ed. R. Patel), 401–424. CRC Press.

8 Trelles, J.A., Fernandez, M., Lewkowicz, E.S. et al. (2003). Purine nucleoside synthesis from uridine using immobilised *Enterobacter gergoviae* CECT 875 whole cell. *Tetrahedron Lett.* 44: 2605–2609.

9 Fernandez-Lucas, J., Fresco-Taboada, A., Acebal, C. et al. (2011). Enzymatic synthesis of nucleoside analogues using immobilized 2′-deoxyribosyltransferase from *Lactobacillus reuteri*. *Appl. Microbiol. Biotechnol.* 91 (2): 317–327.

10 Luo, W., Liu, Y., Zhu, X. et al. (2011). Cloning and characterization of purine nucleoside phosphorylase in *Escherichia coli* and subsequent ribavirin biosynthesis using immobilized recombinant cells. *Enzyme Microb. Technol.* 48 (6–7): 438–444.

11 Mikhailopulo, I.A. and Miroshnikov, A.I. (2013). Some recent findings in the biotechnology of biologically important nucleosides. *Biotechnol. Acta* 6 (4).

12 De Benedetti, E.C., Rivero, C.W., and Trelles, J.A. (2015). Development of a nanostabilized biocatalyst using an extremophilicmicroorganism for ribavirin biosynthesis. *J. Mol. Catal. B: Enzym.* 121: 90–95.

13 Bzowska, A., Kulikowska, E., and Shugar, D. (2000). Purine nucleoside phosphorylases: properties, functions, and clinical aspects. *Pharmacol. Ther.* 88 (3): 349–425.

14 Fresco-Taboada, A., de la Mata, I., Arroyo, M., and Fernandez-Lucas, J. (2013). New insights on nucleoside 2′-deoxyribosyltransferases: a versatile biocatalyst for one-pot one-step synthesis of nucleoside analogs. *Appl. Microbiol. Biotechnol.* 97 (9): 3773–3785.

15 Okuyama, K., Shibuya, S., Hamamoto, T., and Noguchi, T. (2003). Enzymatic synthesis of 2′-deoxyguanosine with nucleoside deoxyribosyltransferase-II. *Biosci. Biotechnol., Biochem.* 67 (5): 989–995.

16 Gupta, M. and Nair, V. (2006). Adenosine deaminase in nucleoside synthesis. A review. *Collect. Czech. Chem. Commun.* 71: 769–787.

17 Angajala, G., Pavan, P., and Subashini, R. (2016). Lipases: an overview of its current challenges and prospectives in the revolution of biocatalysis. *Biocatal. Agr. Biotechnol.* 7 (Suppl. C): 257–270.

18 Klibanov, A.M. (2001). Improving enzymes by using them in organic solvents. *Nature* 409 (6817): 241–246.
19 Kren, V. and Thiem, J. (1997). Glycosylation employing bio-systems: from enzymes to whole cells. *Chem. Soc. Rev.* 26 (6): 463–473.
20 Santaniello, E., Ciuffreda, P., and Alessandrini, L. (2005). Synthesis of modified purine nucleosides and related compounds mediated by adenosine deaminase (ADA) and adenylate deaminase (AMPDA). *Synthesis* 2005 (4): 509–526.
21 De Clercq, E. (2013). Antivirals: past, present and future. *Biochem. Pharmacol.* 85 (6): 727–744.
22 Trelles, J.A. and Lapponi, M.J. (2018). Immobilization techniques applied to the development of biocatalysts for the synthesis of nucleoside analogue derivatives. *Curr. Pharm. Des.* 23 (45): 6879–6897.
23 Li, F., Maag, H., and Alfredson, T. (2008). Prodrugs of nucleoside analogues for improved oral absorption and tissue targeting. *J. Pharm. Sci.* 97 (3): 1109–1134.
24 Johnson, Z.L., Lee, J.H., Lee, K. et al. (2014). Structural basis of nucleoside and nucleoside drug selectivity by concentrative nucleoside transporters. *eLife* 3: e03604.
25 Périgauda, C., Gosselina, G., and Imbacha, J.L. (1992). Nucleoside analogues as chemotherapeutic agents: a review. *Nucleosides Nucleotides Nucleic Acids* 11 (2–4): 903–945.
26 Bergman, A.M., Kuiper, C.M., Voorn, D.A. et al. (2004). Antiproliferative activity and mechanism of action of fatty acid derivatives of arabinofuranosylcytosine in leukemia and solid tumor cell lines. *Biochem. Pharmacol.* 67 (3): 503–511.
27 De Clercq, E. and Field, H.J. (2006). Antiviral prodrugs – the development of successful prodrug strategies for antiviral chemotherapy. *Br. J. Pharmacol.* 147 (1): 1–11.
28 Wang, J.X., Sun, X., and Zhang, Z.R. (2002). Enhanced brain targeting by synthesis of 3′,5′-dioctanoyl-5-fluoro-2′-deoxyuridine and incorporation into solid lipid nanoparticles. *Eur. J. Pharm. Biopharm.* 54 (3): 285–290.
29 Han, H., de Vrueh, R.L., Rhie, J.K. et al. (1998). 5′-Amino acid esters of antiviral nucleosides, acyclovir, and AZT are absorbed by the intestinal PEPT1 peptide transporter. *Pharm. Res.* 15 (8): 1154–1159.
30 Ferrero, M. and Gotor, V. (2000). Biocatalytic selective modifications of conventional nucleosides, carbocyclic nucleosides, and *C*-nucleosides. *Chem. Rev.* 100 (12): 4319–4348.
31 Moris, F. and Gotor, V. (1993). A useful and versatile procedure for the acylation of nucleosides through an enzymic reaction. *J. Org. Chem.* 58 (3): 653–660.
32 Gao, W.L., Li, N., and Zong, M.H. (2013). Enzymatic regioselective acylation of nucleosides in biomass-derived 2-methyltetrahydrofuran: kinetic study and enzyme substrate recognition. *J. Biotechnol.* 164 (1): 91–96.
33 Yoo, C.B. and Jones, P.A. (2006). Epigenetic therapy of cancer: past, present and future. *Nat. Rev. Drug Discovery* 5 (1): 37–50.

34 Herdewijn, P. (2008). *Modified Nucleosides in Biochemistry Biotechnology and Medicine*. Weinheim: Wiley-VCH.

35 Mizuma, T., Ohta, K., Hayashi, M., and Awazu, S. (1992). Intestinal active absorption of sugar-conjugated compounds by glucose transport system: implication of improvement of poorly absorbable drugs. *Biochem. Pharmacol.* 43 (9): 2037–2039.

36 Zhang, Y. and Knapp, S. (2016). Glycosylation of nucleosides. *J. Org. Chem.* 81 (6): 2228–2242.

37 Blažek, J., Jansa, P., Baszczyňski, O. et al. (2012). An enzymatic glycosylation of nucleoside analogues using β-galactosidase from *Escherichia coli*. *Bioorg. Med. Chem.* 20 (9): 3111–3118.

38 Efimtseva, E.V. and Mikhailov, S.N. (2004). Disaccharide nucleosides. *Russ. Chem. Rev.* 73: 401–414.

39 Niu, G. and Tan, H. (2015). Nucleoside antibiotics: biosynthesis, regulation, and biotechnology. *Trends Microbiol.* 23 (2): 110–119.

40 Sun, R. and Wang, L. (2013). Inhibition of *Mycoplasma pneumoniae* growth by FDA-approved anticancer and antiviral nucleoside and nucleobase analogs. *BMC Microbiol.* 13: 184.

41 Struthers, H., Viertl, D., Kosinski, M. et al. (2010). Charge dependent substrate activity of C3′ and N3 functionalized, organometallic technetium and rhenium-labeled thymidine derivatives toward human thymidine kinase 1. *Bioconjugate Chem.* 21 (4): 622–634.

42 Alberto, R., N'Dongo, H.P., Clericuzio, M. et al. (2009). Functionalized thymidine derivatives as carriers for the γ-emitter technetium tricarbonyl moiety. *Inorg. Chim. Acta* 362 (13): 4785–4790.

43 Kassis, A.I., Adelstein, S.J., and Mariani, G. (1996). Radiolabeled nucleoside analogs in cancer diagnosis and therapy. *Q. J. Nucl. Med.* 40 (3): 301–319.

44 Boutureira, O., Matheu, M. I., Díaz, Y., Castillón, S. (2013). Synthesis of C-Nucleosides. In: *Chemical Synthesis of Nucleoside Analogues* (ed. P. Merino), 263–316. Hoboken, NJ: Wiley.

45 Trelles, J.A. and Rivero, C.W. (2013). Whole cell entrapment techniques. In: *Immobilization of Enzymes and Cells*, 3rde (ed. M.J. Guisan), 365–374. Totowa, NJ: Humana Press.

46 Zucca, P., Fernandez-Lafuente, R., and Sanjust, E. (2016). Agarose and its derivatives as supports for enzyme immobilization. *Molecules* 21 (11).

47 Gotovtsev, P.M., Yuzbasheva, E.Y., Gorin, K.V. et al. (2015). Immobilization of microbial cells for biotechnological production: modern solutions and promising technologies. *Appl. Biochem. Microbiol.* 51 (8): 792–803.

48 Sheldon, R.A. and van Pelt, S. (2013). Enzyme immobilisation in biocatalysis: why, what and how. *Chem. Soc. Rev.* 42 (15): 6223–6235.

49 Michelini, E. and Roda, A. (2012). Staying alive: new perspectives on cell immobilization for biosensing purposes. *Anal. Bioanal. Chem.* 402 (5): 1785–1797.

50 Klein, J. and Ziehr, H. (1990). Immobilization of microbial cells by adsorption. *J. Biotechnol.* 16 (1–2): 1–15.

51 Guisan, J.M. (2013). New opportunities for immobilization of enzymes. *Methods Mol. Biol.* 1051: 1–13.

52 Hung, C.P., Lo, H.F., Hsu, W.H. et al. (2008). Immobilization of *Escherichia coli* novablue gamma-glutamyltranspeptidase in Ca-alginate-κ-carrageenan beads. *Appl. Biochem. Biotechnol.* 150 (2): 157–170.

53 Shin, H.J. (2012). Agarose-gel-immobilized recombinant bacterial biosensors for simple and disposable on-site detection of phenolic compounds. *Appl. Microbiol. Biotechnol.* 93 (5): 1895–1904.

54 Haraguchi, K. (2007). Nanocomposite hydrogels. *Curr. Opin. Solid State Mater. Sci.* 11: 47, 47–54, 54.

55 Cappa, V.A., Rivero, C.W., Sambeth, J.E., and Trelles, J.A. (2016). Bioproduction of floxuridine using nanostabilized biocatalysts. *Chem. Eng. Technol.* 39 (9): 1723–1730.

56 Orive, G., Hernández, R.M., Gascón, A.R., and Pedraz, J.L. (2006). Encapsulation of cells in alginate gels. In: *Immobilization of Enzymes and Cells* (ed. J.M. Guisan), 345–355. Totowa, NJ: Humana Press.

57 Górecka, E. and Jastrzębska, M. (2011). Immobilization techniques and biopolymer carriers – a review. *Biotechnol. Food Sci.* 75 (1–2): 27–34.

58 Brena, B., Gonzalez-Pombo, P., and Batista-Viera, F. (2013). Immobilization of enzymes: a literature survey. *Methods Mol. Biol.* 1051: 15–31.

59 Gupta, M.N. and Mattiasson, B. (1992). Unique applications of immobilized proteins in bioanalytical systems. *Methods Biochem. Anal.* 36: 1–34.

60 Datta, S., Christena, L.R., and Sriramulu Rajaram, Y.R. (2013). Enzyme immobilization: an overview on techniques and support materials. *3 Biotech* 3: 1–9.

61 Chang, M.-Y. and Juang, R.-S. (2007). Use of chitosan–clay composite as immobilization support for improved activity and stability of β-glucosidase. *Biochem. Eng. J.* 35 (1): 93–98.

62 Zhou, C. and Wu, Q. (2011). A novel polyacrylamide nanocomposite hydrogel reinforced with natural chitosan nanofibers. *Colloids Surf., B* 84 (1): 155–162.

63 Paul, D.R. and Robeson, L.M. (2008). Polymer nanotechnology: nanocomposites. *Polymer* 49 (15): 3187–3204.

64 An, N., Zhou, C.H., Zhuang, X.Y. et al. (2015). Immobilization of enzymes on clay minerals for biocatalysts and biosensors. *Appl. Clay Sci.* 114: 283–296.

65 Li, S., Meng Lin, M., Toprak, M.S. et al. (2010). Nanocomposites of polymer and inorganic nanoparticles for optical and magnetic applications. *Nano Rev.* 1: doi: 10.3402/nano.v1i0.5214.

66 Holy, A. and Votruba, I. (1987). *Nucleic Acids Symp. Ser.* 18: 69–72.

67 Ewing, D.F., Holý, A., Votruba, I. et al. (1992). Synthesis of 4- and 5-amino-1-(2-deoxy-D-erythro-pentofuranosyl)imidazole nucleosides by chemical and biotransformation methods. *Carbohydr. Res.* 216 (Suppl. C): 109–118.

68 Prasad, A.K., Trikha, S., and Parmar, V.S. (1999). Nucleoside synthesis mediated by glycosyl transferring enzymes. *Bioorg. Chem.* 27 (2): 135–154.

69 Trelles, J.A., Fernández-Lucas, J., Condezo, L.A., and Sinisterra, J.V. (2004). Nucleoside synthesis by immobilised bacterial whole cells. *J. Mol. Catal. B: Enzym.* 30 (5): 219–227.

70 Bickerstaff, G.F. (1997). Immobilization of enzymes and cells. In: *Immobilization of Enzymes and Cells* (ed. G.F. Bickerstaff), 1–11. Totowa, NJ: Humana Press.

71 Fernández-Lucas, J., Condezo, L.A., Martinez-Lagos, F., and Sinisterra, J.V. (2007). Synthesis of 2′-deoxyibosylnucleosides using new 2′-deoxyribosyltransferase microorganism producers. *Enzyme Microb. Technol.* 40 (5): 1147–1155.

72 Rivero, C.W., De Benedetti, E.C., Lozano, M.E., and Trelles, J.A. (2015). Bioproduction of ribavirin by green microbial biotransformation. *Process Biochem.* 50 (6): 935–940.

73 Nóbile, M., Médici, R., Terreni, M., Lewkowic, E.S., Iribarren, A.M. (2012). Use of *Citrobacter koseri* whole cells for the production of arabinonucleosides: a larger scale approach among others. *Process Biochem.* 47: 2182–2188.

74 Pianko, S. and McHutchison, J.G. (2000). Treatment of hepatitis C with interferon and ribavirin. *J. Gastroenterol. Hepatol.* 15 (6): 581–586.

75 Ferenci, P., Brunner, H., Nachbaur, K. et al. (2001). Combination of interferon induction therapy and ribavirin in chronic hepatitis C. *Hepatology* 34 (5): 1006–1011.

76 Borden, K.L. and Culjkovic-Kraljacic, B. (2010). Ribavirin as an anti-cancer therapy: acute myeloid leukemia and beyond? *Leuk. Lymphoma* 51 (10): 1805–1815.

77 Pettersson, F., Yau, C., Dobocan, M.C. et al. (2011). Ribavirin treatment effects on breast cancers overexpressing eIF4E, a biomarker with prognostic specificity for luminal B-type breast cancer. *Clin. Cancer Res.* 17 (9): 2874–2884.

78 Liu, P., Sharon, A., and Chu, C.K. (2008). Fluorinated nucleosides: synthesis and biological implication. *J. Fluorine Chem.* 129 (9): 743–766.

79 Rivero, C.W., Britos, C.N., Lozano, M.E. et al. (2012). Green biosynthesis of floxuridine by immobilized microorganisms. *FEMS Microbiol. Lett.* 331 (1): 31–36.

80 Britos, C.N., Cappa, V.A., Rivero, C.W. et al. (2012). Biotransformation of halogenated 2′-deoxyribosides by immobilized lactic acid bacteria. *J. Mol. Catal. B: Enzym.* 79: 49–53.

81 Cappa, V.A., Rivero, C.W., Britos, C.N. et al. (2014). An efficient biocatalytic system for floxuridine biosynthesis based on *Lactobacillus animalis* ATCC 35046 immobilized in Sr-alginate. *Process Biochem.* 49: 1169–1175.

82 Cappa, V.A. and Trelles, J.A. (2017). One-pot biosynthesis of idoxuridine using nanostabilized lactic acid bacteria. *Process Biochem.* 62 (Suppl. C): 169–173.

83 Trelles, J.A., Valino, A.L., Runza, V. et al. (2005). Screening of catalytically active microorganisms for the synthesisof 6-modified purine nucleosides. *Biotechnol. Lett.* 27 (11): 759–763.

84 De Benedetti, E.C., Rivero, C.W., Britos, C.N. et al. (2012). Biotransformation of 2,6-diaminopurine nucleosides by immobilized *Geobacillus stearothermophilus*. *Biotechnol. Progr.* 28 (5): 1251–1256.

85 Beutler, E. (1992). Cladribine (2-chlorodeoxyadenosine). *Lancet* 340 (8825): 952–956.

86 Robak, T., Jamroziak, K., Gora-Tybor, J. et al. (2010). Comparison of cladribine plus cyclophosphamide with fludarabine plus cyclophosphamide as first-line therapy for chronic lymphocytic leukemia: a phase III randomized study by the Polish Adult Leukemia Group (PALG-CLL3 Study). *J. Clin. Oncol.* 28 (11): 1863–1869.

87 Sigal, D.S., Miller, H.J., Schram, E.D., and Saven, A. (2010). Beyond hairy cell: the activity of cladribine in other hematologic malignancies. *Blood* 116 (16): 2884–2896.

88 Holowiecki, J., Grosicki, S., Giebel, S. et al. (2012). Cladribine, but not fludarabine, added to daunorubicin and cytarabine during induction prolongs survival of patients with acute myeloid leukemia: a multicenter, randomized phase III study. *J. Clin. Oncol.* 30 (20): 2441–2448.

89 Mikhailopulo, I.A., Zinchenko, A.I., Kazimierczuk, Z. et al. (1993). Synthesis of 2-chloro-2′-deoxyadenosine by microbiological transglycosylation. *Nucleosides Nucleotides Nucleic Acids* 12 (3–4): 417–422.

90 De Benedetti, E.C., Rivero, C.W., and Trelles, J.A. (2015). Development of a nanostabilized biocatalyst using an extremophilic microorganism for ribavirin biosynthesis. *J. Mol. Catal. B: Enzym.* 121: 90–95.

91 Casas-Godoy, L., Duquesne, S., Bordes, F. et al. (2012). Lipases: an overview. In: *Lipases and Phospholipases: Methods and Protocols* (ed. G. Sandoval), 3–30. Totowa, NJ: Humana Press.

92 Gotor-Fernández, V., Brieva, R., and Gotor, V. (2006). Lipases: useful biocatalysts for the preparation of pharmaceuticals. *J. Mol. Catal. B: Enzym.* 40 (3): 111–120.

93 Palomo, J.M., Muñoz, G., Fernández-Lorente, G. et al. (2002). Interfacial adsorption of lipases on very hydrophobic support (octadecyl–Sepabeads): immobilization, hyperactivation and stabilization of the open form of lipases. *J. Mol. Catal. B: Enzym.* 19–20: 279–286.

94 Li, X.F., Lu, Z.H., Zhao, G.L. et al. (2012). A facile whole-cell biocatalytic approach to regioselective synthesis of monoacylated 1-beta-D-arabinofuranosylcytosine: influence of organic solvents. *Bioresour. Technol.* 114: 6–11.

95 Li, X.F., Zhu, Z., Zhao, G.L. et al. (2012). A novel biocatalytic approach to acetylation of 1-beta-D-arabinofuranosylcytosine by *Aspergillus oryzae* whole cell in organic solvents. *Appl. Microbiol. Biotechnol.* 93 (1): 143–150.

96 Yang, M.Y., Wu, H., Lu, Z.H. et al. (2014). Regioselective synthesis of cytarabine monopropionate by using a fungal whole-cell biocatalyst in nonaqueous medium. *Bioorg. Med. Chem. Lett.* 24 (15): 3377–3380.

97 Panero, J., Trelles, J., Rodano, V. et al. (2006). Microbial hydrolysis of acetylated nucleosides. *Biotechnol. Lett.* 28 (14): 1077–1081.

98 Tamalampudi, S., Talukder, M.R., Hama, S. et al. (2008). Enzymatic production of biodiesel from Jatropha oil: a comparative study of immobilized-whole cell and commercial lipases as a biocatalyst. *Biochem. Eng. J.* 39 (1): 185–189.

99 Xiao, M.A.N., Qi, C., and Obbard, J.P. (2011). Biodiesel production using *Aspergillus niger* as a whole-cell biocatalyst in a packed-bed reactor. *GCB Bioenergy* 3 (4): 293–298.

100 Hemachander, C., Bose, N., and Puvanakrishnan, R. (2001). Whole cell immobilization of *Ralstonia pickettii* for lipase production. *Process Biochem.* 36 (7): 629–633.

101 Yang, M., Wu, H., Lian, Y. et al. (2014). Influence of organic solvents on catalytic behaviors and cell morphology of whole-cell biocatalysts for synthesis of 5′-arabinocytosine laurate. *PLoS One* 9 (8): e104847.

102 Yang, M., Wu, H., Lian, Y. et al. (2014). Using ionic liquids in whole-cell biocatalysis for the nucleoside acylation. *Microb. Cell Fact.* 13: 143.

103 Sinha, V.R. and Kumria, R. (2001). Colonic drug delivery: prodrug approach. *Pharm. Res.* 18 (5): 557–564.

104 Ye, M., Yan, L.-Q., Li, N., and Zong, M.-H. (2012). Facile and regioselective enzymatic 5′-galactosylation of pyrimidine 2′-deoxynucleosides catalyzed by β-glycosidase from bovine liver. *J. Mol. Catal. B: Enzym.* 79 (Suppl. C): 35–40.

105 Yan, L.-Q., Li, N., and Zong, M.-H. (2014). First enzymatic galactosylation of acyclic nucleoside drugs by β-galactosidase: synthesis of water-soluble β-D-galactosidic prodrugs. *Biotechnol. Bioprocess Eng.* 19 (4): 586–591.

106 Yan, L.-Q., Li, N., and Zong, M.-H. (2013). First and facile enzymatic synthesis of β-fucosyl-containing disaccharide nucleosides through β-galactosidase-catalyzed regioselective glycosylation. *J. Biotechnol.* 164 (2): 371–375.

107 Ye, M., Yu, C.-Y., Li, N., and Zong, M.-H. (2011). Highly regioselective glucosylation of 2′-deoxynucleosides by using the crude β-glycosidase from bovine liver. *J. Biotechnol.* 155 (2): 203–208.

108 Youzhi Zhou, J.C., Zhang, J., Ke, L., and He, B. (2016). Precisely regulated galactosylation of nucleoside analogues in aqueous hydrophilic solvents catalyzed by solvent-stable β-galactosidase. *RSC Adv.* 6.

109 Kakefuda, A., Shuto, S., Nagahata, T. et al. (2010). Nucleosides and nucleotides. 132. Synthesis and biological evaluations of ring-expanded oxetanocin analogues: Purine and pyrimidine analogues of 1,4-anhydro-2-deoxy-d-arabitol and 1,4-anhydro-2-deoxy-3-hydroxymethyl-d-arabitol. *Tetrahedron* 50 (34): 10167–10182.

110 Vince, R. and Brownell, J. (1990). Resolution of racemic carbovir and selective inhibition of human immunodeficiency virus by the (−) enantiomer. *Biochem. Biophys. Res. Commun.* 168 (3): 912–916.

111 Médici, R., Lewkowicz, E., S., and Iribarren, A., M. (2009). *Arthrobacter oxydans* as a biocatalyst for purine deamination. *FEMS Microbiol. Lett.* 289 (1): 20–26.

112 Medici, R., Taverna, M., Lewkowicz, E. et al. (2008). Coupled biocatalysts applied to the synthesis of nucleosides. *Nucleic Acids Symp. Ser. (Oxf.)* 52: 541–542.

113 O'Hagan, D., Schaffrath, C., Cobb, S. L., Hamilton, J. T. G., and Murphy, C. D. Nature 2002, 416, 279.

114 Sanada, M., Miyano, T., Iwadare, S. et al. (1986). Biosynthesis of fluorothreonine and fluoroacetic acid by the thienamycin producer, *Streptomyces cattleya*. *J. Antibiot.* 39 (2): 259–265.

115 Deng, H., Cobb, S.L., McEwan, A.R. et al. (2006). The fluorinase from *Streptomyces cattleya* is also a chlorinase. *Angew. Chem. Int. Ed.* 45 (5): 759–762.

116 Martarello, L., Schaffrath, C., Deng, H. et al. (2003). The first enzymatic method for C–18F bond formation: the synthesis of 5′-[18F]-fluoro-5′-deoxyadenosine for imaging with PET. *J. Labelled Compd. Radiopharm.* 46 (13): 1181–1189.

117 Médici, R., Ignacio Garaycoechea, J., Andrés Dettorre, L. et al. (2011). Biocatalysed halogenation of nucleobase analogues. *Biotechnol. Lett.* 33 (10): 1999–2003.

118 Galarion, L.H. and Rivera, W. (2016). Isolation and characterization of marine sediment bacteria capable of biocatalyzing bromination of indole. *J. Sci. Ind. Res.* 75 (6): 359–364.

5

Enzymatic Synthesis of Nucleic Acid Derivatives by Immobilized Enzymes

Jesús Fernández-Lucas[1] *and Miguel Arroyo*[2]

[1] *Universidad Europea de Madrid, Facultad de Ciencias Biomédicas, Departamento de Farmacia y Biotecnología, Urbanización "El Bosque," Calle Tajo s/n, 28670 Villaviciosa de Odón, Spain*
[2] *Universidad Complutense de Madrid, Facultad de Ciencias Biológicas, Departamento de Bioquímica y Biología Molecular, c/José Antonio Novais 2, 28040 Madrid, Spain*

5.1 Introduction

Nucleic acid derivatives (NADs) are important molecules that display antiviral and anticancer activity due to their capacity for inhibiting DNA synthesis. Moreover, NADs can be also used as starting materials for the synthesis of oligonucleotides for antisense therapy [1]. Such compounds have been formerly synthesized by traditional organic chemistry, which often requires time-consuming multistep processes, including protection and deprotection reactions on the heterocycle base and/or the pentose moiety to allow the modification of natural nucleosides and nucleotides [2]. As an alternative, biocatalysis has become an increasingly handy tool in the industrial production of NADs. Actually, NADs synthesis employing whole cell or isolated enzymes is considered an ecofriendly and sustainable approach since many steps can be reduced in the process, and mild reaction conditions are usually employed [3]. Moreover, chemo-, enantio-, and regioselectivity of enzymes on such complex molecules have demonstrated to be excellent features for the preparation of NADs. Several types of enzymatic reactions can be carried out to obtain different nucleoside and nucleotide analogues, such as acylation, deacylation, transglycosylation, halogenation, and deamination reactions. Microbial hydrolases, glycosyltransferases, kinases, aldolases, and isomerases are some examples of useful enzymes for this purpose. From an industrial point of view, these enzymes should tolerate harsh conditions, including high temperatures, extreme pH values, high substrate and product concentrations, and the presence of organic cosolvents. In such conditions, the robustness of enzymes can be increased by the most appropriate immobilization technique that would allow the preparation of suitable industrial biocatalysts to be used repeatedly, leading to a cost-effective process [4].

Enzymatic and Chemical Synthesis of Nucleic Acid Derivatives, First Edition.
Edited by Jesús Fernández-Lucas and María-José Camarasa Rius.

5.2 Immobilized Glycosyltransferases

Some glycosyltransferases have demonstrated to be very efficient for the enzymatic synthesis of natural and modified nucleosides with antiviral and antitumoral activity. In general, nucleoside phosphorylases (NPs) and nucleoside 2′-deoxyribosyltransferases (NDTs) have been used in nucleosides synthesis by mediating the transfer of glycosyl residues to acceptor bases. Nevertheless, recycling and improved stability of these enzymes must be accomplished prior to their application in industrial bioreactors. Among all immobilization approaches, attachment to solid supports (natural or synthetic) has been the preferred choice to reach these objectives. Supports are selected on the basis of their mechanical strength, high enzyme loading capacity, physical stability, nontoxicity and cheap cost. In this context, epoxy-activated synthetic carriers such as Eupergit-C® and Sepabeads® as well as ionic and aldehyde-activated agaroses have been identified as suitable supports for the immobilization of both NPs and NDTs.

5.2.1 Immobilized Nucleoside Phosphorylases

Nucleosides can be prepared via a one-pot enzymatic transglycosylations catalyzed by NPs [5]. In a simultaneous cascade, a first NP catalyzes the phosphorolysis of a nucleoside leading to a sugar-1-phosphate, while a second NP catalyzes the synthesis of a new nucleoside in the presence of a second nucleobase. NPs are multimeric enzymes that can be divided into two main structural groups: NPI and NPII. Enzymes from the NPI group share a conserved structural fold and include purine nucleoside phosphorylases (PNPs, EC 2.4.2.1) and uridine phosphorylases (UPs, EC 2.4.2.3). On the other hand, members of the NPII group are structurally different from the NPI group, and include thymidine phosphorylases (TP, EC 2.4.2.4) and pyrimidine nucleoside phosphorylases (PyNPs, EC 2.4.2.2). NPs from bacteria are generally preferred for the synthesis of nucleoside analogues due to their broader substrate specificity, and many of them have been cloned and expressed. Nowadays, application of immobilized NPs in one-pot transglycosylations has demonstrated to be as efficient as the application of immobilized whole cells for nucleoside production [6].

5.2.1.1 Stabilization of Nucleoside Phosphorylases by Immobilization

NPs are multimeric enzymes that may be easily deactivated during their immobilization, likely due to dissociation of their protein subunits or their uncorrected assembly. Different immobilization strategies have been tried in order to prevent such events and thus stabilize these biocatalysts [4, 7]. Likewise, robust immobilized enzymes are obviously required under extreme conditions for protein stability such as the presence of organic cosolvents or high pH and temperature values. Actually, in the enzymatic synthesis of some purine nucleosides, such conditions are needed to ensure the solubility of guanine, which is hardly dissolved in aqueous buffer [8]. At the beginning, a simple ionic attachment to an ion-charged support was enough to afford the required stabilization at very high temperatures, such as in the case of PyNP and PNP from *Bacillus stearothermophilus* adsorbed on DEAE-Toyopearl 650 M resin [9] and the co-immobilization of recombinant UP and PNP from *Escherichia coli* on DEAE-52 [10]. Nevertheless, an enzyme adsorption on different ionic supports followed by a post-immobilization technique (such as cross-linking with a poly-aldehyde macromolecule) has currently become an interesting approach to significantly increase the stability of NPs. In this sense, immobilization of trimeric UP from *Bacillus subtilis* was successfully achieved in epoxy-activated Sepabeads EC-EP3 coated with poly(ethyleneimine) (PEI, whose molecular weight was 20 000 Da), and finally treated with a 20% oxidized dextran (Dx), affording a covalent multipoint cross-linking between the enzyme subunits and the support. The obtained immobilized derivative (UP-Sep-PEI-Dx) was highly stable even under drastic experimental conditions, maintaining its activity for almost six hours at pH 10 and 45 °C unlike the free enzyme that was quickly deactivated [11]. Similarly, homodimeric thymidine phosphorylase from *E. coli* (*Ec*TP) was also immobilized by the same procedure, allowing the preparation of an active biocatalyst, which resulted up to sixfold and threefold more stable than the free enzyme and its adsorbed (but not cross-linked) counterpart, respectively, at pH 10 and 37 °C [12]. In addition, covalent multipoint attachment to highly activated supports has also contributed to a strong stabilization of some NPs, despite its detrimental effect on enzyme activity. As an example, tetrameric PNP from *B. subtilis* was covalently immobilized on glyoxyl-agarose in the presence of stabilizing additives ($MgCl_2$ and inosine) as well as a surface-active agent (Triton X-100), rendering an immobilized biocatalyst that kept 44% of its activity after 24 hours at pH 10 and 45 °C, whereas the free PNP was completely inactivated in a few minutes [11]. Likewise, recombinant UP and PNP from *E. coli* were co-immobolized on epoxy-activated Sepabeads EC-EP, and the obtained biocatalyst showed no reduction of activity after its storage for more than one year [13]. Finally, several multimeric NPs from different microbial origin were covalently linked to aldehyde-activated agarose, allowing the preparation of immobilized biocatalysts that were highly stable at pH 10, as well as in the presence of methanol or acetonitrile as cosolvent [14].

5.2.1.2 Synthesis of Nucleosides Catalyzed by Immobilized Nucleoside Phosphorylases

Many immobilized biocatalysts have been successfully employed in the one-pot synthesis of several natural and unnatural nucleosides under unfriendly reaction conditions. When substrates and/or products are poorly soluble in water, enzymatic transglycosylations may be carried out at very high temperatures (up to 60–70 °C) and therefore, robust immobilized biocatalysts must be employed to catalyze such reactions instead of their free counterparts. As a first example, immobilized crude preparation from *B. stearothermophilus* containing PyNP and PNP was used for the continuous production of 5-methyluridine (Figure 5.1), a starting compound for the synthesis of 3′-azido-2′,3′-dideoxythymidine (AZT, zidovudine, or retrovir). Half-lives of the immobilized enzymes activity were 47 and 4.5 days, at 60 and 70 °C, respectively, whereas the half-life of the crude enzymes activity was 14 hours at 70 °C [9].

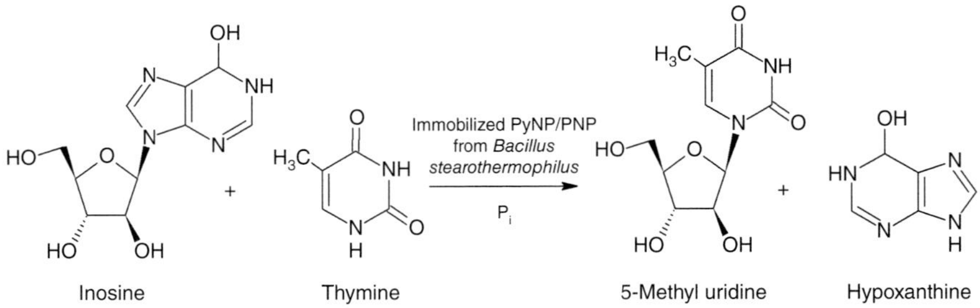

Figure 5.1 Enzymatic synthesis of 5-methyluridine catalyzed by immobilized PyNP and PNP from *Bacillus stearothermophilus* JTS 859. One milliliter of the immobilized biocatalyst was packed into a column reactor (7 × 26 mm) 0.75 mM inosine and thymine and 20 mM potassium phosphate (pH 7.0) was fed 20 ml h^{-1} continuously at 60 °C for 17 days. Source: Hori et al. 1991 [9]. Reproduced with permission from Elsevier.

At high temperatures, co-immobilized UP and PNP from *E. coli* have been used in the synthesis of antileukemic 506U78, a prodrug of *ara*-G (9-β-D-arabinofuranosyl guanine) [10]. Since 506U78 is poorly water soluble, the enzymatic transglycosylation from *ara*-uridine and 6-methoxy guanine was carried out at 55 °C (Figure 5.2) in order to favor product solubilization. The immobilized biocatalyst was thermostable and could be recycled several times in bioreactors. After biocatalyst removal, a crude product was crystallized from hot aqueous base (pH 11). Likewise, co-immobilized UP and PNP from *E. coli* were also successfully employed in the synthesis of natural nucleosides (2′-deoxyadenosine, 2′-deoxyguanosine) and of nucleoside analogues of pharmaceutical interest such as mizoribine, *ara*-A (vidarabine), *ara*-DAMP, and fludarabine-des-phosphate at 60 °C (Figure 5.2). Product yields between 70% and 85% were achieved in a few hours (two to six hours depending on enzyme specificity). In addition, the immobilized biocatalyst also retained its activity in the presence of 40% dimethylsulfoxide (DMSO), an organic cosolvent that was needed in the preparation of fludarabine-des-phosphate. Moreover,

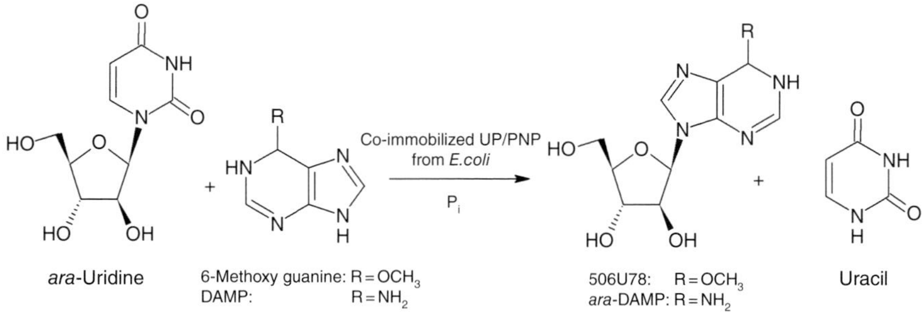

Figure 5.2 Enzymatic transglycosylations catalyzed by immobilized UP and PNP from *E. coli*. Synthesis of 506U78 was carried out in 10 mM phosphate buffer pH 7.4 at 55 °C [10], whereas synthesis of β-D-arabinofuranosyl-2,6-diaminopurine (*ara*-DAMP) in 30 mM phosphate buffer pH 7.0 at 60 °C. Source: Zuffi et al. 2004 [13]. Reproduced with permission from T&F.

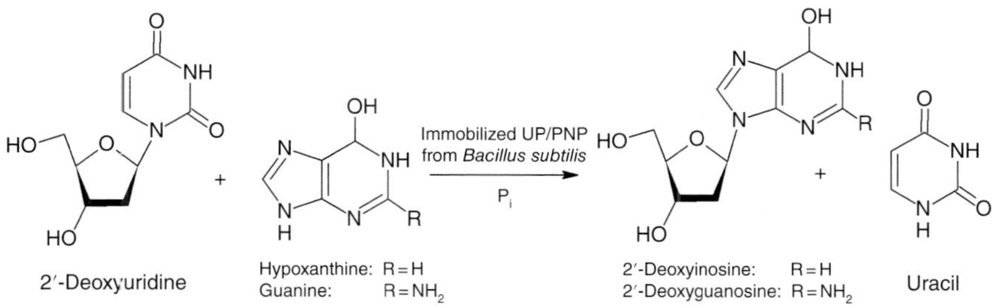

Figure 5.3 Enzymatic syntheses of 2′-deoxyguanosine and 2′-deoxyinosine catalyzed by immobilized UP and PNP from *B. subtilis*. 2′-Deoxyuridine (50 mM) was used as sugar donor, whereas guanine or hypoxanthine (25 mM) were used as sugar acceptors. Both reactions were performed in 10 mM carbonate pH 10 and 45 °C [15].

30 consecutive batch reactions could be performed with the same biocatalyst showing the same good yields [13].

Enzymatic transglycosylations at very alkaline conditions and/or in the presence of a water-miscible organic cosolvent (methanol, acetonitrile, etc.) have been also established as other alternatives to efficiently increase bioconversion yields by enhancing substrate and/or product solubility, although highly stabilized enzymes are required in such harsh conditions. As mentioned before, different immobilized NPs have been prepared in order to achieve the desired quaternary structure stabilization of these enzymes. For instance, immobilized UP and PNP from *B. subtilis* were employed at pH 10 in the synthesis of natural purine nucleosides (Figure 5.3), achieving 92% yield in 24 hours in the synthesis of 2′-deoxyguanosine, and 85% yield in 5 hours in the synthesis of 2′-deoxyinosine [15]. After biocatalyst removal by filtration, the desired product can be easily isolated by precipitation from the reaction medium. In this sense, 2′-deoxyguanosine can be precipitated from the reaction mixture as a white solid at pH 5.5 and 4 °C, with a 97% HPLC purity [11].

In the presence of an organic cosolvent such as acetonitrile, immobilized microbial NPs were successfully employed in the one-pot synthesis of *ara*-A and 2′,3′-dideoxyinosine (ddI, didanosine) achieving 74% yield at 48 hours and 44% yield at 24 hours, respectively (Figure 5.4) [14, 16].

Immobilized *Ec*TP or immobilized pyrimidine nucleoside phosphorylase from *B. subtilis* (*Bs*PyNP) were successfully used for the one-pot synthesis of floxuridine (5-fluoro-2′-deoxyuridine) starting from 2′deoxyuridine or thymidine as the sugar donor and 5-fluorouracil as the base acceptor (Figure 5.5), yielding product conversions of 73% (*Ec*TP) and 76% (*Bs*PyNP), respectively (about 9 $g\,l^{-1}$).

The immobilized biocatalysts could be recycled and employed in the synthesis of other 5-halogenated pyrimidine 2′-deoxyribonucleosides in fully aqueous medium [17].

5.2.2 Immobilized Nucleoside 2′-Deoxyribosyltransferases

NDTs (EC 2.4.2.6) catalyze the exchange of a purine or pyrimidine base of a 2′-deoxyribonucleoside (donor) to a free purine or pyrimidine base (acceptor).

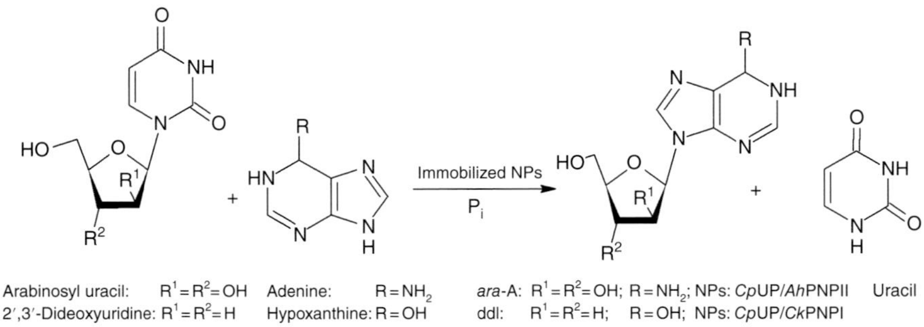

Figure 5.4 Enzymatic synthesis of *ara*-A and ddI catalyzed by immobilized NPs. Reactions were carried out at room temperature, in 50 mM phosphate pH 7.5 containing 10% (v/v) of acetonitrile. Donor concentration was 10 mM and acceptor concentration was 5 mM [14]. Immobilized NPs; *Cp*UP: uridine phosphorylase from *Clostridium perfringens*; *Ah*PNPII: purine nucleoside phosphorylase type II from *Aeromonas hydrophila*; *Ck*PNP: purine nucleoside phosphorylase type I from *Citrobacter koseri* [14, 16].

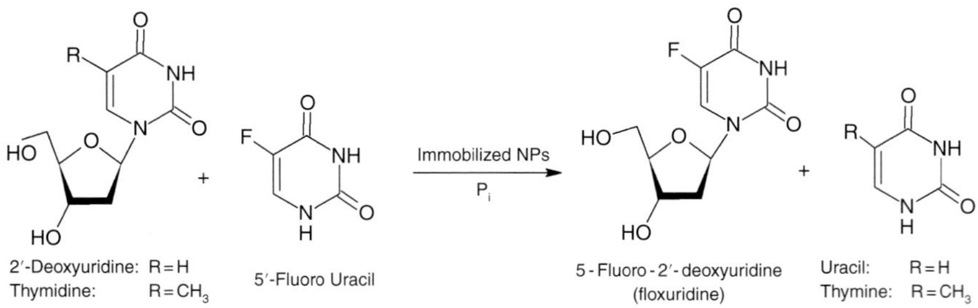

Figure 5.5 Enzymatic synthesis of 5-fluoro-2′-deoxyuridine (floxuridine) catalyzed by different immobilized NPs. Best results were afforded in10 mM phosphate pH 7.5 and nucleoside/nucleobase 3 : 1 molar ratio. Examples of immobilized NPs: *Ec*TP, thymidine phosphorylase from *E. coli*; *Bs*PyNP, pyrimidine nucleoside phosphorylase from *Bacillus subtilis*. Source: Serra et al. 2013 [17]. Reproduced with permission from Elsevier.

According to their substrate specificities, NDTs have been divided into two groups: NDT type I (PDTs), specific for purine bases; and NDT type II (NDT), specific for both purine and pyrimidine bases as donor and/or acceptor [18–20]. The main source of NDTs for nucleoside enzymatic synthesis comes from different microbial species of *Lactobacillus*, although quite interesting enzymes from extremophiles have been described as well [21]. As a rule of thumb, immobilized NDTs clearly offer some advantages compared to immobilized NPs in the synthesis of nucleoside analogues. On one hand, some transglycosylations catalyzed by NPs need the performance of both UP and PNP enzymes ("two-step" reactions), and this could actually be a disadvantage compared to the "one-step" action of NDTs [20]. On other hand, NPs have sometimes been immobilized separately using different methods and supports in order to achieve the best performance for each enzyme [11, 15].

5.2.2.1 Stabilization of Nucleoside 2′-Deoxyribosyltransferases by Immobilization

Since NDTs are also multimeric enzymes like NPs, different immobilization strategies have been attempted to stabilize their quaternary protein structure, including binding to ionic-exchangers and hydrophobic supports and covalent multipoint attachment to activated supports (such as epoxy-activated and aldehyde-activated carriers) as well as ionic adsorption to supports functionalized with PEI and followed by cross-linking with aldehyde-dextran. The first immobilized NDT biocatalysts were prepared with the crude enzyme from *Lactobacillus leichmannii* that was linked by hydrophobic interaction on octyl-sepharose or by covalent attachment to poly(acrylamide-co-*N*-acryloxysuccinimide) [22]. Although crude enzyme extracts might reduce costs in the preparation of immobilized enzymes, biocatalysts with higher specific activity, and, therefore, greater industrial potential, can be obtained only when pure enzymes are employed during immobilization. The following examples are immobilized NDTs biocatalysts that have been prepared from purified enzyme solutions. For instance, immobilization of NDT from *Lactobacillus reuteri* (*Lr*NDT) was carried out by covalent attachment to epoxy-activated Sepabeads EC-EP303 [23]. An increased thermal stability was shown by the best immobilized *Lr*NDT biocatalyst since its half-life at 60 °C was 17.9 hours whereas the half-life of the free enzyme was 17.9 minutes at the same temperature. Moreover, the immobilized enzyme activity was enhanced in the presence of different water-miscible organic cosolvents at 10–30% (v/v) [24], paving the way for the application of this biocatalyst in the synthesis of natural nucleosides and analogues using purine bases with low solubility in aqueous buffer. Recently, magnetic supports for enzyme immobilization has emerged as an alternative approach that renders easy separation of biocatalysts from the reaction medium that can be employed in fluidized bed reactors that are stabilized by applying a magnetic field. In this sense, *Lr*NDT was covalently linked to glutaraldehyde-activated magnetic chitosan, allowing the preparation of a robust immobilized biocatalyst. In fact, thermal stability was enhanced compared to immobilized *Lr*NDT on Sepabeads EC-EP, since the half-life of the new biocatalyst at 60 °C was 56.3 hours, whereas full activity was observed after storage at 40 °C for 144 hours [25]. Recently, other NDTs have been immobilized by other methods different from the covalent attachment to a support. For instance, NDT from *Bacillus psychrosaccharolyticus* (*Bp*NDT) was adsorbed onto PEI-functionalized agarose, and subsequent cross-linked with aldehyde–dextran (20 000 kDa and 70% oxidation degree), [21]. Likewise, active immobilized biocatalysts were obtained by ionic adsorption of NDT from *Lactobacillus animalis* (*La*NDT) on DEAE-sepharose and Q-agarose whose activity were retained for 64 and 24 hours, respectively [26]. Moreover, a higher stabilization of *La*NDT was achieved by an oriented covalent immobilization of the His-tagged recombinant enzyme onto an hetero-functional support that was activated with nickel-chelates and glyoxyl groups [27], Firstly, His-tagged *La*NDT was immobilized by employing its metal-binding affinity, through the Ni^{2+} cations on the support, allowing the preparation of the immobilized derivative Ni^{2+}-*La*NDT. Afterward, an intense multipoint covalent attachment involving surface lysine residues of the enzyme was accomplished at alkaline pH, and finally, reversible Schiff's base adducts were reduced to stable primary amino bonds, allowing the preparation of the final immobilized biocatalyst (Ni^{2+}-Gx-*La*NDT). Despite covalent immobilization affecting enzyme activity (Ni^{2+}-Gx-*La*NDT derivative displayed only 50% of the initial activity, whereas Ni^{2+}-*La*NDT showed no loss of activity), improved storage stability was obtained since Ni^{2+}-Gx-*La*NDT retained full activity for more than 78 days at 4 °C, whereas Ni^{2+}-*La*NDT was inactive after 65 days. Moreover, multipoint attachment prevented enzyme leakage at 50 °C.

5.2.2.2 Synthesis of Nucleosides Catalyzed by Immobilized 2′-Deoxyribosyltransferases

The described immobilized NDTs have been successfully employed in the one-pot "one-step" synthesis of many natural and unnatural nucleosides (Figure 5.6). Immobilized *Lr*NDT on Sepabeads EC-EP could be used in one-pot transglycosylations at very mild conditions of pH and temperature and also under unfriendly reaction conditions that include the presence of organic cosolvents [23, 24]. As a matter of fact, immobilized *Lr*NDT was successfully used in the enzymatic synthesis of several nucleoside analogues with therapeutic activity (2,6-diaminopurine-2′-deoxyriboside, 5-fluoro-2′-deoxyuridine, 5-chlorodeoxyuridine, 5-bromo-2′-deoxyuridine, 5-iodo-2′-deoxyuridine, 5-ethyl-2′-deoxyuridine, 5-aza-2′-deoxycytidine, 2′-fluoro-2′-deoxycytidine,

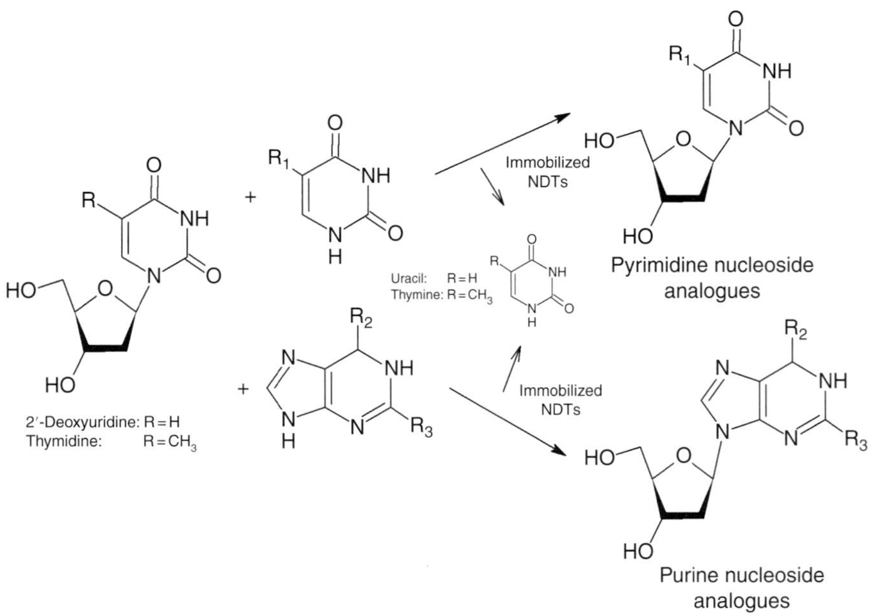

Figure 5.6 Enzymatic synthesis of nucleoside analogues catalyzed by immobilized nucleoside 2′-deoxyribosyltransferases (NDTs): NDT from *Lactobacillus reuteri* (*Lr*NDT) [23]; NDT from *Bacillus psychrosaccharolyticus* (*Bp*NDT)(21); and NDT from *Lactobacillus animalis* (*La*NDT) [26, 27].

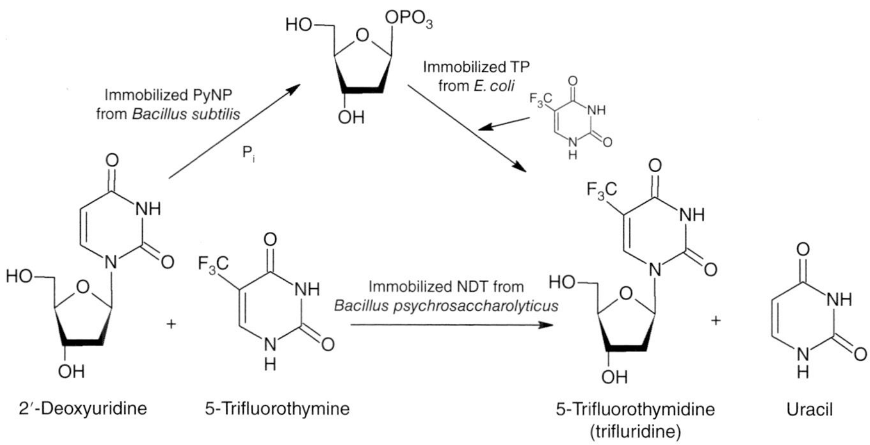

Figure 5.7 Two approaches for the enzymatic synthesis of 5-trifluorothymidine (trifluridine) employing immobilized nucleoside 2′-deoxyribosyltransferase [21, 28] and immobilized nucleoside phosphorylases [17].

and 5-fluoro-2′-deoxycytidine), and could be recycled at least 26 times retaining most of its activity.

Interestingly, immobilized *Lr*NDT on magnetic chitosan beads demonstrated to be also active in the synthesis of some nucleoside analogues as well as the arabinosyl-nucleosides vidarabine (*ara*-A) and cytarabine (*ara*-C) [25]. Furthermore, this is the first report that describes the enzymatic synthesis of these arabinosyl-nucleosides catalyzed by an immobilized NDT, although the bioconversion yields were not very high (14–17% at 72 hours). In this case, the immobilized biocatalyst could be easily recovered and recycled for 30 consecutive batch reactions with negligible loss of catalytic activity. In addition, immobilized *Bp*NDT was successfully used in the preparation of therapeutic nucleoside analogues, obtaining 64.4% product yield at two hours for trifluridine synthesis, and 76.2% conversion yield of decytabine at three hours [21, 28]. This immobilized NDT derivative could be recycled for at least 30 consecutive cycles with 25% loss of activity, and could be an alternative to the use of NPs in the "two-step" enzymatic synthesis of trifluridine (Figure 5.7).

Interestingly, immobilized *La*NDT by ionic adsorption on DEAE-agarose was satisfactorily used to obtain different halogenated pyrimidine and purine 2′-deoxyribonucleosides (including the antimetabolite floxuridine) with yields above 50% at short reaction times [26]. The developed biocatalyst was also able to accept 2′,3′-dideoxyribose and arabinose moieties as substrates, obtaining the anti-human immunodeficiency virus (HIV) compounds 5-fluoro-2′,3′-dideoxyuridine (FddUrd) and 5-fluorouracil-arabinonucleoside (ara5FUra) in a one-step reaction. In addition, immobilized His-tagged *La*NDT on hetero-functional support after multipoint attachment (Ni^{2+}-Gx-*La*NDT) could be reused for more than 300 hours of successive batches in floxuridine synthesis at 30 °C. On the contrary, the non-covalent attached enzyme (Ni^{2+}-*La*NDT) was inactivated after 22 hours in the same conditions [27].

5.2.3 Immobilized Nucleobase Phosphoribosyltransferases

Phosphoribosyltransferases (PRTs) are enzymes that catalyze the reversible transfer of the 5-phosphoribosyl group from 5-phospho-α-D-ribosyl-1-pyrophosphate (PRPP) to purine and pyrimidine nucleobases or their derivatives in the presence of Mg^{2+}. According to substrate specificity, PRTs can be classified as: (i) 6-aminopurine phosphoribosyltransferases (6-aminopurine PRTs, EC 2.4.2.7), which are strictly specific for 6-aminopurines such as adenine, 2-fluoroadenine, or 2-chloroadenine [29, 30]; (ii) 6-oxopurine phosphoribosyltransferases (6-oxo PRTs, EC 2.4.2.8), which show variable specificity against 6-oxopurines such as hypoxanthine, xanthine, and guanine among others (HPRT, GPRT, XPRT, HGPRT, or HGXPRT) and different 6-oxo and 6-mercaptopurine derivatives [29, 31]; (iii) uracil phosphoribosyltransferases (UPRTs, EC 2.4.2.9), which are highly specific for uracil, but are also able to recognize some 5-halogenated derivatives [29, 32]; and (iv) orotate phosphoribosyltransferases (OPRTs, EC 2.4.2.10), which are able to use orotic acid and other pyrimidine derivatives as substrates [29]. Despite their great potential as biocatalysts, only a few examples of immobilized PRTs have been reported in the literature. In this sense, an interesting multi-enzymatic cascade system for the synthesis of nucleoside analogues using cross-linked enzyme aggregates (CLEAs) has been developed [33]. In this system, an engineered hypoxanthine phosphoribosyltransferase (8B3PRT) was co-aggregated in the presence of ammonium sulfate with several enzymes for adenosine-5′-triphosphate (ATP) and PRPP regeneration (ribokinase pyruvate kinase, PRPP synthetase, and adenylate kinase), as well as bovine serum albumin (BSA), and further cross-linked with glutaraldehyde in order to obtain an efficient immobilized biocatalyst, which was successfully used in the synthesis of a great variety of nucleotide analogues. More recently, hypoxanthine–guanine–xanthine phosphoribosyltransferase from *Thermus thermophilus* HB8 (*Tt*HGXPRT) was covalently attached to glutaraldehyde-activated MagReSyn®Amine magnetic iron oxide porous microparticles [34]. Two different approaches allowed the preparation of highly active biocatalysts: (i) an enzyme immobilization through its N-terminal residue at pH 8.5 allowing the preparation of M*Tt*HGXPRT1-3 derivative and (ii) a multipoint covalent immobilization through its surface lysine residues at pH 10 (M*Tt*HGXPRT4-5 derivative). As observed, thermal stability and reusability of a magnetic M*Tt*HGXPRT3 derivative was higher than its M*Tt*HGXPRT4-5 counterpart, and consequently the first biocatalyst was used in the synthesis of ribonucleoside monophosphates with application as food additives [34].

In this sense, M*Tt*HGXPRT3 was successfully employed in the enzymatic production of several dietary nucleotides, such as inosine-5′-monophosphate (IMP) and guanosine-5′-monophosphate (GMP) (Figure 5.8), which show immunostimulant and taste enhancer effects.

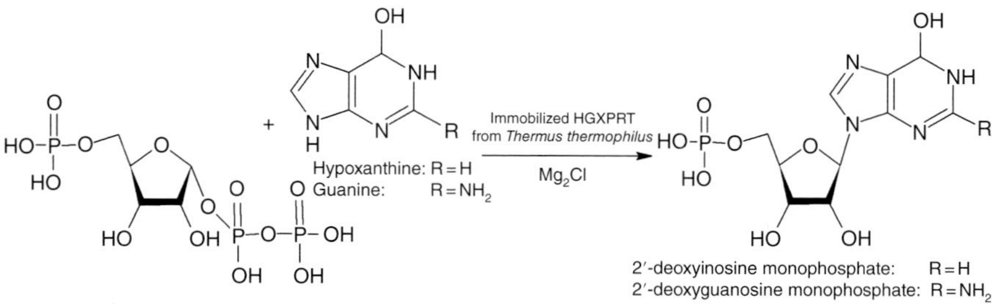

Figure 5.8 Enzymatic synthesis of dietary nucleoside-5′-monophosphates employing immobilized hypoxanthine–guanine–xanthine phosphoribosyltransferase from *Thermus thermophilus* HB8 (*Tt*HGXPRT) onto MagReSyn® aldehyde microspheres. Reaction conditions: 12 mM tris/HCl pH 8.0 and 60 °C. Source: del Arco et al. 2018. [34]. ©2018 by MDPI (http://www.mdpi.org). Reproduction is permitted for noncommercial purposes.

5.3 Immobilized Nucleoside Oxidase

A crude extract of *Stenotrophomonas maltophilia* exhibiting nucleoside oxidase activity was immobilized on Eupergit C beads. The immobilized biocatalyst was able to convert the 5′-hydroxyl groups of different unnatural nucleosides to their corresponding carboxylic acids [35], including the nucleoside 1-[2-chloro-6-[(2,2-diphenylethyl)amino]-9*H*-purin-9-yl]-1-deoxy-β-D-ribofuranuronic acid (Figure 5.9), a key intermediate in the synthesis of a novel group of compounds with broad anti-inflammatory properties. It was found that the same batch of enzyme could be re-used for five cycles.

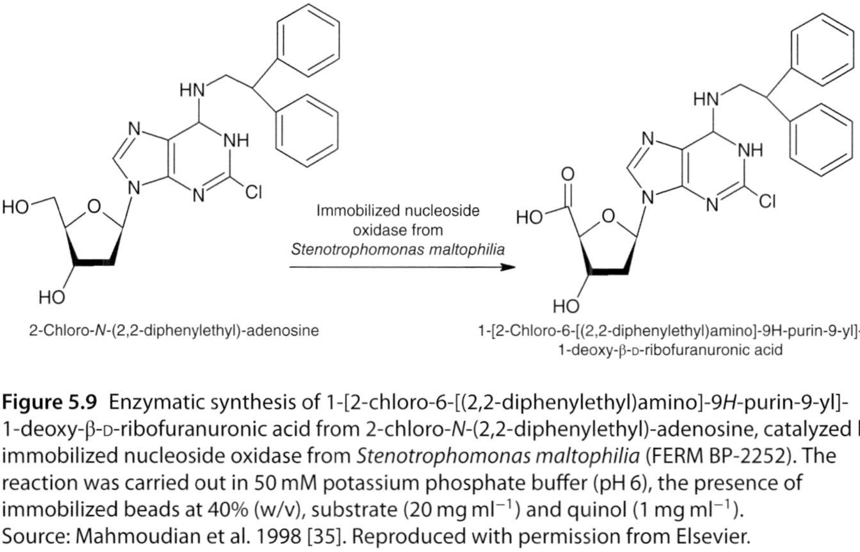

Figure 5.9 Enzymatic synthesis of 1-[2-chloro-6-[(2,2-diphenylethyl)amino]-9*H*-purin-9-yl]-1-deoxy-β-D-ribofuranuronic acid from 2-chloro-*N*-(2,2-diphenylethyl)-adenosine, catalyzed by immobilized nucleoside oxidase from *Stenotrophomonas maltophilia* (FERM BP-2252). The reaction was carried out in 50 mM potassium phosphate buffer (pH 6), the presence of immobilized beads at 40% (w/v), substrate (20 mg ml^{-1}) and quinol (1 mg ml^{-1}). Source: Mahmoudian et al. 1998 [35]. Reproduced with permission from Elsevier.

5.4 Immobilized Hydrolases

5.4.1 Immobilized Lipases

Regioselective acylation represents a common method of protecting the hydroxyl group in nucleosides and has proven to be the most successful strategy for the preparation of nucleoside derivatives with higher oral bioavailability [3, 36]. Microbial lipases have demonstrated to be an excellent "green" alternative to the traditional chemical modification of nucleosides due to their high regioselectivity at mild conditions in reaction systems that may even include the presence of water-miscible organic cosolvents or ionic liquids [37, 38]. The main immobilized microbial lipases used for regioselective acylation of nucleosides are the commercial immobilized preparations of lipases from *Thermomyces lanuginosus* (Lipozyme® TL IM) [39], *Mucor miehei* (Lipozyme), *Burkholderia cepacia* (PSL-C), and lipase B from *Candida antarctica* (Novozym 435, CAL-B). Herein, some representative examples of lipase-catalyzed regioselective acylations of nucleosides are shown. For instance, an efficient synthesis of 5′-*O*-benzoyl-2′-deoxynucleosides has been performed through the regioselective acylation of the parent nucleosides catalyzed by immobilized CAL-B with vinyl benzoate at 60 °C in tetrahydrofurane (THF) as solvent [40]. The protected monomers are important precursors for the synthesis of therapeutically useful oligonucleotides and modified nucleosides. In a similar strategy, immobilized lipase from *Burkholderia cepacia* (PSL-C) mediated floxuridine benzoylation (Figure 5.10), obtaining an excellent conversion (>99%) and great regioselectivity at 3′-position (92%) in anhydrous THF with the addition of 1-butyl-2,3-dimethylimidazolium hexafluorophosphate ($[C_4MIm]PF_6$) [41].

The same preferential acylation of the secondary hydroxy group of floxuridine has been carried out by PSL-C with vinyl crotonate, yielding

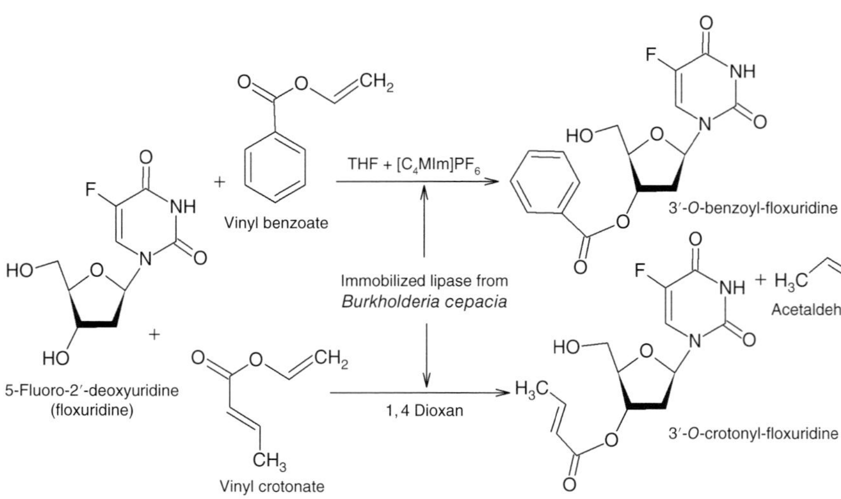

Figure 5.10 Regioselective acylations of floxuridine catalyzed by immobilized lipase from *Burkholderia cepacia* (PSL-C) [41, 42].

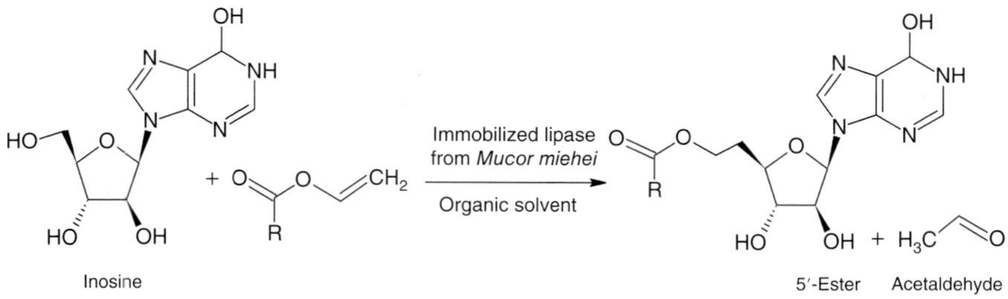

Figure 5.11 Synthesis of inosine esters catalyzed by immobilized lipase from *Mucor miehei* in organic solvents. Source: Wang et al. 2005 [44]. Reproduced with permission from Elsevier.

3′-*O*-crotonyl-floxuridine [42]. PSL-C was also described to provide the 3′-*O*-crotonylated derivatives of other nucleosides, whereas immobilized CAL-B afforded the 5′-*O*-crotonylated compounds. As a consequence, the combination of both immobilized lipases was successfully used to achieve the diacylated nucleosides [43]. Regioslective acylations catalyzed by immobilized lipases with wide range of vinyl esters in nonconventional media have been also reported for the preparation of several nucleoside derivatives. Enzymatic synthesis of fatty acid inosine esters was carried out by immobilized lipase from *M. miehei* (Lipozyme) in acetone (Figure 5.11), obtaining 5′-*O*-acyl derivatives that are more lipophilic than inosine and thus suitable for potentially pharmaceutical application [44].

In a similar approach, a regioselective acylation at the 5′-OH position of *ara*-C was catalyzed by immobilized CAL-B (Novozym 435) with vinyl fatty acid esters (laurate or stearate) in binary organic solvents [45, 46], or vinyl propionate in several ionic liquids [47] (Figure 5.12). Likewise, PSL-C displayed a 3′-regioselectivity of 80% toward 3′-hydroxyl of 6-azauridine employing stearate vinyl ester [48].

Immobilized CAL-B (Novozym 435) was also used in the regioselective undecylenoylation of 5′-hydroxyl of purine nucleosides in THF. The obtained nucleoside esters could serve as starting materials for the synthesis of functional polymeric prodrugs as well as antifungal agents without bad smell [49]. Such regioselective undecylenoylation was extended to other nucleosides such as

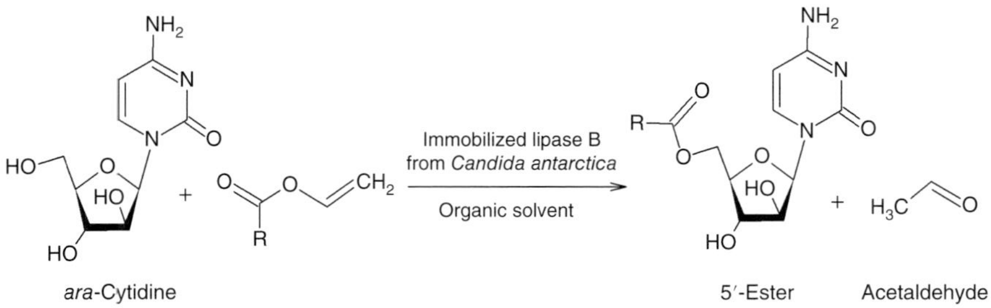

Figure 5.12 Regioselective acylation of *ara*-C catalyzed by immobilized lipase B from *Candida antarctica* [45–47]. Highest 5′-regioselectivity (>99.9%) and conversion (97%) was obtained with the mixture hexane–pyridine (1 : 3).

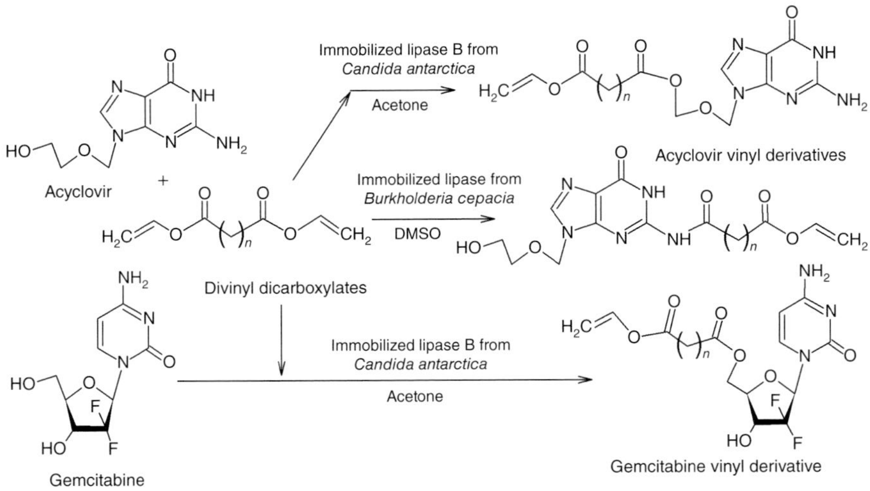

Figure 5.13 Enzymatic synthesis of vinyl derivatives of acyclovir and gemcitabine for the preparation of polymerizable prodrugs [52, 53].

8-chloroadenosine and its analogues, but employing immobilized lipase from *Penicillium expansum* in the biomass-derived solvent MeTHF [50]. Several pyrimidine nucleosides with therapeutic activity such as floxuridine, idoxuridine, stavudine, and zidovudine, were modified with this immobilized lipase [51]. As mentioned before, nucleoside undecylenoylated esters can be very useful for the preparation of polymerizable prodrugs, but other nucleoside esters and amides can be prepared by lipase-catalyzed acylations. A facile control of the acylation at the primary hydroxyl of acyclovir was achieved employing immobilized CAL-B in pyridine or acetone, whereas PSL-C (PS Amano) in DMSO selectively acylated the amino group [52] (Figure 5.13). The obtained acyclovir vinyl derivatives were used to obtain five polymerizable prodrugs.

A similar approach was developed to synthetize four polymerizable vinyl gemcitabine esters. In this sense, selective transesterification of gemcitabine was achieved with divinyl dicarboxylates using immobilized CAL-B in acetone [53], reaching the highest yield (93%) when the molar ratio of divinyl adipate to gemcitabine was 4 : 1. The same immobilized enzyme (Novozym 435) was also used in the regioselective coupling of α-lipoic acid to the 5′-hydroxyl group of gemcitabine [54] (Figure 5.14).

Synthesis of macromolecular prodrugs has been extended to the preparation of a ribavirin acrylate by a chemoenzymatic approach using Novozym 435 [55]. Finally, immobilized lipases have been broadened to other applications such as the enzymatic separation of furanosyl and pyranosyl nucleosides [56], and the preparation of key building blocks for the synthesis of oligonucleotides known as UNAs (unlocked nucleic acids) [57] and LNAs (lock nucleic acids) [58].

5.4.2 Immobilized Proteases

A stable immobilized derivative of *B. subtilis* protease N on activated agarose was successfully used in the regioselective hydrolysis of acetylated nucleosides, giving

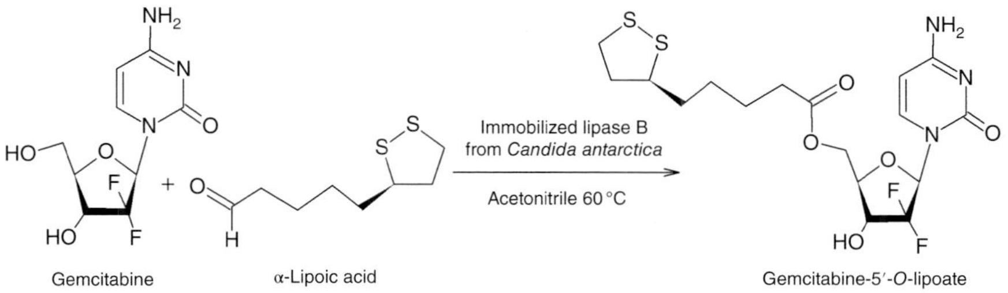

Figure 5.14 Enzymatic synthesis of gemcitabine-5′-*O*-lipoate catalyzed by immobilized lipase B from *Candida antarctica*. This compound shows higher activity compared to its parent counterpart against non-small cell lung cancer. Source: Geromichalou et al. 2015 [54]. Reproduced with permission from Elsevier.

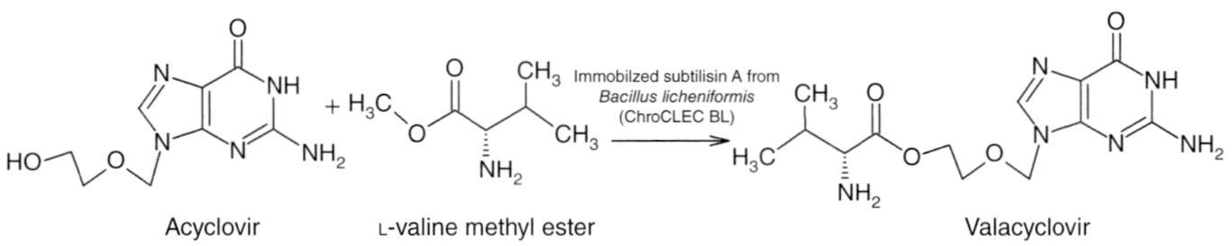

Figure 5.15 Enzymatic synthesis of valacyclovir catalyzed by immobilized subtilisin A from *Bacillus licheniformis*. Source: McClean et al. 2011 [60]. Reproduced with permission from Elsevier.

new intermediates for the synthesis of capecitabine [59]. Cross-linked enzyme crystals of subtilisin A from *Bacillus licheniformis* (Chiro CLEC-BL) was used in the synthesis of enantiopure valacyclovir at high rates employing L-valine methyl ester as acyl donor and *t*-butanol as cosolvent [60] (Figure 5.15). The same immobilized preparation has been used for the aminoacylation of the 3′-hydroxyl group of lobucavir using the binary mixture of acetone and dimethylformamide (DMF, 70 : 30) as solvent and L-valine *p*-nitrophenyl ester as acyl donor [36].

5.4.3 Immobilized Esterases

Cholesterol esterase from Boehringer immobilized on Accurel PP was used for the enzymatic resolution of the racemic butyrate ester of 2′,3′-dideoxy-5-fluoro-3′-thiacytidine (Figure 5.16), resulting in the cleavage of the (−) isomer to the corresponding alcohol (emtricitabine, FTC) with a 31% conversion yield and 98% enantioselectivity [61].

5.4.4 Immobilized Deaminases

Cytidine deaminase from *E. coli* was covalently immobilized on epoxy-activated Eupergit C in order to carry out the enzymatic resolution of a racemic mixture of (±) 2′,3′-dideoxy-3′-thiacytidine. The (+) isomer was selectively deaminated from the racemic mixture (Figure 5.17), and after a two-column purification process, optically pure (−) 2′,3′-dideoxy-3′-thiacytidine (lamivudine, 3TC) was obtained with a 76% yield [62].

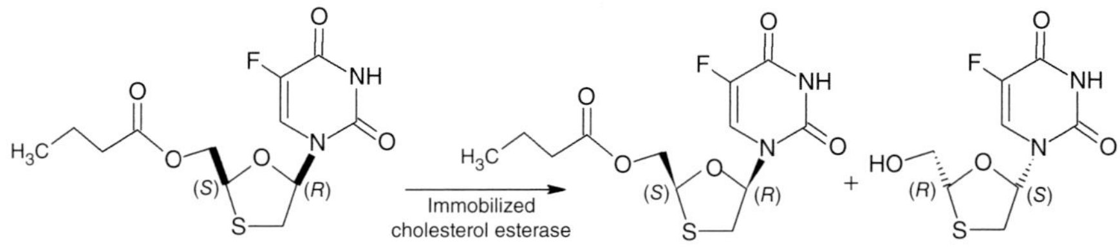

Figure 5.16 Enzymatic resolution of racemic (±) 2′,3′-dideoxy-5-fluoro-3′-thiacytidine butyrate using immobilized cholesterol esterase yielding the (−) isomer alcohol (FTC, emtricitabine). Reaction conditions: 5 mM potassium phosphate pH 7.0, 30 °C. Source: Osborne et al. 2006 [61]. Reproduced with permission from ACS.

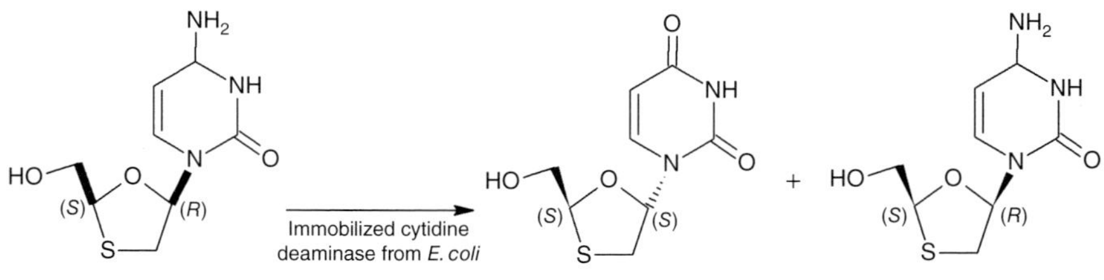

Figure 5.17 Synthesis of lamivudine (3TC) by enzymatic resolution of racemic (±) 2′,3′-dideoxy-5-fluoro-3′-thiacytidine catalyzed by an immobilized cytidine deaminase from *E. coli*. Source: Mahmoudian et al. 1993 [62]. Reproduced with permission from Elsevier.

5.4.5 Immobilized *S*-Adenosylhomocysteine Hydrolases

S-Adenosylhomocysteine (SAH) has been described as an effective sedative, a good sleep modulator, and a new anticonvulsant. SAH can be synthesized from adenosine and homocysteine by using a microbial *S*-adenosylhomocysteine hydrolase (SAHase, EC 3.3.1.1). In this sense, recombinant SAHase from *Corynebacterium glutamicum* ATCC 13032 was covalently bound to epoxy-activated Eupergit C, and the obtained biocatalyst was suitable for synthesizing SAH (Figure 5.18), as well as other important *S*-nucleosidyl-homocysteine compounds [63].

5.5 Immobilized Phosphopentomutases

Microbial phosphopentomutases (PPMs, EC 5.4.2.7) transfer the phosphate group from position C5 of α-D-ribose-5-phosphate (R5P) to position C1, yielding α-D-ribose-1-phosphate (R1P). R1P serves as a substrate for purine or pyrimidine NPs that accept a free base as a second substrate to form a nucleoside by phosphate displacement. In this sense, phosphopentomutase from *E. coli* ATCC 4157 was covalently linked to glyoxyl-activated agarose in the presence of 10% (v/v) glycerol in order to enhance enzyme stability at the alkaline conditions needed for enzyme binding.

This immobilized biocatalyst was efficiently used in ribavirin synthesis from 1,2,4-triazole-3-carboxamide (TCA) and R5P in the presence of a microbial

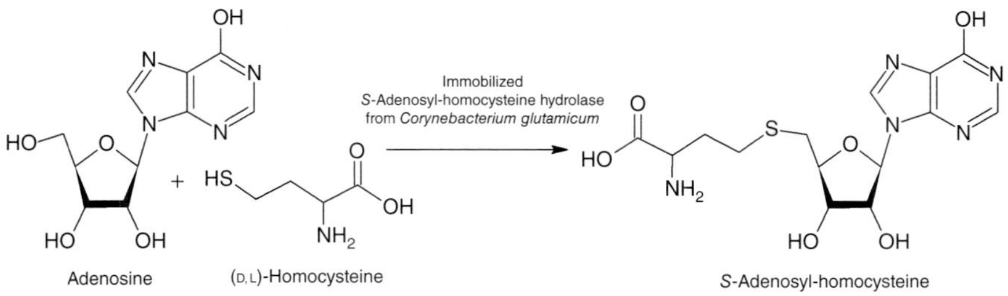

Figure 5.18 Enzymatic synthesis of *S*-adenosyl-homocysteine by immobilized CgSAHase from *Corynebacterium glutamicum* ATCC 13032. Reaction was carried out in 50 mM phosphate buffer pH 7.6 containing adenosine (5–200 μM), (D,L)-homocysteine (10–400 μM) and 2 g (wet weight) of immobilized enzyme in a final volume of 5 ml. Source: Lozada-Ramirez et al. 2012 [63]. Reproduced with permission from Springer Nature.

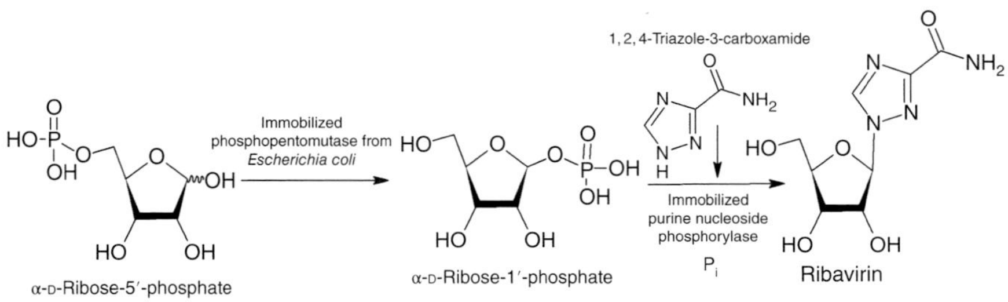

Figure 5.19 Enzymatic synthesis of ribavirin catalyzed by a multi-enzymatic system containing immobilized phosphopentomutase and immobilized PNP. Reaction conditions: 25 mM phosphate pH 7.0 at 25 °C. Source: Rivero et al. 2017 [64]. Reproduced with permission from Elsevier.

purine NP (from Sigma) which was immobilized on agarose activated with cyanogen bromide (Figure 5.19) [64].

5.6 Immobilized Deoxyribonucleoside Kinases

Deoxyribonucleoside kinases (dNKs) catalyze the transfer of a phosphate group from a donor nucleotide (such as ATP) to the 5′-hydroxyl group of a nucleoside, thus forming the corresponding 5′-mononucleotide. This enzymatic reaction can be useful in biocatalysis for the phosphorylation of nucleosides as a greener alternative to the chemical process based on $POCl_3$ which lacks regioselectivity, resulting in mixtures of mono- and dinucleotides. Solubility of poorly water-soluble nucleosides with therapeutic activity can be significantly enhanced by phosphorylation, and the resulting nucleotides may be considered as prodrugs of their parent nucleosides such as vidarabine monophosphate (araA-MP, used for the systemic treatment of *Herpex simplex* viruses) and fludarabine monophosphate (FaraA-MP, used against leukemias and lymphomas). In this sense, fruit fly (*Drosophila melanogaster*) deoxyribonucleoside kinase (*Dm*dNK; EC 2.7.1.145) was adsorbed on a solid ion exchange support and further cross-linked with aldehyde dextran, allowing the preparation of a highly stable biocatalyst at pH 10

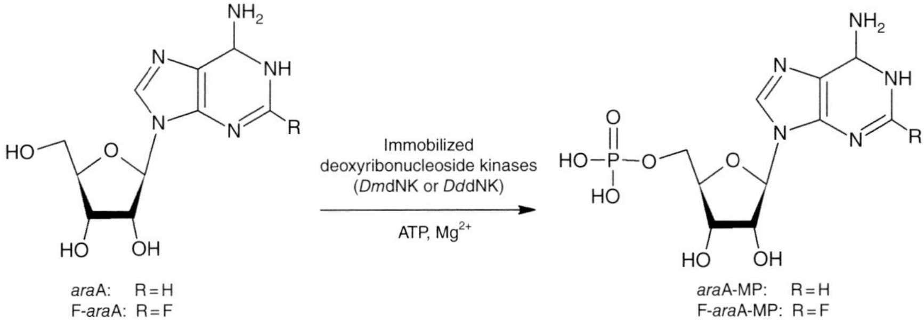

Figure 5.20 Enzymatic 5′-phosphorylation of araA and F-araA catalyzed by immobilized deoxyribonucleoside kinases (dNKs). *Dm*dNK, deoxyribonucleoside kinase from *Drosophila melanogaster*; *Dd*dAK, deoxyadenosine kinase from *Dictyostelium discoideum*; ATP, adenosine-5′-triphosphate. Reaction conditions: 50 mM ammonium acetate pH 8; $MgCl_2$ 2 mM; 37 °C [65, 66].

[65]. Similarly, deoxyadenosine kinase from *Dictyostelium discoideum* (*Dd*dAK) has been immobilized by ionic interaction on derivatized Sepabeads EC-EP with PEI, and further cross-linked with Dx [66]. Both immobilized biocatalysts based on *Dm*dNK and *Dd*dAK were successfully employed in the preparative synthesis of araA-MP and FaraA-MP (Figure 5.20). For the synthesis of these high-added value products, the use of stoichiometric ATP could be justified. However, for the synthesis of other monophosphates, ADP produced during the reaction should be recycled by using other enzymes such as pyruvate, acetate, or polyphosphate kinases.

References

1 Fernandez-Lucas, J. (2015). Multienzymatic synthesis of nucleic acid derivatives: a general perspective. *Appl. Microbiol. Biotechnol.* 99: 4615–4627.

2 Boryski, J. (2008). Reactions of transglycosylation in the nucleoside chemistry. *Curr. Org. Chem.* 12: 309–325.

3 Li, N., Smith, T.J., and Zong, M.H. (2010). Biocatalytic transformation of nucleoside derivatives. *Biotechnol. Adv.* 28: 348–366.

4 Garcia-Galan, C., Berenguer-Murcia, A., Fernandez-Lafuente, R., and Rodrigues, R.C. (2011). Potential of different enzyme immobilization strategies to improve enzyme performance. *Adv. Synth. Catal.* 353: 2885–2904.

5 Lewkowicz, E.S. and Iribarren, A.M. (2006). Nucleoside phosphorylases. *Curr. Org. Chem.* 10: 1197–1215.

6 Trelles, J.A., Fernandez-Lucas, J., Condezo, L.A., and Sinisterra, J.V. (2004). Nucleoside synthesis by immobilised bacterial whole cells. *J. Mol. Catal. B:Enzym.* 30: 219–227.

7 Fernandez-Lafuente, R. (2009). Stabilization of multimeric enzymes: strategies to prevent subunit dissociation. *Enzyme Microb. Technol.* 45: 405–418.

8 Yokozeki, K. and Tsuji, T. (2000). A novel enzymatic method for the production of purine-2′-deoxyribonucleosides. *J. Mol. Catal. B: Enzym.* 10: 207–213.

9 Hori, N., Watanabe, M., Sunagawa, K. et al. (1991). Production of 5-methyluridine by immobilized thermostable purine nucleoside phosphorylase and pyrimidine nucleoside phosphorylase from *Bacillus stearothermophilus* JTS-859. *J. Biotechnol.* 17: 121–131.

10 Mahmoudian, M. (2000). Biocatalytic production of chiral pharmaceutical intermediates. *Biocatal. Biotransform.* 18: 105–118.

11 Rocchietti, S., Ubiali, D., Terreni, M. et al. (2004). Immobilization and stabilization of recombinant multimeric uridine and purine nucleoside phosphorylases from *Bacillus subtilis*. *Biomacromolecules* 5: 2195–2200.

12 Serra, I., Serra, C.D., Rocchietti, S. et al. (2011). Stabilization of thymidine phosphorylase from *Escherichia coli* by immobilization and post immobilization techniques. *Enzyme Microb. Technol.* 49: 52–58.

13 Zuffi, G., Ghisotti, D., Oliva, I. et al. (2004). Immobilized biocatalysts for the production of nucleosides and nucleoside analogues by enzymatic transglycosylation reactions. *Biocatal. Biotransform.* 22: 25–33.

14 Serra, I., Ubiali, D., Piskur, J. et al. (2013). Developing a collection of immobilized nucleoside phosphorylases for the preparation of nucleoside analogues: enzymatic synthesis of arabinosyladenine and 2,3-dideoxyinosine. *ChemPlusChem* 78: 157–165.

15 Ubiali, D., Rocchietti, S., Scaramozzino, F. et al. (2004). Synthesis of 2′-deoxynucleosides by transglycosylation with new immobilized and stabilized uridine phosphorylase and purine nucleoside phosphorylase. *Adv. Synth. Catal.* 346: 1361–1366.

16 Serra, I., Ubiali, D., Albertini, A.M. et al. (2010). Microbial nucleoside phosphorylases as efficient biocatalysts for the synthesis of antiviral and antitumoral nucleosides. *J. Biotechnol.* 150: S408.

17 Serra, I., Bavaro, T., Cecchini, D.A. et al. (2013). A comparison between immobilized pyrimidine nucleoside phosphorylase from *Bacillus subtilis* and thymidine phosphorylase from *Escherichia coli* in the synthesis of 5-substituted pyrimidine 2′-deoxyribonucleosides. *J. Mol. Catal. B: Enzym.* 95: 16–22.

18 Mikhailopulo, I.A. (2007). Biotechnology of nucleic acid constituents – state of the art and perspectives. *Curr. Org. Chem.* 11: 317–333.

19 Mikhailopulo, I.A. and Miroshnikov, A.I. (2011). Biologically important nucleosides: modern trends in biotechnology and application. *Mendeleev Commun.* 21: 57–68.

20 Fresco-Taboada, A., de la Mata, I., Arroyo, M., and Fernandez-Lucas, J. (2013). New insights on nucleoside 2′-deoxyribosyltransferases: a versatile biocatalyst for one-pot one-step synthesis of nucleoside analogs. *Appl. Microbiol. Biotechnol.* 97: 3773–3785.

21 Fresco-Taboada, A., Serra, I., Fernandez-Lucas, J. et al. (2014). Nucleoside 2′-deoxyribosyltransferase from psychrophilic bacterium *Bacillus psychrosaccharolyticus* – preparation of an immobilized biocatalyst for the enzymatic synthesis of therapeutic nucleosides. *Molecules* 19: 11231–11249.

22 Hicks, N. and Hutchinson, D.W. (1994). Synthesis of nucleoside analogs using immobilized *N*-deoxyribosyltransferases. *Biocatalysis* 11: 1–7.

23 Fernandez-Lucas, J., Fresco-Taboada, A., Acebal, C. et al. (2011). Enzymatic synthesis of nucleoside analogues using immobilized 2′-deoxyribosyltransferase from *Lactobacillus reuteri*. *Appl. Microbiol. Biotechnol.* 91: 317–327.

24 Fernandez-Lucas, J., Fresco-Taboada, A., de la Mata, I., and Arroyo, M. (2012). One-step enzymatic synthesis of nucleosides from low water-soluble purine bases in non-conventional media. *Bioresour. Technol.* 115: 63–69.

25 Fernandez-Lucas, J., Harris, R., Mata-Casar, I. et al. (2013). Magnetic chitosan beads for covalent immobilization of nucleoside 2′-deoxyribosyltransferase: application in nucleoside analogues synthesis. *J. Ind. Microbiol. Biotechnol.* 40: 955–966.

26 Britos, C.N., Lapponi, M.J., Cappa, V.A. et al. (2016). Biotransformation of halogenated nucleosides by immobilized *Lactobacillus animalis* 2′-*N*-deoxyribosyltransferase. *J. Fluorine Chem.* 186: 91–96.

27 Méndez, M.B., Rivero, C.W., López-Gallego, F. et al. (2018). Development of a high efficient biocatalyst by oriented covalent immobilization of a novel recombinant 2′-*N*-deoxyribosyltransferase from *Lactobacillus animalis*. *J. Biotechnol.* 270: 39–43.

28 Fresco-Taboada, A., Serra, I., Arroyo, M. et al. (2016). Development of an immobilized biocatalyst based on *Bacillus psychrosaccharolyticus* NDT for the preparative synthesis of trifluridine and decytabine. *Catal. Today* 259: 197–204.

29 del Arco, J. and Fernández-Lucas, J. (2017). Purine and pyrimidine phosphoribosyltransferases: a versatile tool for enzymatic synthesis of nucleoside-5′-monophosphates. *Curr. Pharm. Des.* 23: 6898–6912.

30 del Arco, J., Martínez, M., Donday, M. et al. (2018). Cloning, expression and biochemical characterization of xanthine and adenine phosphoribosyltransferases from *Thermus thermophilus* HB8. *Biocatal. Biotransform.* 36: 216–223.

31 del Arco, J., Cejudo-Sanches, J., Esteban, I. et al. (2017). Enzymatic production of dietary nucleotides from low-soluble purine bases by an efficient, thermostable and alkali-tolerant biocatalyst. *Food Chem.* 237: 605–611.

32 del Arco, J., Acosta, J., Pereira, H.M. et al. (2018). Enzymatic production of non-natural nucleoside-5′-monophosphates by a thermostable uracil phosphoribosyltransferase. *ChemCatChem* 10: 439–448.

33 Scism, R.A. and Bachmann, B.O. (2010). Five-component cascade synthesis of nucleotide analogues in an engineered self-immobilized enzyme aggregate. *ChemBioChem* 11: 67–70.

34 del Arco, J., Martínez-Pascual, S., Clemente-Suárez, V.J. et al. (2018). One-pot, one-step production of dietary nucleotides by magnetic biocatalysts. *Catalysts* 8: 184.

35 Mahmoudian, M., Rudd, B.A.M., Cox, B. et al. (1998). A versatile procedure for the generation of nucleoside 5′-carboxylic acids using nucleoside oxidase. *Tetrahedron* 54: 8171–8182.

36 Iglesias, L.E., Lewkowicz, E.S., Medici, R. et al. (2015). Biocatalytic approaches applied to the synthesis of nucleoside prodrugs. *Biotechnol. Adv.* 33: 412–434.

37 Kumar, V. and Malhotra, S.V. (2008). Synthesis of nucleoside-based antiviral drugs in ionic liquids. *Bioorg. Med. Chem. Lett.* 18: 5640–5642.

38 Kumar, V., Parmar, V.S., and Malhotra, S.V. (2010). Structural modifications of nucleosides in ionic liquids. *Biochimie* 92: 1260–1265.

39 Fernandez-Lafuente, R. (2010). Lipase from *Thermomyces lanuginosus*: uses and prospects as an industrial biocatalyst. *J. Mol. Catal. B: Enzym.* 62: 197–212.

40 García, J., Fernández, S., Ferrero, M. et al. (2004). A mild, efficient and regioselective enzymatic procedure for 5′-*O*-benzoylation of 2′-deoxynucleosides. *Tetrahedron Lett.* 45: 1709–1712.

41 Li, N., Ma, D., and Zong, M.H. (2008). Enhancing the activity and regioselectivity of lipases for 3′-benzoylation of floxuridine and its analogs by using ionic liquid-containing systems. *J. Biotechnol.* 133: 103–109.

42 Zhao, Z., Zong, M., and Li, N. (2009). Efficient regioselective synthesis of 3′-*O*-crotonylfloxuridine catalysed by *Pseudomonas cepacia* lipase. *Biotechnol. Appl. Biochem.* 52: 45–51.

43 Diaz-Rodriguez, A., Fernandez, S., Lavandera, I. et al. (2005). Novel and efficient regioselective enzymatic approach to 3′-, 5′- and 3′,5′-di-*O*-crotonyl 2′-deoxynucleoside derivatives. *Tetrahedron Lett.* 46: 5835–5838.

44 Wang, N., Wu, Q., Wu, W.B. et al. (2005). Lipase-catalyzed regioselective synthesis of lipophilic inosine ester derivatives. *J. Mol. Catal. B Enzym.* 35: 14–18.

45 Li, X.F., Zong, M.H., and Zhao, G.L. (2010). Efficient synthesis of 5′-*O*-laurate of 1-beta-D-arabinofuranosylcytosine via highly regioselective enzymatic acylation in binary solvent mixtures. *Bioorg. Med. Chem. Lett.* 20: 4125–4127.

46 Li, X.F., Zong, M.H., and Zhao, G.L. (2010). Highly regioselective enzymatic synthesis of 5′-*O*-stearate of 1-beta-D-arabinofuranosylcytosine in binary organic solvent mixtures. *Appl. Microbiol. Biotechnol.* 88: 57–63.

47 Li, X.F., Lou, W.Y., Smith, T.J. et al. (2006). Efficient regioselective acylation of 1-beta-D-arabinofuranosylcytosine catalyzed by lipase in ionic liquid containing systems. *Green Chem.* 8: 538–544.

48 Wang, Z.Y., Bi, Y.H., and Zong, M.H. (2012). Regioselectivity reversal in acylation of 6-azauridine catalyzed by *Burkholderia cepacia* lipase. *Biotechnol. Lett.* 34: 55–59.

49 Gao, W.L., Li, N., and Zong, M.H. (2011). Highly regioselective synthesis of undecylenic acid esters of purine nucleosides catalyzed by *Candida antarctica* lipase B. *Biotechnol. Lett.* 33: 2233–2240.

50 Gao, W.L., Liu, H., Li, N., and Zong, M.H. (2012). Regioselective enzymatic undecylenoylation of 8-chloroadenosine and its analogs with biomass-based 2-methyltetrahydrofuran as solvent. *Bioresour. Technol.* 118: 82–88.

51 Gao, W.L., Li, N., and Zong, M.H. (2013). Enzymatic regioselective acylation of nucleosides in biomass-derived 2-methyltetrahydrofuran: kinetic study and enzyme substrate recognition. *J. Biotechnol.* 164: 91–96.

52 Li, X., Wu, Q., Lv, D.S., and Lin, X.F. (2006). Controllable synthesis of polymerizable ester and amide prodrugs of acyclovir by enzyme in organic solvent. *Bioorg. Med. Chem.* 14: 3377–3382.

53 Zhou, G.J., Li, X., Teng, M.X. et al. (2012). Synthesis and characterization of saccharide-functionalized polymer-gemcitabine conjugates based on chemoenzymatic selective strategy. *J. Appl. Polym. Sci.* 124: 1840–1847.

54 Geromichalou, E., Sayyad, N., Kyriakou, E. et al. (2015). Regioselective chemical and rapid enzymatic synthesis of a novel redox – antiproliferative molecular hybrid. *Eur. J. Med. Chem.* 96: 47–57.

55 Kryger, M.B., Wohl, B.M., Smith, A.A., and Zelikin, A.N. (2013). Macromolecular prodrugs of ribavirin combat side effects and toxicity with no loss of activity of the drug. *Chem. Commun.* 49: 2643–2645.

56 Maity, J., Shakya, G., Singh, S.K. et al. (2008). Efficient and selective enzymatic acylation reaction: separation of furanosyl and pyranosyl nucleosides. *J. Org. Chem.* 73: 5629–5632.

57 Singh, S.K., Reddy, L.C., Srivastava, S. et al. (2012). Selective biocatalytic acylation studies on 5′-*O*-(4,4′-dimethoxytrityl)-2′,3′-secouridine: an efficient synthesis of UNA monomer. *Nucleosides Nucleotides Nucleic Acids* 31: 831–840.

58 Singh, S.K., Sharma, V.K., Bohra, K. et al. (2011). Biocatalytic deacylation studies on tetra-*O*-acyl-beta-D-xylofuranosyl nucleosides: synthesis of xylo-LNA monomers. *J. Org. Chem.* 76: 7556–7562.

59 Bavaro, T., Cattaneo, G., Serra, I. et al. (2016). Immobilization of neutral protease from *Bacillus subtilis* for regioselective hydrolysis of acetylated nucleosides: application to capecitabine synthesis. *Molecules* 21: 1621.

60 McClean, K., Preston, C., Spence, D. et al. (2011). Biocatalytic synthesis of valaciclovir using commercial enzymes. *Tetrahedron Lett.* 52: 215–218.

61 Osborne, A.P., Brick, D., Ruecroft, G., and Taylor, I.N. (2006). Immobilization of cholesterol esterase for use in multiple batch biotransformations to prepare (−)-FTC (emtricitabine). *Org. Process Res. Dev.* 10: 670–672.

62 Mahmoudian, M., Baines, B.S., Drake, C.S. et al. (1993). Enzymatic production of optically pure (2′R-*cis*)-2′-deoxy-3′-thiacytidine (3TC, lamivudine): a potent anti-HIV agent. *Enzyme Microb. Technol.* 15: 749–755.

63 Lozada-Ramirez, J.D., Sanchez-Ferrer, A., and Garcia-Carmona, F. (2012). Enzymatic synthesis of *S*-adenosylhomocysteine: immobilization of recombinant *S*-adenosylhomocysteine hydrolase from *Corynebacterium glutamicum* (ATCC 13032). *Appl. Microbiol. Biotechnol.* 93: 2317–2325.

64 Rivero, C.W., De Benedetti, E.C., Gallego, F.L. et al. (2017). Biosynthesis of an antiviral compound using a stabilized phosphopentomutase by multipoint covalent immobilization. *J. Biotechnol.* 249: 34–41.

65 Serra, I., Conti, S., Piskur, J. et al. (2014). Immobilized *Drosophila melanogaster* deoxyribonucleoside kinase (Dm dNK) as a high performing biocatalyst for the synthesis of purine arabinonucleotides. *Adv. Synth. Catal.* 356: 563–570.

66 Serra, I., Ubiali, D., Piskur, J. et al. (2017). Immobilization of deoxyadenosine kinase from *Dictyostelium discoideum* (*Dd*dAK) and its application in the 5′-phosphorylation of arabinosyladenine and arabinosyl-2-fluoroadenine. *ChemistrySelect* 2: 5403–5408.

6

Synthesis of Nucleic Acid Derivatives by Multi-Enzymatic Systems

Qingbao Ding[1,2]

[1] *Rutgers University, Department of Microbiology, Biochemistry and Molecular Genetics, New Jersey Medical School, 205 South Orange Avenue, Newark, NJ 07103, USA*
[2] *East China University of Science and Technology, State Key Laboratory of Bioreactor Engineering, 130 Meilong Road, Shanghai 200237, China*

6.1 Multi-Enzymatic Systems in Biosynthesis

Enzymes are natural catalysts that increase the rate of chemical reactions. Over the last few decades, thousands of enzymes have been developed to produce chemicals. These biocatalytic reactions may involve one enzyme for a specific substrate or two or more enzymes that carry out a series of reactions to yield an expected product. When combinations of several enzymes are used, the process may become more complex, and some otherwise challenging syntheses can be achieved. In recent years, the terms "cascade reactions" or "tandem reactions" have been used to refer to multi-enzymatic reactions [1–4].

The ideal multi-enzymatic systems are living cells. The oldest application of multi-enzymatic systems is the use of these living cells, especially whole cells of microorganisms (e.g. brewing of alcohol and vinegar and industrial fermentation of hundreds of chemicals). There are many advantages to using cellular multi-enzymatic systems, such as inexpensive raw material. In addition, these systems do not require isolation and stabilization of enzymes or addition of cofactors. However, the complexity of cellular systems makes the synthetic process difficult to control, and maximum conversion of the substrate cannot be attained because some of the substrate is used to support cell growth.

Unlike *in vivo* multi-enzymatic systems, *in vitro*, reactions can be easily optimized by changing enzyme and substrate concentrations, varying pH or temperature, or adding cofactors and solvents. Generally, the substrate is converted only to the desired product, and there are no side metabolites, so the purity of the final product *in vitro* is much higher than that in living cells, and the process is very simple.

Despite the many advantages of *in vitro* systems, there remain some limitations. Only a few enzymes can be chosen for this process, so the products can only be produced from specific precursors. Enzymes need to be obtained in sufficient quantities and must be functionally stable under the reaction

Enzymatic and Chemical Synthesis of Nucleic Acid Derivatives, First Edition.
Edited by Jesús Fernández-Lucas and María-José Camarasa Rius.

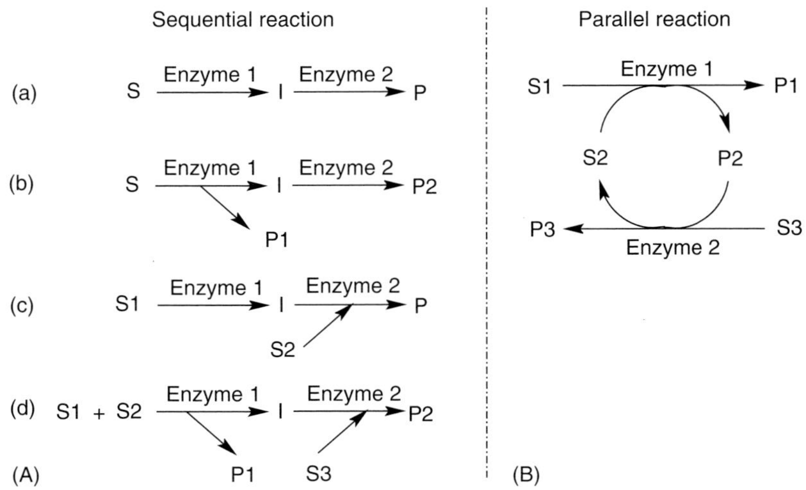

Figure 6.1 Scheme of different multi-enzymatic reactions. S, substrate; I, intermediate; P, product.

conditions. Cofactor-dependent enzymes require cofactor regeneration from inexpensive substrates to reduce costs.

As we know, a multi-enzymatic reaction refers to a biocatalytic process in which two or more transformations are involved and are catalyzed by two or more enzymes. Two kinds of multi-enzymatic systems have been proposed based on types of reaction intermediates (Figure 6.1) [5].

(1) Sequential reactions
This multi-enzyme reaction system can be divided into two subtypes.
   (i) *Linear multi-enzymatic systems* (Figure 6.1A(a)): In these systems, each transformation occurs in tandem, and each intermediate product is the substrate for the following reaction. The intermediate can be separated from the reaction medium or used directly in other reactions. In this system, unstable or toxic intermediates do not accumulate and affect the entire process, thus leading to better yields. It is easy to drive these reversible reactions to completion.
   (ii) *Branched multi-enzymatic systems* (Figure 6.1A(c-d)): In these systems, in addition to the main product, another product is produced, or another substrate is added to the reaction. More complex systems can involve additional substrates and products. The principle behind these systems is that the reaction equilibria can be shifted to obtain the desired product.

(2) Parallel multi-enzymatic systems (Figure 6.1B)
In these systems, two products are simultaneously produced from two different substrates by two distinct biocatalytic reactions. Normally, two different reactions are coupled via cofactors or cosubstrates. Cofactor regeneration is usually used in this system, such as NADPH or NADH regeneration in redox reactions and NTP regeneration in energy transfer or phosphate transfer. NADPH, NADH, and NTP are much more expensive than the substrates

or products in the reactions, and coupling the reactions can greatly reduce the cost.

Of course, current multi-enzymatic systems are not limited to the above classification because these systems may contain several steps that have been reasonably combined [6, 7]. A cost-effective process requires maximal conversion of all substrates used in the reaction to the desired product but without accumulation of toxic intermediates or undesired side products, which must be easily removable from the mixture.

In a multi-enzymatic system, all enzymes can be added simultaneously (one-pot) into the reaction mixture or sequentially into a reaction vessel. These enzymes can be in solution (homogeneous catalysis) or immobilized (heterogeneous catalysis).

6.2 General Overview of Multi-Enzymatic Synthesis of Nucleic Acid Derivatives

Natural nucleosides, nucleotides, and their derivatives play a variety of roles in biochemistry, biology, and medicine. These molecules are commonly used for the synthesis of antiviral and antitumoral agents as well as for the synthesis of oligonucleotides and glycoconjugates. Traditionally, these types of compounds are synthesized by multistep chemical methods [8, 9] that include protection or deprotection of heterocyclic bases or pentose. Unexpected side products or undesired racemic mixtures significantly decrease the total yield and efficiency. Toxic chemical reagents and organic solvents must sometimes be employed in these reactions, which can be harmful to the environment.

Therefore, many efforts have been devoted to synthesizing these compounds with biological reactions catalyzed by whole cells or enzymes [10–13]. Biocatalytic synthesis has many advantages, such as very mild reaction conditions, high stereo- or regioselectivity, and eco-friendliness. Many enzymatic or chemoenzymatic syntheses of nucleosides or nucleotides have been developed to substitute chemical synthesis.

Many enzymes are used individually or in combination to biosynthesize nucleosides or nucleotides [14]: (i) glycosyltransferases (GTs), such as nucleoside phosphorylase (NPase), nucleoside 2′-deoxyribosyltransferase (NDTase), and phosphoribosyltransferase (PRTase); (ii) deaminases, such as adenylate deaminase (AMP deaminase, AMPDase), adenosine deaminase (ADase), and cytidine deaminase (CDase); (iii) kinases, such as deoxynucleoside kinase (dNKase), nucleotide kinase (NtKase), acetate kinase (ACKase), and polyphosphate kinase (PPKase); and (iv) halogenases, such as fluorinase and chlorinase. Some other enzymes, like oxygenase, isomerase, and aldolase, are also used for the biosynthesis of nucleosides and nucleotides.

Although many enzymes have been successfully used for the biosynthesis of nucleosides, nucleotides, and their derivatives, some complex reactions cannot be completed with a single enzyme. For example, ATP is required for the phosphorylation of nucleosides to nucleotides, but ATP is much more expensive than the nucleotide products.

For more efficient synthesis, the use of two or more enzymes can offer several advantages, especially for much more complex synthesis schemes. The use of two or more enzymes can shift the equilibrium of reversible processes toward the desired direction and eliminate problems with product or substrate inhibition. In addition, side products can be removed to increase the ease of separation, and the use of expensive substrates can be avoided.

Over the last two decades, many multi-enzymatic processes have been developed to replace multistep chemical procedures. In this chapter, different methods of *in vitro* multi-enzymatic synthesis of nucleosides, nucleotides, and their analogues are described.

6.3 Multi-Enzymatic Synthesis of Nucleosides and Their Derivatives

Nucleosides and their derivatives are of great interest to medical and biochemical researchers studying antiviral and anticancer therapies. Traditionally, nucleosides are readily synthesized by chemical condensation of a protected base and ribose. Unfortunately, there are many drawbacks to this method, for example, this method involves multiple steps and toxic catalysts, offers low yields, and results in the production of optical isomers. Therefore, the replacement of chemical synthesis by biosynthesis is encouraged. As described above, some enzymes have been successfully used for the synthesis of nucleosides and nucleoside analogues, and there are great advantages to using combinations of these enzymes.

6.3.1 Multi-Enzymatic Synthesis of Nucleosides and Their Analogues by Nucleoside Phosphorylase

NPases mainly include purine nucleoside phosphorylase (PNPase), uridine phosphorylase (UPase), thymidine phosphorylase (TPase), and pyrimidine nucleoside phosphorylase (PyNPase). NPases catalyze the reversible phosphorolysis of nucleosides to form a base and ribose-1-phosphate (R1P) and participate in the metabolism of nucleic acids [15, 16]. NPases have been widely used in the biosynthesis of nucleosides in many studies [17]. Here, we provide some examples, focusing on the combinatorial use of NPases since 2010 (Table 6.1).

Table 6.1 shows that several purine nucleoside analogues (PNAs) are prepared by combinatorial use of NPases. In most cases, PNAs are not easy to chemically synthesize because the purine base is more unstable than the pyrimidine base, leading to greater protection of the active group in the purine ring, which requires many steps. Unlike PNAs, pyrimidine nucleoside analogues (PyNAs) are easier to synthesize chemically and require only a few steps. For example, dideoxyuridine (ddU), arabinoside uracil (ara-U), deoxyuridine (dUR), and thymidine are generally used as starting materials in PNA synthesis as special ribosyl donors.

Among these nucleosides, 5-methyluridine (5-MUR) and thymidine are of great interest to researchers because these nucleosides can be used as raw

Table 6.1 Multi-enzymatic synthesis of nucleosides by combination with NPase.

Substrates					
Nucleoside	**Base**	**Enzymes**	**Product**	**Yield (%)**	**References**
Guanosine	Thymine			85	[18, 19]
Ara-U	(*R*)-4 or (*S*)-4[a)]	UPase PNPase	(*R*)-5 or (*S*)-5[b)]	65 or 46	[20]
araU	Adenine	UPase PNPase	Vidarabine	80	[21]
Uridine	TCA	UPase PNPase	Rabivirin	80	[22]
Uridine Thymidine Ara-U	2,6-Dichloropurine or 2-F-6-Cl-purine	PyNPase PNPase	2,6-Dihalourine riboside or deoxyriboside or arabinoside	60–85	[23, 24]
Thymidine	Adenine	TPase PNPase	2′-Deoxyadenosine	96	[25]
	6-Methyladenine		6-Methyl-2′-deoxyadenosine	95	
Thymidine or 2′-deoxyuridine	2,6-Diamonpurine	TPase PNPase	2,6-Diaminopurine deoxynucleoside	90	[26, 27]
Uridine		UPase PNPase	2,6-Diaminopurine nucleoside		
2′-Deoxyguanosine	5-Methyl-6(5)-azapyrimidine	PNPase TPase	5-Methyl-6(5)-azapyrimidine deoxynucleoside	80	[28]
2′,3′-Dideoxyuridine	Hypoxanthine	UPase PNPase	2′,3′-Dideoxyinosine	30	[29]
Uridine	5,6-Disubstituted benzimidazole	UPase PNPase	5,6-Disubstituted benzimidazole riboside	54–98	[30]
Thymidine		TPase PNPase	5,6-Disubstituted benzimidazole deoxyriboside	51–85	
Uridine or deoxyuridine	5-Substituted 4,6-difluorobenzimidazoles	UPase PNPase	5-Substituted 4,6-difluorobenzimidazoles ribo- or deoxriboside	60–90	[31]
Uridine or deoxyuridine		PyNPase PNPase			

TCA, 1,2,4-triazole-3-carboxamide.

a) (*R*)-4, (3*R*)-4-[6-(2-Amino-purin-6-yl)amino-hexanoyl]-7,8-difluoro-3-methyl-3,4-dihydro-2*H*-[1,4]benzoxazine; (*S*)-4, (3*S*)-4-[6-(2-Amino-purin-6-yl)amino-hexanoyl]-7,8-difluoro-3-methyl-3,4-dihydro-2*H*-[1,4]benzoxazine.

b) (3*R*)-4-{6-[2-Amino-9-(β-D-arabinofuranosyl)purin-6-yl]amino-hexanoyl}-7,8-difluoro-3-methyl-3,4-dihydro-2*H*-[1,4]benzoxazine; (3*S*)-4-{6-[2-Amino-9-(β-D-arabinofuranosyl)purin-6-yl]amino-hexanoyl}-7,8-difluoro-3-methyl-3,4-dihydro-2*H*-[1,4]benzoxazine.

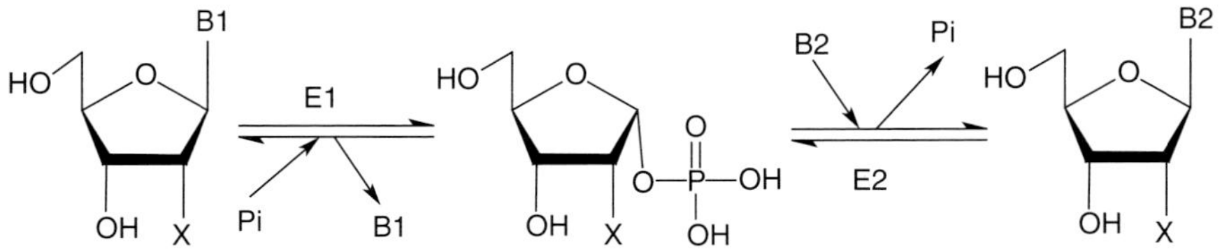

Figure 6.2 Mechanism of the biosynthesis of NAs by NPase. B1, B2: purine or pyrimidine base; X: H or OH; E1, E2: NPase (if B1 and B2 are the same type, then E1 and E2 can be the same).

materials for the synthesis of AZT [32], an anti-HIV medicine. However, direct biosynthesis of thymidine from another deoxyriboside is not advantageous because it is expensive. However, high yields can be achieved in the production of 5-MUR from guanosine with an NPase, which is a cheap nucleoside that can be produced by fermentation. The side product, guanine, has very low solubility in water, which drives the guanosine phosphorolysis forward, increasing R1P production, and increased R1P levels then increase the formation of the nucleoside.

The mechanism of biosynthesis of nucleosides by NPases involves two steps that are catalyzed by a single NPase (Figure 6.2). If B1 and B2 are the same type of base (i.e. if both are a purine or a pyrimidine base), only one kind of NPase can be used, otherwise two types of NPases are required, usually catalyzing the interconversion between PNAs and PyNAs. The interconversion between PNAs and PyNAs can make the subsequent separation easier due to the different chemical characteristics of these molecules.

6.3.2 Transglycosylation Coupled with Xanthine Oxidase

Transglycosylation catalyzed by NPases is reversible, which may lead to lower conversion in some reactions. Therefore, removing the side-product, i.e. the base, may greatly shift the equilibrium toward the product nucleoside. Inosine has higher solubility in water and lower cost and can be used as raw material to biosynthesize other nucleosides. Unlike guanosine, low conversion is achieved when inosine is used as a ribosyl donor because hypoxanthine has much higher solubility in water than guanine. Hori et al. [33] described a method for the biosynthesis of 5-MUR catalyzed by NPase, in which hypoxanthine was converted to urate by xanthine oxidase (XODase) (Figure 6.3); urate is not an NPase substrate. The yields of 5-MUR in the presence and absence of xanthine oxidase were 76 and 33%, respectively, when the initial concentrations of inosine, thymine, and phosphate were 5 mM each. After optimization of the reaction conditions, the yield of 5-MUR obtained was 94% when 40 mM inosine was used. A series of other uracil nucleoside analogues were synthesized with high efficiency by this method [34].

Similarly, Pal and Nair [35] have also shown that the yield of thymidine obtained from 2′-deoxyinosine (2′-dIR) increased when the reaction was catalyzed by whole cells of *Bacillus stearothermophilus* and XODase at a temperature of 55 °C (Figure 6.3). In this reaction, 2′-deoxyadenosine (2′-dAR) was used as starting material because 2′-dAR is more universal than 2′-dIR. In this

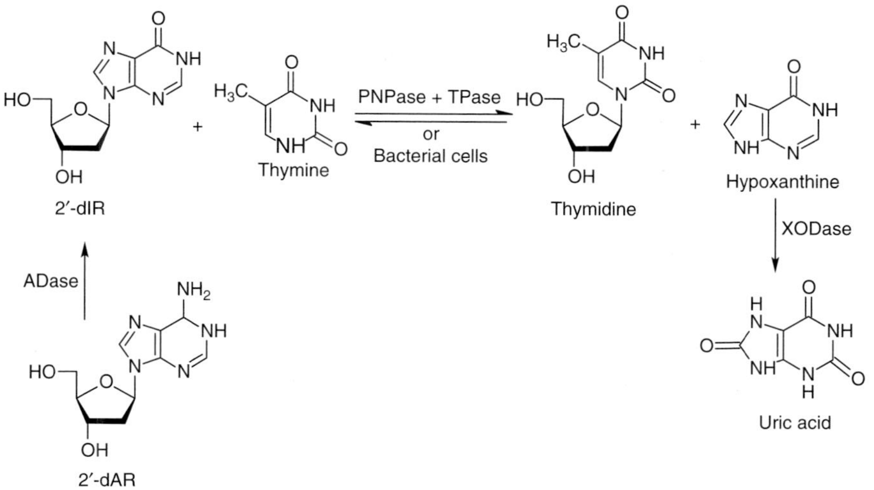

Figure 6.3 Multi-enzymatic synthesis of thymidine in the presence of XODase.

reaction, 2′-dAR was first deaminated to 2′-dIR by ADase. Some other purine deoxynucleosides have also been prepared by this method with good yields [36].

6.3.3 Transglycosylation Reactions Coupled with Deamination

ADase and adenylic acid deaminase (AMPDase) can specifically catalyze the hydrolytic deamination of 6-aminopurine nucleosides while leaving the amino group of the 2-amino base moiety unaltered [37]. In addition, this enzyme shows broad substrate specificity, and the use of this enzyme can be extended to carbocyclonucleosides or acyclonucleosides. The properties of ADase and AMPDase and several examples of the biosynthesis of nucleosides by these enzymes have been described in detail in a review by Santaniello et al. [37].

Reactions catalyzed by NPases or NDTase can be coupled with deamination, which improves the biosynthesis of nucleosides. ADase and AMPDase have been successfully used in these reactions for the synthesis of guanosine or inosine derivatives due to the poor solubility of guanine or hypoxanthine. Guanine is too insoluble in water to be used as a substrate of PNPase or NDTase, leading to very low yields in the preparation of guanosine derivatives. However, some guanine analogues, such as 2,6-diamino purine (DAP) and 2-amino-6-chloro purine (2A6ClP), exhibit good solubility in water. These analogues can be substituted for guanine in transglycosylation to first form guanine nucleoside derivatives. Then, these guanine nucleoside derivatives can be easily deaminated by using ADase or AMPDase to produce high yields of guanosine derivatives.

The natural deoxynucleoside 2′-deoxyguanosine (dGR) is a raw material in the synthesis of dGTP or other derivatives. dGR is not stable in acid and is difficult to chemically synthesize. Because of the insolubility of guanine, low yields are obtained from direct transglycosylation by PNPase [38] or NDTase [39]. However, DAP is highly soluble in water. First, 2,6-diamino purine riboside (DAPR)

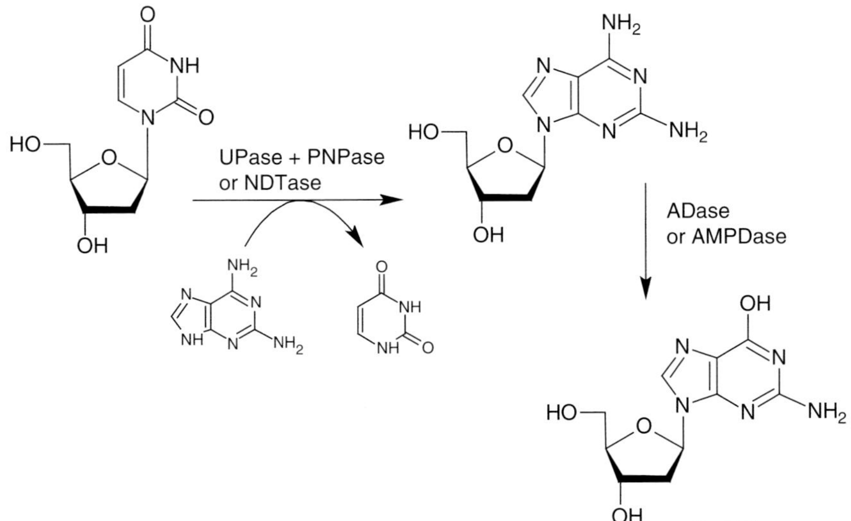

Figure 6.4 Multi-enzymatic synthesis of 2′-dGR by combining reactions catalyzed by NPase or NDTase and ADase (or AMPDase).

is synthesized by PNPase or NDTase, and then, DAPR is easily deaminated by ADase or AMPDase to obtain high yields of dGR (Figure 6.4).

9-β-D-arabinofuranosylguanine (araG) can also be prepared by a similar method (Figure 6.5) [40]. 2,6-Diamino purine arabinofuranoside (araDAP) was first synthesized at a 76% yield at 60 °C by araU from *Enterobacter gergoviae*. Then, the reaction mixture containing araDAP was centrifuged, and cells of *Arthrobacter oxydans* or commercial ADase were added. Under optimum conditions, the yield of araG was as high as 60% when *A. oxydans* was used, while it was 100% when commercial ADase was used.

Another guanosine analogue, 3′-dGR, can also be prepared with a combination of an NPase and deaminase (Figure 6.6) [41, 42], while 3′-deoxycytidine (dCR) was deaminated to 3′-dUR because 3′-dCR is not the substrate of an NPase and is easily produced from cytidine [43]. This step was catalyzed by CDase. 3′-dUR was converted to 2,6-diaminopurine 3′-deoxyriboside (3′-DAPdR), and the latter was deaminated to 3′-dGR. Under optimized conditions (incubation of the substrates at 52 °C for 26 hours), the yield of 3′-DAPdR attained was 72% when the reaction was catalyzed by *Escherichia coli* BMT-4D/1A. The deamination yield attained by using commercial ADase reached 85% when starting with partially purified 3′-DAPdR. A similar method has also been applied to the biosynthesis of 2′-fluoro-2′-deoxyguanosine (2′-F-dGR) [44] to obtain an overall yield of 72–79%.

6.3.4 ADase in Combination with Lipase

It has been reported that the 5′-OH group of adenosine is essential for the activity of ADase [45], which was also confirmed by deamination of 2′- and

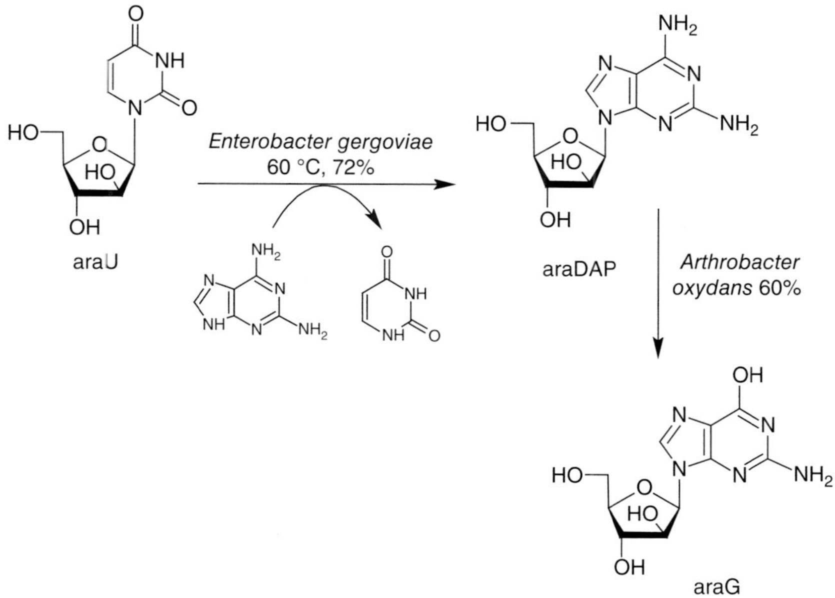

Figure 6.5 Multi-enzymatic synthesis of araG.

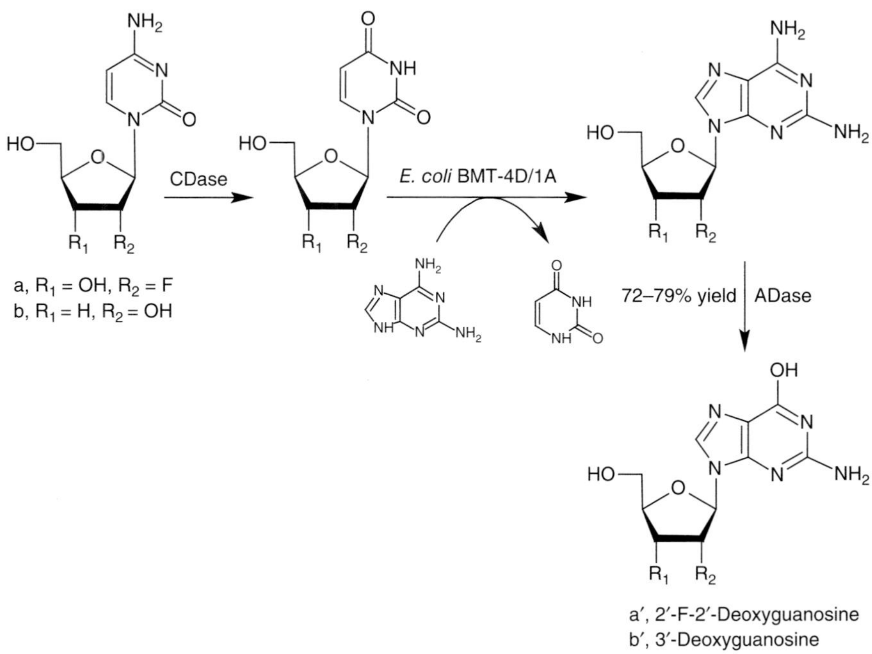

Figure 6.6 Multi-enzymatic synthesis of 2′-F-2′-deoxyguanosine and 3′-deoxyguanosine.

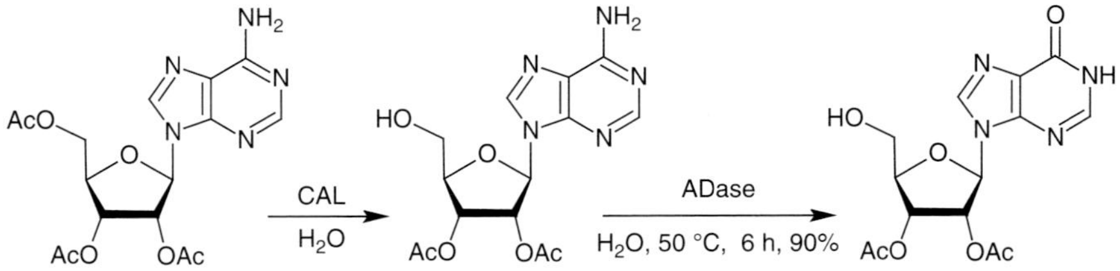

Figure 6.7 Multi-enzymatic synthesis of protected inosine by a combination of CAL and ADase.

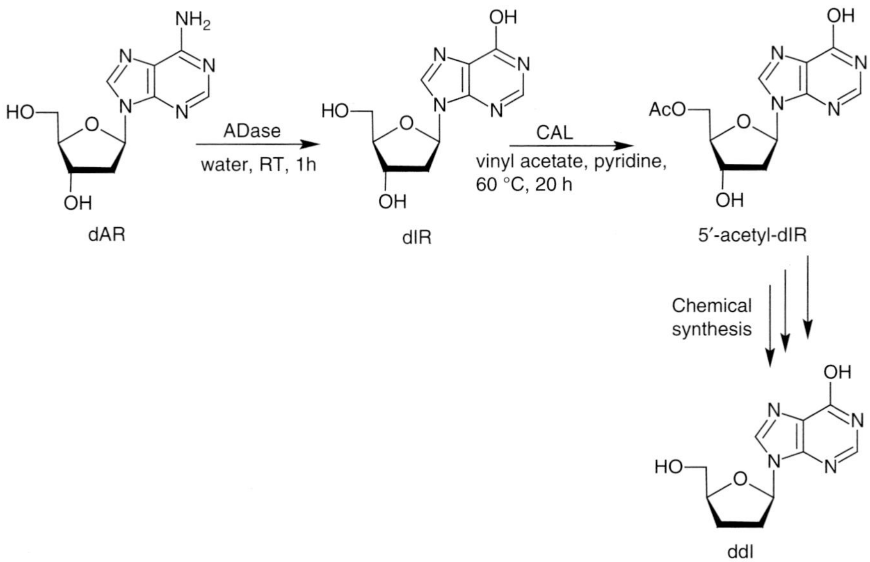

Figure 6.8 Chemoenzymatic synthesis of ddI.

3′-deoxyadenosine acetates [46]. Adenosine with a 5′-acetate group cannot be deaminated by ADase. However, adenosine 2′,3′-diacetate, obtained by selective hydrolysis of adenosine triacetate by lipase (CAL), can be easily deaminated by ADase (six hours, 90%) (Figure 6.7).

Ciuffreda et al. [47] described a chemoenzymatic process to produce 2′,3′-dideoxyinosine (ddI) (Figure 6.8), another anti-HIV medicine. 2′-dIR was efficiently acetylated by CAL in pyridine or tetrahydrofuran at 60 °C to yield the 5′-acetate dIR, which then was chemically converted into ddI. The advantage of this process is that it does not require complex purification of intermediates. The total yield of ddI obtained from dIR was 68%.

6.3.5 Esterification of Nucleosides

araG exhibits activity against acute lymphoblastic leukemia [48]; however, the poor solubility of this molecule limits its usage. One prodrug of araG, nelarabine, is eight times more water soluble than araG but still exhibits poor solubility. 5′-Acetate nelarabine appears more soluble and bioavailable. The

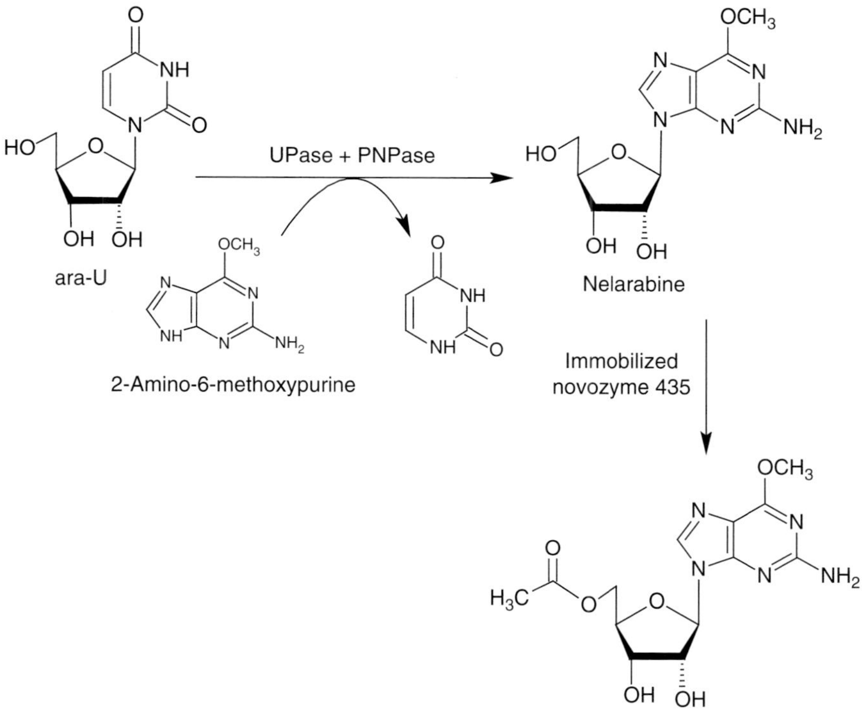

Figure 6.9 Multi-enzymatic synthesis of 5′-acetate nelarabine by a combination of transglycosylation and selective esterification.

multi-enzymatic synthesis of 5′-acetate nelarabine by a combination of transglycosylation and selective esterification is shown in Figure 6.9 [49, 50]. In this process, 2-amino-6-methoxypurine and araU were used as raw materials, and PNPase and UPase were used as catalysts. To improve solubility, nelarabine was selectively esterified with isopropenyl acetate at the 5′ position by immobilized lipase (Novozyme 435) in solvent, and a molar conversion of 99% was attained.

Doxifluridine (5′-deoxy-5-fluorouridine, DFUR) was previously developed for the treatment of some cancers because this compound is selectively converted to 5-FU by human TPase in cancer cells [51]. Direct biosynthesis of DFUR does not occur due to the lack of a 5-deoxyribosyl donor. In one chemoenzymatic approach for the synthesis of DFUR (Figure 6.10), 2′,3′,5′-tri-*O*-acetyluridine was employed by Bavaro et al. as starting material, and two biocatalysts were involved [52]. In the first step, the substrate, 2′,3′,5′-tri-*O*-acetyluridine, was regioselectively hydrolyzed by *Pseudomonas fluorescens* lipase (PFL), and after four chemical steps, 5′-deoxyuridine (5′-dUR) was obtained. Then, a one-enzyme, one-pot transglycosylation catalyzed by immobilized UPase from *Bacillus subtilis* afforded DFUR with approximately 40% conversion. Alternatively, TPase from *E. coli*, immobilized on PEI (polyethylenimine)-activated Sepabeads® and stabilized with 20% oxidized dextran, has also been used [53].

Nucleoside drugs may exhibit low oral bioavailability and have serious side effects. The use of alkoxycarbonylated nucleosides may reduce side effects.

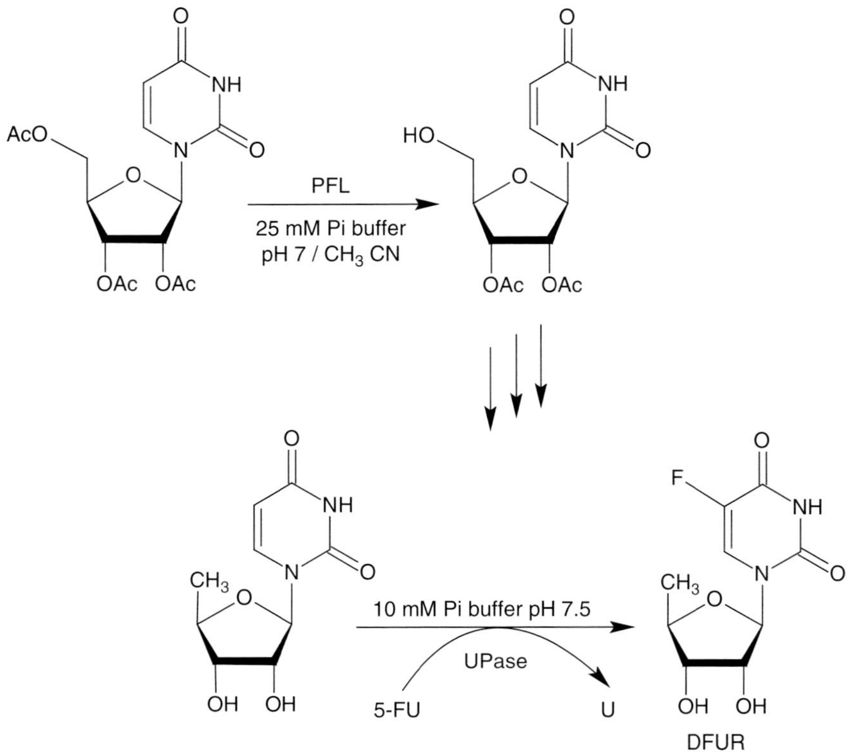

Figure 6.10 Chemoenzymatic synthesis of Doxifluridine.

A series of novel conjugates between *N*-heterocyclic compounds (Figure 6.11) and ribavirin were generated by linking these compounds together via dicarboxylates [54]. *Mucor javanicus* lipase (amino lipase M, MJL) was used for the enzymatic Markovnikov addition of *N*-heterocyclic compounds to divinyl dicarboxylates. The intermediates a′–d′ were obtained with 12–83% isolated yields. The products of the first step were subjected to enzymatic acylation by CAL-B in acetone, furnishing the end products a–d at 41–93% isolated yields.

6.3.6 Multi-Enzymatic Synthesis of Fluorine Nucleosides

In the bacterium *Streptomyces cattleya*, the enzyme fluorinase was found to catalyze the conversion of fluoride ion and *S*-adenosylmethionine (SAM) to 5′-fluoro-5′-deoxyfluoroadenosine (5′-FDA) (Figure 6.12) [55]. Reports of fluorination are rare, and organofluorine compounds are known to be toxic [56, 57]. This enzyme provides a new biotechnological method for the preparation of fluorine nucleosides [58, 59].

The reaction catalyzed by fluorinase is reversible. For efficient conversion, the equilibrium of the reaction must be shifted toward 5′-FDA synthesis, which can be achieved via a multi-enzymatic synthesis. Deng et al. reported a multi-enzymatic system to synthesize ^{18}F-fluoronucleosides (Figure 6.12) [60].

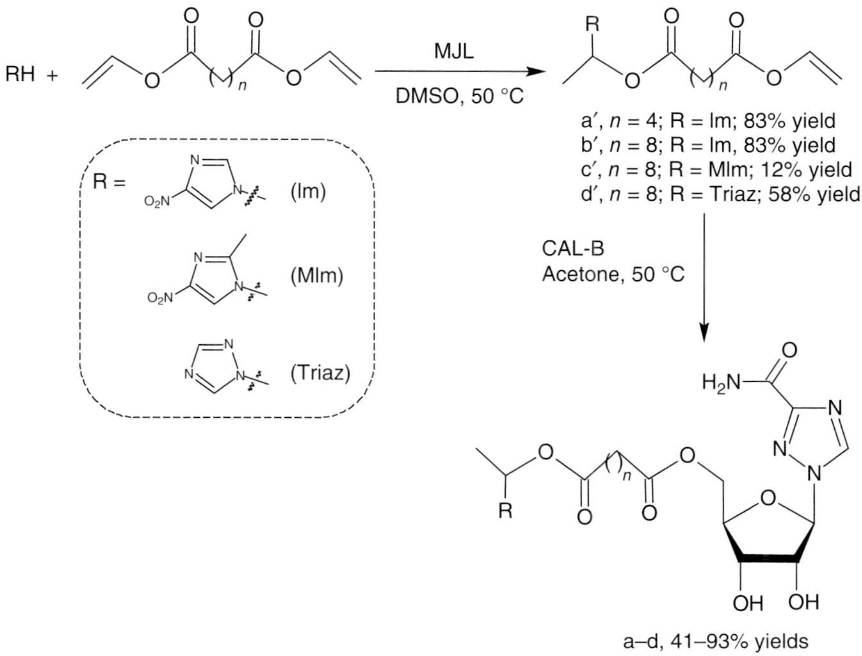

Figure 6.11 Two-lipase-catalyzed sequential synthesis of ribavirin derivatives.

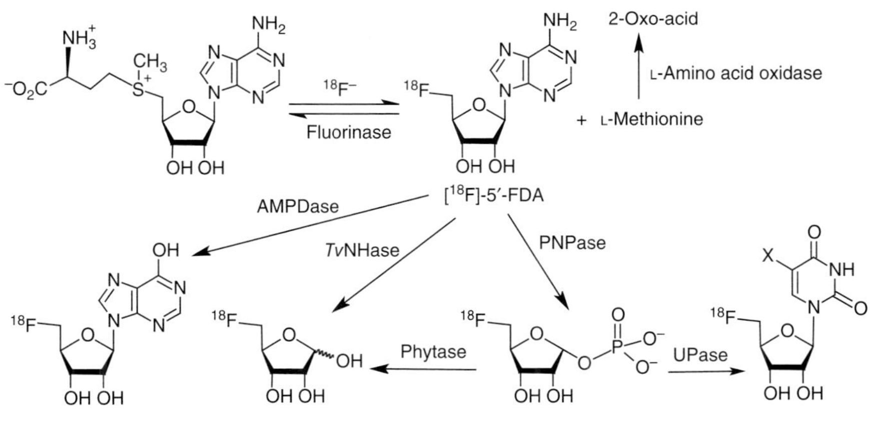

Figure 6.12 Fluorinase coupled with other enzymes for the synthesis of various ^{18}F-labeled compounds.

Compounds containing the F-18 isotope are utilized in positron emission tomography due to the long half-life of this isotope. The enzymatic synthesis of fluorine nucleosides by fluorinase is an attractive method because this method involves rapid and clean synthesis.

The method involved coupling of fluorinase to L-amino acid oxidase, AMPDase, or PNPase and phytase. The resultant product, ^{18}F-5′-FDA, accumulated in the reaction mixture, while L-methionine was removed by the action of the oxidase. The radiochemical yield (RCY) of ^{18}F-5′-FDA was greater than 90% at 35 °C with oxidase. AMPDase could deaminate ^{18}F-5′-FDA to form another fluorinated nucleoside, ^{18}F-5′-deoxyinosine (^{18}F-5′-FDI), with an RCY of 75%. ^{18}F-5′-FDA was also phosphorylyzed by PNPase to form R1P and then converted into fluororibose (5-FDR) with an RCY of 45% after four hours. However, when not used in combination other enzymes, fluorinase did not yield any radiochemical product.

Onega et al. [61] improved the above method by substituting PNPase with nucleoside hydrolase (NHase) to obtain high yields of ^{18}F-5-FDR. In a previous work, human breast cancer cells showed good uptake and very limited metabolism of ^{18}F-5-FDR over two hours. Unlike PNPase, NHase directly catalyzes the hydrolysis of nucleosides without phosphate [62, 63]. NHase from *Trypanosoma vivax* (*Tv*NHase) was used to efficiently hydrolyze the *N*-glycosidic bonds of inosine, adenosine, and guanosine. The one-pot reaction with fluorinase and *Tv*NHase was inefficient. However, once the biosynthesis of ^{18}F-5′-FDA (RCY of 97%) was complete and fluorinase was denatured by heat and removed from the reaction mixture, a high RCY (80%) of ^{18}F-5-FDR was obtained after a two-hour reaction with 20 mg ml^{-1} *Tv*NHase.

The intermediate 5-FDRP is an important compound for the biosynthesis of fluorinated nucleosides by using NPase. Margit et al. reported a series of fluorinated nucleosides that were prepared from 5-FDRP [64]. In this base-swap reaction, PNPase (from *S. cattleya*) and PyNPase (from *B. stearothermophilus*) or TPase (commercial) were used. Interestingly, TPase was confirmed to be efficient enough for the biosynthesis of 2′-nondeoxy nucleosides instead of PyNPase although K_M values indicate that TPase is selective toward 2′-deoxy substrates ($K_M = 35$ mM versus 2.2 mM).

Deng et al. [65] have also shown that fluorinase from *S. cattleya* had activity toward chloride ions in addition to fluoride ions. Removal of L-methionine, which is also produced in the halide substitution reaction and inhibits the reverse reaction, and formation of 5′-chloro-5′-deoxyadenosine (5′-ClDA) can be readily detected by HPLC-ES-MS against a reference compound. The crystal structure of fluorinase with 5′-ClDA shows that the protein structure underwent no significant change in response to the chloride atom. When fluorinase is used in combination with AMPDase, 5′-chloro-5′-deoxyinosine (5′-ClDI) is produced, which is easily detected by HPLC [66]. However, no organohalogen products were detected when Br^- or I^- were used in these reactions.

6.3.7 Multi-Enzymatic Synthesis of Nucleosides via R5P

Phosphomutases (PPMases) can rearrange the positions of phosphates within a substrate molecule [67]. Bacterial PPMases transfer the phosphate group from position C5 of R5P to position C1, yielding R1P, which bridges glucose metabolism and RNA biosynthesis [68]. As previously discussed, R1P serves as a substrate for NPase, which accepts a free base as a second substrate to form a nucleoside. The multi-enzymatic synthesis of nucleosides with PPMase and NPases (Figure 6.13) involves only one free base and the product nucleoside; therefore, the purification of the product nucleoside becomes very simple.

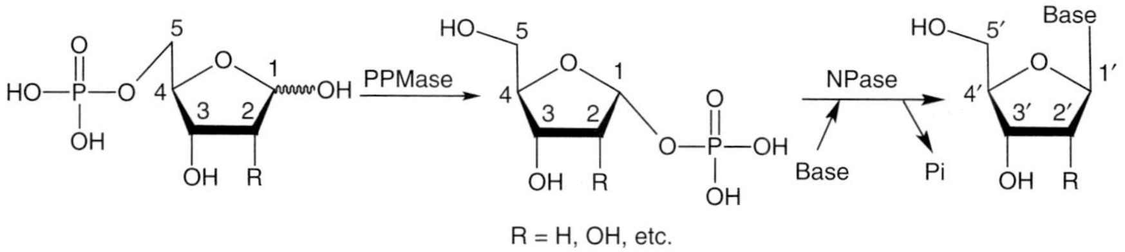

Figure 6.13 Biosynthesis of nucleosides by PPMase and NPase.

PPMase from *E. coli* ATCC 4157 was overexpressed, and it was found that the stability of this enzyme increases in 10% glycerol [69]. This enzyme retained 86% of its initial activity at pH 10 after 18 hours of incubation.

A general strategy for the chemical preparation of furanose 5-phosphates starting from ribose, arabinose, and 2-deoxyribose was developed by Taverna-Porro et al. [70]. A series of ribosyl-modified and base-modified nucleosides were prepared with overexpressed PPMase and commercial NPase; the yield ranged from 7% to 100%, and some arabinosyl nucleosides had relatively low yields. This work showed that PPMase can accept different ribosyl groups.

Stepchenko et al. [71] described a one-pot approach for the synthesis of 2′-deoxynucleosides from 2-deoxyribose by using ribokinase (RKase), PPMase, and PNPase (Figure 6.14). Based on this method, some 8-aza-purine or 8-aza-7-deazapurine 2′-deoxynucleosides were synthesized, depending on the solubility of these bases. Unlike the abovementioned chemical synthesis of the R5P intermediate, the enzymatic synthesis of R5P seems more feasible.

2-Deoxy-D-ribofuranose-5-phosphate (dR5P) can also be synthesized from acetaldehyde and dihydroxyacetone monophosphate via D-glyceraldehyde 3-phosphate by *Klebsiella pneumoniae* B-4-4 by deoxyriboaldolase (DERA)- and triosephosphateisomerase (TPI)-catalyzed reactions (Figure 6.15) [72].

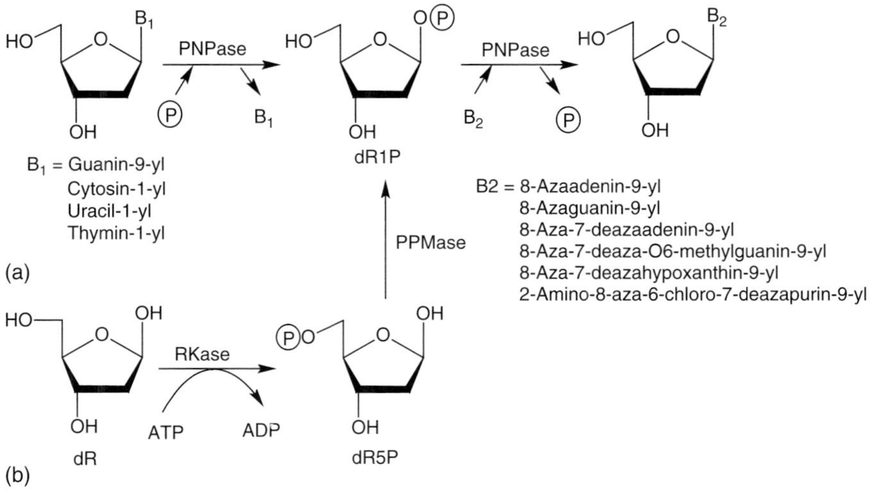

Figure 6.14 Synthesis of deoxynucleosides by the transglycosylation reaction (path a) and multi-enzymatic conversion of 2-deoxy-D-ribose (dR) to deoxynucleosides (path b).

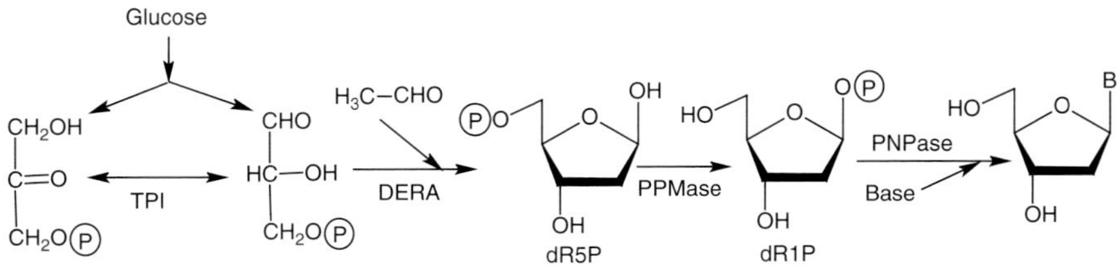

Figure 6.15 Microbial synthesis of 2′-deoxyribonucleosides from triosephosphate, acetaldehyde, and nucleobase.

Under the optimum conditions, 98.7 mM 2-deoxyribose 5-phosphate was produced from 200 mM acetaldehyde and 117 mM dihydroxyacetone phosphate in two hours with a molar yield of 84%. Then, dR5P was converted to 2-deoxy-α-D-ribofuranose-1-phosphate (dR1P) by *E. coli* pTS17/BL21 cells expressing *E. coli* PPMase. In the presence of PNPase and adenine, a mixture of dAR and dIR (in a ratio of 1 : 16) was produced.

Ouwerkerk et al. reported a one-pot, two-step enzymatic reaction involving glycosylation of thymine or uracil (^{13}C and ^{15}N) using dR1P, which was obtained from dR5P, which in turn was obtained from D-glyceraldehyde-3-phosphate (Gla3P) and acetaldehyde [73, 74]. Two enzymes, PPMase and TPase, were used. The yield of thymidine was as high as 61% in the presence of excess acetaldehyde when 0.4 mmol of thymine and 0.1 mmol of dR5P were used as substrates in the presence of 100 units of PPMase and 25 units of TPase.

Scism and Bachmann [75] reported a five-component cascade process for the synthesis of nucleotides and nucleotide analogues (Figure 6.16). Adenylate kinase (AKase), pyruvate kinase (PKase), RKase, phosphoribosyl pyrophosphate synthetase (PPSase), and the engineered hypoxanthine phosphoribosyl transferase 8B3PRT (8B3) were self-immobilized to form a multi-enzyme system, a covalently linked enzyme aggregate or CLEA. In this pathway, PRPP was produced from ribose by RKase and PPSase, and ATP was regenerated by AKase and PKase from phosphoenol pyruvate (PEP). A mutant hypoxanthine phosphoribosyl transferase catalyzes the addition of purine nucleobases to PRPP. 8B3PRT can accept different purine bases, such as 6-chloropurine, 6-thiopurine and purine, and exhibits good conversion.

6.3.8 Other Reactions

A new synthetic strategy for obtaining nucleoside structures by using fructose-1,6-diphosphate aldolase (FDP aldolase) as a catalyst was demonstrated by Liu et al. [76]. FDP aldolase catalyzes the stereospecific condensation of dihydroxyacetone phosphate (DHAP) and D-glyceraldehyde-3-phosphate (G3P) to yield fructose-1,6-diphosphate (Figure 6.17). However, the enzyme accepts a wide variety of aldehydes. FDP aldolase from rabbit muscle can use 3-adenyl-2-hydroxypropanal diethyl acetal (obtained *in situ* from cyclopropylcarbaldehyde diethyl acetal) and formyldiphosphate to furnish interesting ara-related structures. After the reaction catalyzed by FDP aldolase was

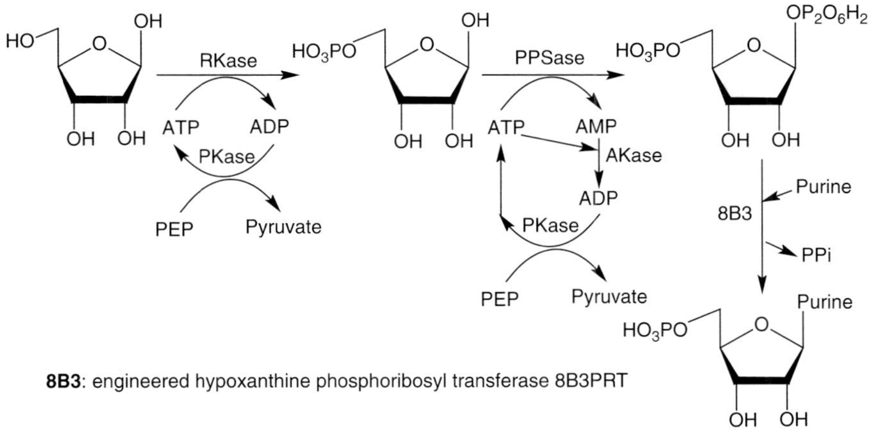

Figure 6.16 Multi-enzymatic synthesis of purine nucleotides.

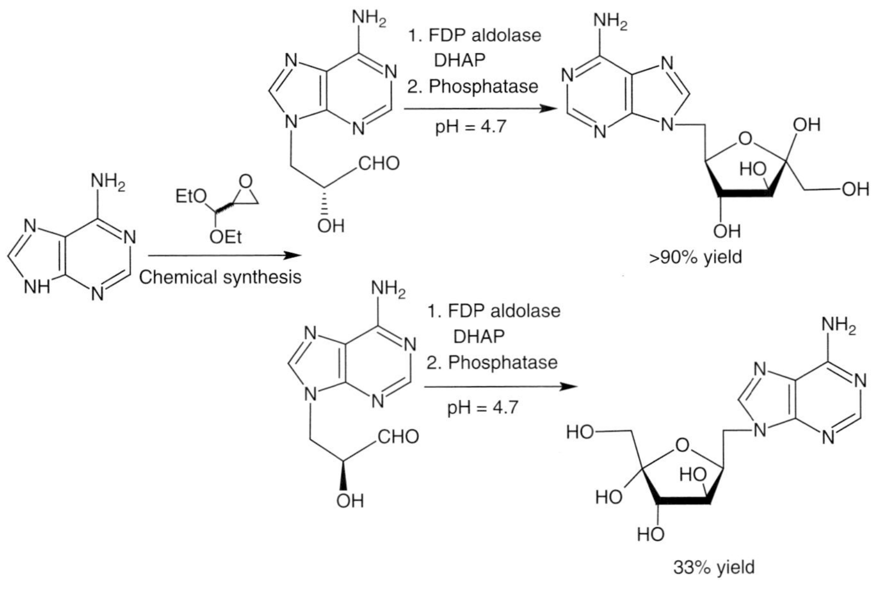

Figure 6.17 Synthesis of adenine nucleoside analogues by FDP aldolase and phosphatase.

complete, the phosphate group was removed by acid phosphates to afford 6-adenyl-6-deoxy-D-fructose at a yield of 20%.

6.4 Multi-Enzymatic Synthesis of Nucleotides and Their Derivatives

Nucleotides and their derivatives are essential molecules in all living cells and include nucleoside 5′-monophosphate (NMP), deoxynucleoside 5′-monophosphate (dNMP), and the di- or tri-phosphates of these molecules

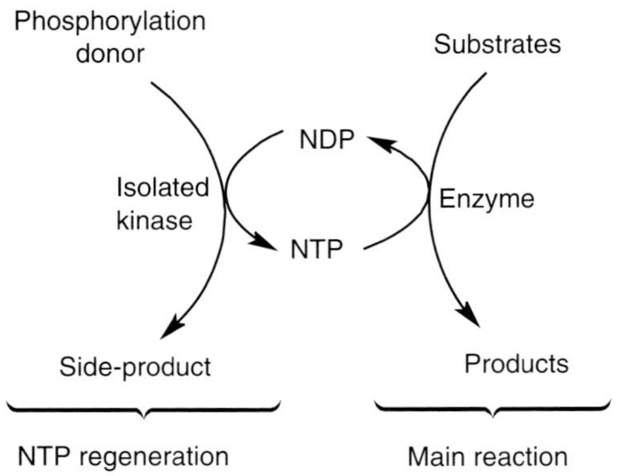

Figure 6.18 NTP regeneration by isolated enzymes.

(NDP, dNDP, NTP, and dNTP). The main nucleotide analogues are sugar-NDPs, which take part in the biosynthesis of polysaccharides.

In the biosynthesis of nucleotides and their analogues, phosphate transfer or energy transfer is generally involved, which makes the reaction very complex and expensive. NTP regeneration systems have been successfully developed to solve these problems. During NTP regeneration, NTP is substituted with another low-cost substrate, and only a very small amount of NTP is used in the reaction as an intermediate to transfer phosphate and energy.

Many methods for NTP regeneration have been developed, and most of these methods are multi-enzymatic reactions (Figure 6.18) with isolated enzymes or whole cells [77]. Here, we discuss only the process catalyzed by isolated enzymes because this process is easy to combine with other reactions.

NTP-regeneration systems with isolated enzymes mainly involve four kinds of kinases: ACKase, PKase, PPKase, and creatine kinase (CKase) [78]. The main phosphate group donors are acetyl phosphate (ACP), PEP, polyphosphate (PolyP), and creatine phosphate (CP). Since isolated enzymes have high NTP-regeneration efficiency, it is very easy to design a novel multi-enzymatic synthesis process combined with an isolated-enzyme NTP-regeneration system for the preparation of bioactive compounds.

6.4.1 Multi-Enzymatic Synthesis of NMPs and dNMPs

Industrially, NMPs can be produced from RNA by nuclease P1, and dNMPs can be produced from DNA. However, a complex separation is required because four kinds of NMPs with similar chemical properties have to be isolated. In addition, for dNMP preparation, the availability of the DNA raw material DNA is very limited. Fortunately, several multi-enzymatic approaches for the synthesis of (d)NMPs were developed from related nucleosides.

Thymidine 5′-monophosphate (TMP) is readily synthesized from thymidine by thymidine kinase (TKase) coupled with ACKase, which catalyzes ATP regeneration (Figure 6.19) [79]. Thymidine is industrially available since it is used as the starting material for the chemical synthesis of zidovudine (ZDV or AZT). The other three 2′-deoxynucleosides, namely, 2′-dAR, 2′-dCR, and 2′-dGR, are not

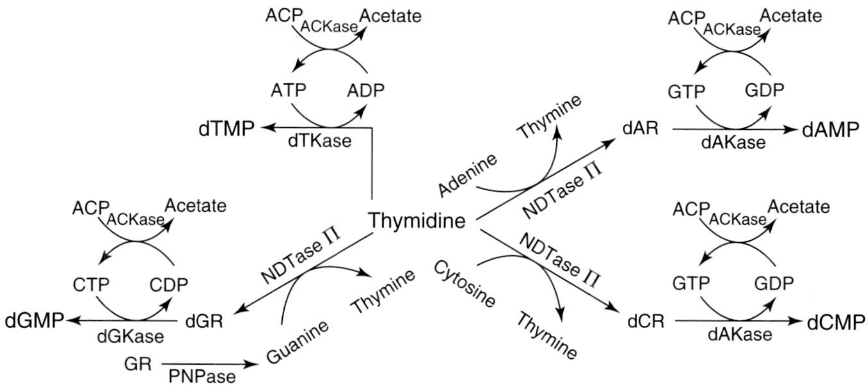

Figure 6.19 Multi-enzymatic synthesis of dNMPs coupled with NTP regeneration by ACKase.

industrially available but can be synthesized from thymidine by using an NPase or NDTase.

A novel one-pot procedure for the synthesis of dAMP or dCMP was designed by Zou et al. [80]. This procedure involved the production of 2′-dAR or 2′-dCR by the enzyme *N*-deoxyribosyltransferase II (NDTase II) from *Lactobacillus delbrueckii* and 2′-deoxynucleoside phosphorylation catalyzed by 2′-dNKase from *B. subtilis* (Figure 6.19). A GTP-regeneration system catalyzed by ACKase from *E. coli* was coupled with a deoxyadenosine kinase (dAKase). Yields of 90.1% for dAMP and 62.3% for dCMP were obtained from 10 mM thymidine. dGMP could also be prepared by a similar method, but the yield was very low because of the insolubility of guanine. Li et al. [81] improved this method (Figure 6.19) by substituting guanine with guanosine. PNPase increased the yield of 2′-dGR, which in turn increased the yield of dGMP to 76%.

Qian et al. [82] designed a process for the synthesis of UMP and CMP from uridine and cytidine, respectively, using a recombinant uridine-cytidine kinase (UCKase) and ACKase. The yield of UMP (3 hours) or CMP (10 hours) was as high as 97%. This method is efficient enough to replace RNA hydrolysis.

6.4.2 Multi-Enzymatic Synthesis of NTPs and dNTPs

NTPs and dNTPs are fundamental reagents in modern molecular biology. NTPs and dNTPs cannot be extracted from living cells although these molecules are widely distributed in all living cells. Both NTPs and dNTPs are easily generated from NDPs and dNDPs by several kinases, such as ACKase, PPKase, and PKase. The inexpensive phosphate donors ACP, PolyP, and PEP can be chemically prepared from very cheap compounds. However, NDPs and dNDPs are relatively expensive, and are usually generated from NMPs and dNMPs, respectively, which is catalyzed by nucleoside monophosphate kinase (NMPKase).

Haynie et al. [83] described a multi-enzymatic synthesis of a mixture of NTPs from RNA. A mixture of NMPs was obtained by the hydrolysis of RNA by nuclease P1. Then, 0.34 mol of total NMP was converted to NTP by NMPKase and

ACKase with ACP as the ultimate phosphate donor. The yields were as follows: ATP, 90%; GTP, 90%; CTP, 60%; and UTP, 40%.

Regeneration of ATP from PolyP and AMP by polyphosphate AMP phosphotransferase (PAPase) from *Acinetobacter johnsonii* and AKase was described by Resnick and Zehnder [84]. AMP and 2′-dAMP were efficiently phosphorylated to ADP and 2′-dADP, respectively, which were converted to ATP and 2′-dATP, respectively, by ADKase. However, this multi-enzymatic system had lower efficiency toward other NTPs. Kameda et al. [85] improved this ATP regeneration and showed that this system could regenerate GTP from GMP.

A multi-enzymatic method for the synthesis of ribavirin 5′-triphosphate (RTP) was reported by Kim and Whotesides [86], who used the PEP/PKase system for ATP regeneration. The yield of RTP was 93%, with 97% purity. Ribavirin 5′-diphosphate (RDP) was not detected in the reaction, implying that the conversion of ribavirin 5′-monophosphate (RMP) to RDP by adenylate kinase is the rate-limiting step in the sequence; the conversion of RDP to RTP by PKase is probably rapid.

Bao and Ryu presented a strategy for the total integrated biosynthesis of four dNTP components, dATP, dGTP, dCTP, and dTTP, using dNMPs as the starting material and the four dNMP kinases (Figure 6.20) [87]. A dATP regeneration system catalyzed by PKase was used for dNTP synthesis. dNMP was first converted to dNDP by dNMPKase, and then, dNDP was converted to dNTP by PKase.

5-Fluorouridine 5′-triphosphate (5F-UTP) and 5-fluorocytidine 5′-triphosphate (5F-CTP) were also successfully produced by a multi-enzymatic system (Figure 6.21) [88]. For 5F-UTP synthesis, 5F-UMP was first prepared by using uracil phosphoribosyl transferase (UPRTase). Then, 5F-UMP was converted to 5F-UDP and 5F-UTP with NMPKase and PKase, respectively. Five equivalents of ATP, which was generated from PEP via 3-PGA by PGMase and enolase, were used in this process. Similarly, 5-fluorocytidine was phosphorylated to 5F-CMP by uridine kinase (UKase) and then converted to 5F-CTP. In this process, three equivalents of ATP were required.

6.4.3 Multi-Enzymatic Synthesis of NDP-Sugars and Other NDP Derivatives

Oligosaccharides are required in many types of cell recognition and signal transduction processes. Many large-scale oligosaccharide syntheses are based on biochemical methods that require sugar nucleotides, usually NDP-sugars, as substrates. However, these substrates are too expensive due to limited research

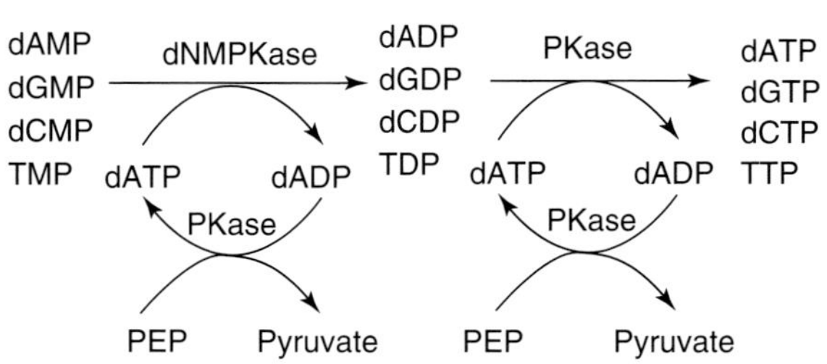

Figure 6.20 Coupled reaction scheme for the total biosynthesis of dNTP from dNMP.

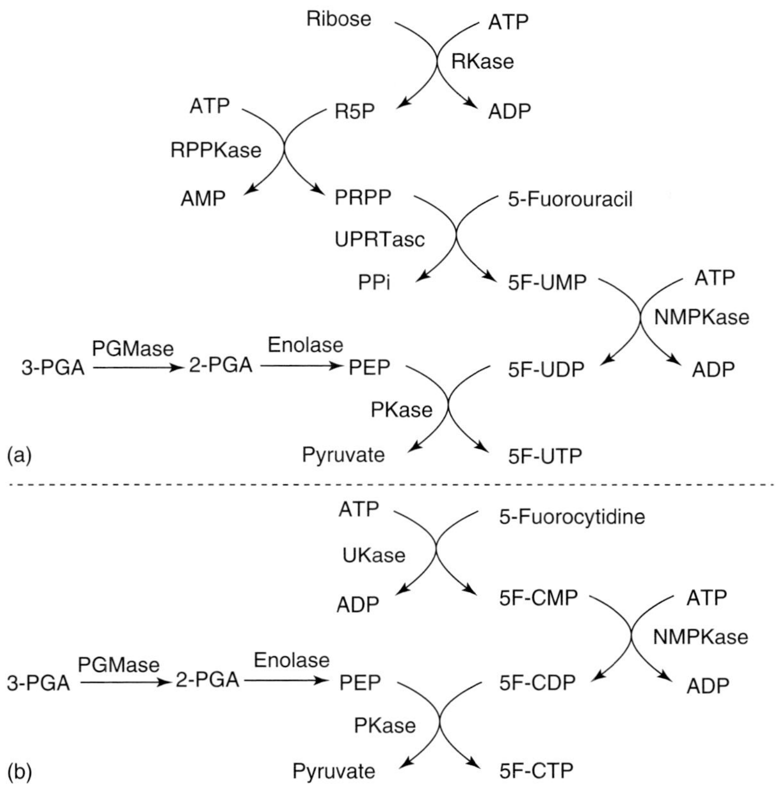

Figure 6.21 Scheme of multi-enzymatic synthesis of (a) 5F-UTP and (b) 5F-CTP.

on oligosaccharides. To overcome this challenge, several enzymatic processes for the synthesis of NDP-sugars have been developed, and most of these processes are based on multi-enzymatic reactions that use cheaper starting substrates, such as NMPs or dNMPs.

In these multi-enzymatic processes, NTPs or dNTPs are generated from NMPs or dNMPs via NDPs or dNDPs by NMPKases or dNMPKases and an NTP regeneration system. The sugar group is mainly derived from sugar-1-phosphate. (d)NDP-sugars can be produced from (d)NTP and sugar-1-phosphate via a reaction catalyzed by NDP-sugar synthase (also called sugar-1-phosphate nucleotide transferase). The preparation of NDP-sugars has been described in detail in some reviews [89–92].

Sucrose synthase (SuSy) is a glycosyltransferase (GT) discovered in plants and in bacteria [93]. This enzyme catalyzes the reversible transfer of a glucosyl moiety between fructose and an NDP (Figure 6.22), which is usually prepared from an NMP coupled with an ATP-regeneration system. The major advantage of this process is that sugar-1-phosphate does not participate in the reaction. Many examples have been described in a review [93], and here, a multi-enzymatic preparation of TDP-glucose is described as an example [94]. TMP is converted to TDP by TMPKase coupled with ATP regeneration from PEP. Then, TDP and

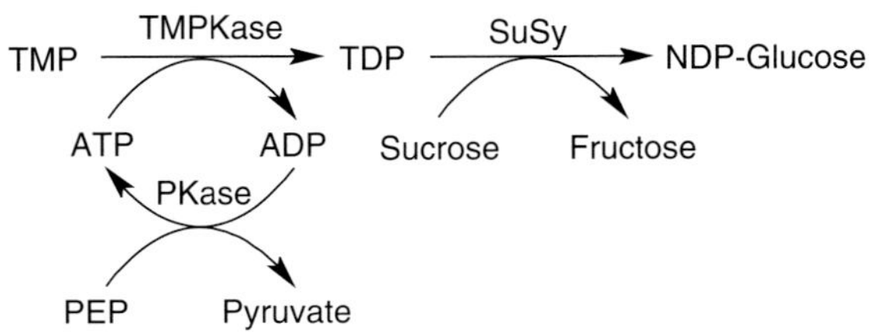

Figure 6.22 Enzymatic synthesis of NDP-glucose by sucrose synthase.

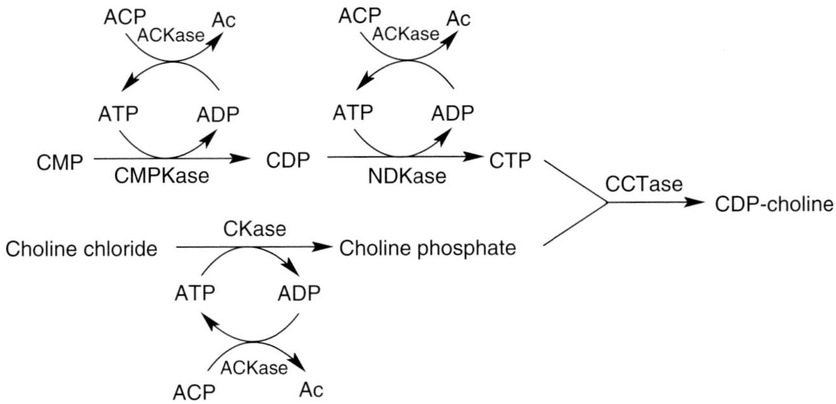

Figure 6.23 Multi-enzymatic synthesis of CDP-choline.

sucrose are converted to TDP-glucose by SuSy via a repetitive batch method; all the enzymes are reused by the removal of 80% of the product solution by ultrafiltration and subsequent addition of fresh substrate solution to start the next batch. After 10 batches, the average yield of TDP-glucose was 94% with TMP as a reference.

One NDP-sugar derivative, cytidine diphosphate choline (CDP-choline, CDP-C), is widely used for treating brain injuries. CDP-C can be produced from CMP and choline via the yeast glycolytic pathway [95–98]. To address production efficiency, Liu et al. [99] recently described a process that integrates an acetate kinase/acetyl phosphate module with *E. coli* strains expressing the enzymes (CMPKase, NDKase, CKase, CCTase) required for converting CMP to CDP-choline, dramatically increasing the CDP-choline titre and productivity (Figure 6.23). Within one hour, 49 mM CDP-choline was produced, with a molar yield of 89.9% and 68.4% with reference to CMP and choline chloride, respectively; the energy utilization efficiency was 79.5%.

6.5 Conclusion

A variety of approaches for multi-enzymatic synthesis, using biological systems or *in vitro* with isolated biocatalysts, have been successfully used for the synthesis of nucleic acid derivatives that are not readily accessible by a single enzyme or by normal organic chemical synthesis. An increasing number of complex processes that use simple and cheap raw materials can be executed due to the development of modern molecular biotechnology. Because these syntheses do not require separation of the intermediates produced in the multi-enzymatic reactions, the usually complex process becomes a very simple "one-pot" process. The shift in the reaction equilibrium is accompanied by higher yields and purities of the products obtained, which increases the utility and applicability of multi-enzymatic synthesis of nucleic acid derivatives.

References

1 Ricca, E., Brucher, B., and Schrittwieser, J.H. (2011). Multi-enzymatic cascade reactions: overview and perspectives. *Adv. Synth. Catal.* 353: 2239–2262.

2 Schrittwieser, J.H., Sattler, J., Resch, V. et al. (2011). Recent biocatalytic oxidation-reduction cascades. *Curr. Opin. Chem. Biol.* 15: 249–256.

3 García-Junceda, E., Lavandera, I., Rother, D., and Schrittwieser, J.H. (2015). (Chemo)enzymatic cascades – nature's synthetic strategy transferred to the laboratory. *J. Mol. Catal. B: Enzym.* 114: 1–6.

4 Schrittwieser, J.H., Velikogne, S., Hall, M., and Kroutil, W. (2018). Artificial biocatalytic linear cascades for preparation of organic molecules. *Chem. Rev.* 118 (1): 270–348.

5 Xue, R. and Woodley, J.M. (2012). Process technology for multi-enzymatic reaction systems. *Bioresour. Technol.* 115: 183–195.

6 Schoffelen, S. and Hest, J.C.M. (2012). Multi-enzyme systems: bring enzymes together in vitro. *Soft Matter* 8: 1736–1746.

7 France, S.P., Hepworth, L.J., Turner, N.J., and Flitsch, S.L. (2017). Constructing biocatalytic cascades: in vitro and in vivo approaches to de novo multi-enzyme pathways. *ACS Catal.* 7: 710–724.

8 Burgess, K. and Cook, D. (2000). Syntheses of nucleoside triphosphates. *Chem. Rev.* 100: 2047–2.59.

9 Qiu, X., Xu, X., and Qing, F. (2010). Recent advances in the synthesis of fluorinated nucleosides. *Tetrahedron* 66: 789–843.

10 Mikhailopulo, I.A. and Miroshnikov, A.I. Biologically important nucleosides: modern trends in biotechnology and application. *Mendeleev Commun.* 21: 57–68.

11 Mikhailopulo, I.A. (2007). Biotechnology of nucleic acid constituents-state of the art and perspectives. *Curr. Org. Chem.* 11: 317–335.

12 Iglesias, L.E., Lewkowicz, E.S., Medici, R. et al. (2015). Biocatalytic approaches applied to the synthesis of nucleoside prodrugs. *Biotechnol. Adv.* 33: 412–434.

13 Lapponi, M.J., Rivero, C.W., Zinni, M.A. et al. (2016). New developments in nucleoside analogues biosynthesis: a review. *J. Mol. Catal.: B Enzym.* 133: 218–233.

14 Fernandez-Lucas, J. (2015). Multienzymatic synthesis of nucleic acid derivatives: a general perspective. *Appl. Microbiol. Biotechnol.* 99: 4651–4627.

15 Giorgelli, F., Bottai, C., Mascia, L. et al. (1997). Recycling of a-D-ribose 1-phosphate for nucleoside interconversion. *Biochim. Biophys. Acta* 1335 (1–2): 16–22.

16 Mascia, L., Cotrufo, T., Cappiello, M., and lpata, P.L. (1999). Ribose 1-phosphate and inosine activate uracil salvage in rat brain. *Biochim. Biophys. Acta* 1472 (1–2): 93–98.
17 Prasad, A.K., Trikha, S., and Parmar, V.S. (1999). Nucleoside synthesis mediated by glycosyl transferring enzymes. *Bioorg. Chem.* 27: 135–154.
18 Zhu, S., Ren, L., Wang, J. et al. (2012). Two-setp efficient synthesis of 5-methyluridine via two thermostable nucleoside phosphorylase from Aeropyrum pernix. *Bioorg. Med. Chem. Lett.* 22: 2102–2104.
19 Zhu, S., Song, D., Gong, C. et al. (2013). Biosynthesis of nucleoside analogues via thermostable nucleoside phosphorylase. *Appl. Microbiol. Biotechnol.* 97: 6769–6778.
20 Eletskaya, B.Z., Konstantinova, I.D., Paramonov, A.S. et al. (2016). Chemoenzymatic arabinosylation of 2-aminopurines bearing the chiral fragment of 7,8-difluoro-3-methyl-3,4-dihydro-2*H*-[1,4]benzoxazines. *Mendeleev Commun.* 26: 6–8.
21 Serra, I., Daly, S., Alcantara, A.R. et al. (2015). Redesigning the synthesis of vidarabine via a multienzymatic reaction catalyzed by immobilized nucleoside phosphorylases. *RSC Adv.* 5: 23569–23577.
22 Rivero, C.W., De Benedetti, E.C., Lozano, M.E., and Trelles, J.A. (2015). Bioproduction of ribavirin by green microbial biotransformation. *Process Biochem.* 50: 935–940.
23 Zhou, X., Mikhailopulo, B.M.N.C., and Neubauer, P. (2015). Immobilization of thermostable nucleoside phosphorylases on MagReSyn® epoxide microspheres and their application for the synthesis of 2,6-dihalogenated purine nucleosides. *J. Mol. Catal.: B Enzym.* 115: 119–127.
24 Zhou, X., Szeker, K., Jiao, L. et al. (2015). Synthesis of 2,6-dihalogenated purine nucleosides by thermostable nucleoside phosphorylases. *Adv. Synth. Catal.* 357: 1237–1244.
25 Liang, S., Li, W., Gao, T. et al. (2010). Enzymatic synthesis of 2′-deoxyadenosine and 6-methylpurine-2′-deoxyriboside by *Escherichia coli* DH5α overexpressing nucleoside phosphorylases from *Escherichia coli* BL21. *J. Biosci. Bioeng.* 110 (2): 165–168.
26 Ding, Q., Ou, L., Wei, D. et al. (2010). Enzymatic synthesis of nucleosides by nucleoside phosphorylase co-expressed in *Escherichia coli*. *J. Zhejiang Univ. Sci. B* 11 (11): 880–888.
27 De Benedetti, E.C., Rivero, C.W., Britos, C.N. et al. (2012). Biotransformation of 2,6-diaminopurine nucleosides by immobilized *Geobacillus stearothermophilus*. *Biotechnol. Progr.* 28: 1251–1256.
28 Stepchenko, V.A., Miroshnikov, A.I., Seelaand, F., and Mikhailopulo, I.A. (2016). Enzymatic synthesis and phosphorolysis of 4(2)-thioxo- and 6(5)-azapyrimidine nucleosides by *E. coli* nucleoside phosphorylases. *Beilstein J. Org. Chem.* 12: 2588–2601.
29 Christoffersen, S., Serra, I., Terreni, M., and Piškur, J. (2010). Nucleoside phosphorylases from *Clostridium Perfringens* in the synthesis of 2′,3′-dideoxyinosine. *Nucleosides Nucleotides Nucleic Acids* 29 (4–6): 445–448.

30 Konstantinova, I.D., Selezneva, O.M., Fateev, I.V. et al. (2013). Chemo-enzymatic synthesis and biological evaluation of 5,6-disubstituted benzimidazole ribo- and 2′-deoxyribonucleosides chemo-enzymatic syntheses and biological evaluation. *Synthesis* 45: 272–280.

31 Kharitonova, M.I., Fateev, I.V., Kayushin, A.L. et al. (2016, 406). Chemoenzymatic synthesis and antiherpes activity of 5-substituted 4,6-difluorobenzimidazoles ribo- and 2′-deoxyribonucleosides. *Synthesis* 48: 394.

32 Warshaw, J.A. and Watanabe, K.A. (1990). 2′-Azido-2′,3′-dideoxypyrimidine nucleosides. Synthesis and antiviral activity against human immunodeficiency virus. *J. Med. Chem.* 33 (6): 1663–1666.

33 Hori, N., Uehara, K., and Mikami, Y. (1991). Effects of xanthine oxidase on synthesis of 5-methyluridine by the ribosyl transfer reaction. *Agric. Biol. Chem.* 55 (4): 1071–1074.

34 Hori, N., Uehara, K., and Mikami, Y. (1992). Enzymatic synthesis of 5-methyluridine from adenosine and thymine with high efficiency. *Biosci. Biotechnol., Biochem.* 56 (4): 580–582.

35 Pal, S. and Nair, V. (1997). Enzymatic synthesis of thymidine using bacterial whole cells and isolated purine nucleoside phosphorylase. *Biocatal. Biotransform.* 15: 147–158.

36 Pal, S. and Nair, V. (1997). Biocatalytic synthesis of base-modified 2′-deoxy-B-D-ribonucleosides with bacterial whole cells. *Biotechnol. Lett.* 19 (4): 349–351.

37 Santaniello, E., Ciuffreda, P., and Alessandrini, L. (2005). Synthesis of modified purine nucleosides and related compounds mediated by adenosine deaminase (ADA) and adenylate deaminase (AMPDA). *Synthesis* 2005 (4): 509–526.

38 Yokozeki, K. and Tsuji, T. (2000). A novel enzymatic method for the production of purine-2′-deoxyribonucleosides. *J. Mol. Catal.: B Enzym.* 10 (1–3): 207–213.

39 Okuyama, K., Shibuya, S., Hamamoto, T., and Noguchi, T. (2003). Enzymatic synthesis of 2′-deoxyguanosine with nucleoside deoxyribosyltransferase-II. *Biosci. Biotechnol. Biochem.* 67 (5): 989–995.

40 Médici, R., Iribarren, A.M., and Lewkowicz, E.S. (2009). Synthesis of 9-b-D-arabinofuranosylguanine by combined use of two whole cell biocatalysts. *Bioorg. Med. Chem. Lett.* 19: 4210–4212.

41 Barai, V.N., Zinchenko, A.I., Eroshevskaya, L.A. et al. (2002). Chemo-enzymatic synthesis of 3-deoxy-β-D-ribofuranosyl purines. *Helv. Chim. Acta* 85: 1893–1900.

42 Barai, V.N., Zinchenko, A.I., Eroshevskaya, L.A. et al. (2003). Chemo-enzymatic synthesis of 3-deoxy-β-D-ribofuranosyl purines and study of their biological properties. *Nucleosides Nucleotides Nucleic Acids* 22 (5–8): 751–753.

43 Sivets, G.G., Klennitskaya, T.V., Zhernosek, E.V., and Mikhailopulo, I.A. (2002). Synthesis of peracylated derivatives of L-ribofuranose from D-ribose and their use for the preparation of β-L-ribonucleosides. *Synthesis* 2002 (2): 0253–0259.

44 Zaitseva, G.V., Zinchenko, A.I., Barai, V.N. et al. (1999). Chemical and enzymatic synthesis and antiviral properties of 2′-deoxy-2′-fluoroguanosine. *Nucleosides Nucleotides* 18 (4–5): 687–688.

45 Bloch, A., Robins, M.J., and McCarthy, J.R. (1967). The role of the 5′-hydroxyl group of adenosine in determining substrate specificity for adenosine deaminase. *J. Med. Chem.* 10 (5): 908–912.

46 Ciuffreda, P., Casati, S., and Santaniello, E. (2000). The action of adenosine deaminase (E.C. 3.5.4.4.) on adenosine and deoxyadenosine acetates: the crucial role of the 5′-hydroxy group for the enzyme activity. *Tetrahedron* 56: 3239–3243.

47 Ciuffreda, P., Casati, S., and Santaniello, E. (1999). Lipase-catalyzed protection of the hydroxy groups of the nucleosides inosine and 2′-deoxyinosine: a new chemoenzymatic synthesis of the antiviral drug 2′,3′-dideoxyinosine. *Bioorg. Med. Chem. Lett.* 9: 1577–1582.

48 Rodriguez, C.O. Jr.,, Stellrecht, C.M., and Gandhi, V. (2003). Mechanisms for T-cell selective cytotoxicity of arabinosylguanine. *Blood* 102: 1842–1848.

49 Herbal, K., Kitteringham, J., Voyle, M., and Whitehead, A.J. (2005). Synthesis of the enantiomer of nelarabine. *Tetrahedron Lett.* 46: 2961–2964.

50 Mahmoudian, M., Eaddy, J., and Dawson, M. (1999). Enzymic acylation of 506U78 (2-amino-9-β-D-arabinofuranosyl-6-methoxy-9*H*-purine), a powerful new anti-leukaemic agent. *Biotechnol. Appl. Biochem.* 29: 229–233.

51 Shimma, N., Umeda, I., Arasaki, M. et al. (2000). The design and synthesis of a new tumor-selective fluoropyrimidine carbamate, capecitabine. *Bioorg. Med. Chem.* 8: 1697–1706.

52 Bavaro, T., Rocchietti, S., Ubiali, D. et al. (2009). A versatile synthesis of 5-functionalized nucleosides through regioselective enzymatic hydrolysis of their peracetylated precursors. *Eur. J. Org. Chem.* 2009 (12): 1967–1975.

53 Serra, I., Bavaro, T., Cecchini, D.A. et al. (2013). A comparison between immobilized pyrimidine nucleoside phosphorylase from *Bacillus subtilis* and thymidine phosphorylase from *Escherichia coli* in the synthesis of 5-substituted pyrimidine 2-deoxyribonucleosides. *J. Mol. Catal.: B Enzym.* 95: 16–22.

54 Liu, B.K., Qian, X.Q., Wu, Q., and Lin, X.F. (2008). Two lipase-catalyzed sequential synthesis of drug derivatives in organic media. *Enzym. Microb. Technol.* 43: 375–380.

55 O'Hagan, D., Schaffrath, C., Cobb, S.L. et al. (2002). Biosynthesis of an organofluorine molecule. *Nature* 416: 279.

56 Harper, D.B. and O'Hagan, D. (1994). The fluorinated natural products. *Nat. Prod. Rep.* 11 (0): 123–133.

57 Wagner, C., El Omari, M., and König, G.M. (2009). Biohalogenation: nature's way to synthesize halogenated metabolites. *J. Nat. Prod.* 72 (3): 540–553.

58 Latham, J., Brandenburger, E., Shepherd, S.A. et al. (2018). Development of halogenase enzymes for use in synthesis. *Chem. Rev.* 118 (1): 232–269.

59 O'Hagan, D. and Deng, H. (2015). Enzymatic fluorination and biotechnological developments of the fluorinase. *Chem. Rev.* 115 (2): 634–649.

60 Deng, H., Cobb, S.L., Gee, A.D. et al. (2006). Fluorinase mediated C–18F bond formation, an enzymatic tool for PET labelling. *Chem. Commun.* 652–654.

61 Onega, M., Domarkas, J., Deng, H. et al. (2010). An enzymatic route to 5-deoxy-5-[18F]fluoro-D-ribose, a [18F]-fluorinated sugar for PET imaging. *Chem. Commun.* 46: 139–141.

62 Vandemeulebroucke, A., De Vos, S., Holsbeke, E.V. et al. (2008). A flexible loop as a functional element in the catalytic mechanism of nucleoside hydrolase from *Trypanosoma vivax*. *J. Biol. Chem.* 283: 22272–22282.

63 Versées, W., Goeminnec, A., Berg, M. et al. Crystal structures of *T. vivax* nucleoside hydrolase in complex with new potent and specific inhibitors. *Biochim. Biophys. Acta* 1794: 953–960.

64 Winkler, M., Domarkas, J., Schweiger, L.F., and O'Hagan, D. (2008). Fluorinase-coupled base swaps: synthesis of [18F]-5′-deoxy-5′-fluorouridines. *Angew. Chem. Int. Ed.* 47: 10141–10143.

65 Deng, H., Cobb, S.L., McEwan, A.R. et al. (2006). The fluorinase from *Streptomyces cattleya* is also a chlorinase. *Angew. Chem. Int. Ed.* 45: 759–762.

66 Margolin, A.L., Borcherding, D.R., Wolf-Kugel, D., and Margolin, N. (1994). AMP deaminase as a novel practical catalyst in the synthesis of 6-oxopurine ribosides and their analogs. *J. Org. Chem.* 59 (24): 7214–7218.

67 Panosian, T.D., Nannemann, D.P., Watkins, G.R. et al. (2011). *Bacillus cereus* phosphopentomutase is an alkaline phosphatase family member that exhibits an altered entry point into the catalytic cycle. *J. Biol. Chem.* 286: 8043–8054.

68 Tozzi, M.G., Camici, M., Mascia, L. et al. (2006). Pentose phosphates in nucleoside interconversion and catabolism. *FEBS J.* 273: 1089–1101.

69 Rivero, C.W., De Benedetti, E.C., Gallego, F.L. et al. (2017). Biosynthesis of an antiviral compound using a stabilized phosphopentomutase by multipoint covalent immobilization. *J. Biotechnol.* 249: 34–41.

70 Taverna-Porro, M., Bouvier, L.A., Pereira, C.A. et al. (2008). Chemoenzymatic preparation of nucleosides from furanoses. *Tetrahedron Lett.* 49: 2642–2645.

71 Stepchenko, V.A., Seela, F., Esipov, R.S. et al. (2012). Enzymatic synthesis of 2′-deoxy-B-D-ribonucleosides of 8-azapurines and 8-aza-7-deazapurines. *Synlett* 23: 1541–1545.

72 Ogawa, J., Saito, K., Sakai, T. et al. (2003). Microbial production of 2-deoxyribose 5-phosphate form acetaldehyde and triosephosphate for the synthesis of 2′-deoxyribonucleosides. *Biosci. Biotechnol. Biochem.* 67: 933–936.

73 Ouwerkerk, N., van Boom, J.H., Lugtenburg, J., and Raap, J. (2000). Chemo-enzymatic synthesis of thymidine ^{13}C-labelled in the 2′-deoxyribose moiety. *Eur. J. Org. Chem.* 2000 (5): 861–866.

74 Ouwerkerk, N., Steenweg, M., De Ruijter, M. et al. One-pot two-step enzymatic coupling of pyrimidine bases to 2-deoxy-D-ribose-5-phosphate. A new strategy in the synthesis of stable isotope labeled deoxynucleosides. *J. Org. Chem.* 67 (5): 1480–1489.

75 Scism, R.A. and Bachmann, B.O. (2010). Five-component cascade synthesis of nucleotide analogues in an engineered self-immobilized enzyme aggregate. *ChemBioChem* 11: 67–70.

76 Liu, K.K.C. and Wong, C. (1992). A new strategy for the synthesis of nucleoside analogues based on enzyme-catalyzed aldol reactions. *J. Org. Chem.* 57 (18): 4789–4791.

77 Endo, T. and Koizumi, S. (2001). Microbial conversion with cofactor regeneration using genetically engineered bacteria. *Adv. Synth. Catal.* 343 (6–7): 521–526.

78 Andexer, J.N. and Richter, M. (2015). Emerging enzymes for ATP regeneration in biocatalytic processes. *ChemBioChem* 16: 380–386.

79 Zhang, J., Qian, Y., Ding, Q., and Ou, L. (2015). Enzymatic manufacture of deoxythymidine 5′-triphosphate with permeable intact cells of *E. coli* coexpressing thymidylate kinase and acetate kinase. *J. Microb. Biotechnol.* 25 (12): 2034–2042.

80 Zou, Z., Ding, Q., Ou, L., and Yan, B. (2013). Efficient production of deoxynucleoside-5′-monophosphates using deoxynucleoside kinase coupled with a GTP-regeneration system. *Appl. Microbiol. Biotechnol.* 97: 9389–9395.

81 Li, Y., Ding, Q., Ou, L., and Zhang, J. (2014). One-pot process of 2′-deoxyguanylic acid catalyzed by a multi-enzyme system. *Biotechnol. Bioprocess Eng.* 20: 37–43.

82 Qian, Y., Ding, Q., Li, Y., and Ou, L. (2014). Phosphorylation of uridine and cytidine by uridine-cytidine kinase. *J. Biotechnol.* 188: 81–87.

83 Haynie, S.L. and Whitesides, G.M. (1990). Preparation of a mixture of nucleoside triphosphates suitable for use in synthesis of nucleotide phosphate sugars from ribonucleic acid using nuclease P1, a mixture of nucleoside monophosphokinases and acetate kinase. *Appl. Biochem. Biotechnol.* 23: 205–220.

84 Resnick, S.M. and Zehnder, A.J.B. (2000). In vitro ATP regeneration from polhyphosphate and AMP by polyphosphate:AMP phosphotransferase and adenylate kinase from *Acinetobacter johnsonii* 210A. *Appl. Environ. Microbiol.* 66: 2045–2051.

85 Kameda, A., Shiba, T., Kawazoe, Y. et al. (2001). A novel ATP regeneration system using polyphosphate-AMP phosphotransferase and polyphosphate kinase. *J. Biosci. Bioeng.* 91 (6): 557–563.

86 Kim, M. and Whitesides, G.M. (1987). Enzyme-catalyzed synthesis of nucleoside triphosphates from nucleoside monophosphates. ATP from AMP and ribavirin 5′-triphosphate from ribavirin 5′-monophosphate. *Appl. Biochem. Biotechnol.* 16: 95–108.

87 Bao, J. and Ryu, D.D.Y. (2007). Total biosynthesis of deoxynucleoside triphosphates using deoxynucleoside monophosphate kinases for PCR application. *Biotechnol. Bioeng.* 98 (1): 1–11.

88 Henning, M., Scott, L.G., Sperling, E. et al. (2007). Synthesis of 5-fluoropyrimidine nucleotides as sensitive NMR probes of RNA structure. *J. Am. Chem. Soc.* 129: 14911–14921.

89 Tsai, T., Wu, C., and Wong, C. (2015). Large-scale enzymatic synthesis of glycans with cofactor regeneration. In: *Glycoscience: Biology and Medicine* (ed. N. Taniguchi et al.), 409–417. Japan: Springer.

90 Thibodeaux, C.J., Melancon, C.E., and Liu, H. (2008). Natural-product sugar biosynthesis and enzymatic glycodiversification. *Angew. Chem. Int. Ed.* 47: 9814–9859.

91 De Bruyn, F., Maertens, J., Beauprez, J. et al. (2015). Biotechnological advances in UDP-sugar based glycosylation of small moleucles. *Biotechnol. Adv.* 33: 288–302.

92 Cai, L. (2012). Recent progress in enzymatic synthesis of sugar nucleotides. *J. Carbohydr. Chem.* 31 (7): 535–552.

93 Schmolzer, K., Gutmann, A., Diricks, M. et al. (2016). Sucrose synthase: a unique glycosyltransferase for biocatalytic glycosylation process development. *Biotechnol. Adv.* 34: 88–111.

94 Elling, L., Rupprath, C., Gunther, N. et al. (2005). An enzyme module system for the synthesis of dTDP-activated deoxysugars from dTMP and sucrose. *ChemBioChem* 6: 1423–1430.

95 Tang, J., Chen, Y., Chen, X. et al. (2010). Production of cytidine 5′-diphosphorylcholine with high utilization of ATP by whole cells of *Saccharomyces cerevisiae*. *Bioresour. Technol.* 101 (22): 8807–8813.

96 Ado, Y., Suzuki, Y., Tadokoro, T. et al. (1979). Regeneration of ATP by immobilized microbial cells and utilizationfor the synthesis of ATP and CDP-choline. *J. Solid-Phase Biochem.* 4 (1): 43–55.

97 Kimura, A., Tatsutomi, Y., Mizushima, N. et al. (1978). Immobilization of glycolysis system of yeasts and production of cytidine diphosphate choline. *Eur. J. Appl. Microbiol. Biotechnol.* 5: 13–16.

98 Kimura, A., Tatsutomi, Y., Matsuno, R. et al. (1981). Some properties of an immobilized glycolysis system of yeast in fermentative phosphorylation of nucleotides. *Eur. J. Appl. Microbiol. Biotechnol.* 11: 78–80.

99 Liu, Y., Wang, J., Xu, C. et al. (2017). Efficient multi-enzyme-catalyzed CDP-choline production driven by an ATP donor module. *Appl. Microbiol. Biotechnol.* 101 (4): 1409–1417.

7

Enzymatic Synthesis Using Polymerases of Modified Nucleic Acids and Genes

Elena Eremeeva and Piet Herdewijn

KU Leuven, Rega Institute for Medical Research, Medicinal Chemistry, Herestraat 49, 3000 Leuven, Belgium

7.1 Introduction

Deoxyribonucleic acid (DNA) and ribonucleic acid (RNA) are key molecules of life on Earth. All organisms from bacteria to humans use the same molecular structure to encode their vital functions. Despite the fact that natural biopolymers are composed of only four letters of the genetic alphabet (G, C, A, and T or U), they can offer a variety of possible sequences and structures. These complex systems are the product of a long-term evolutionary process that highly optimized them to accurately perform their major biological functions such as information storage and processing. However, the traditional notions of the composition of nucleic acids have changed when the first non-canonical 5-methylcytosine was found in *Mycobacterium tuberculosis* in 1925 [1] and more than 30 years later, the N^6-methyladenine (N^6MeA) was discovered in bacterial DNA [2]. Evidently, nature helped evolve additional letters and broadly used them to expand its capabilities since the range of chemical and structural diversity of unmodified nucleic acids are limited [3]. Moreover, the DNA of certain bacteriophages contains no deoxycytidines, deoxythymidines, or deoxyadenosines, but 5-hydroxymethyldeoxycytidine [4], deoxyuridine [5], or 2,6-diaminopurine (DAP) [6] instead. This further provides evidence that the natural evolution stepped beyond the four-letter genetic code.

In the last 60 years, numerous chemically modified nucleic acid monomers have been produced and incorporated into oligomeric sequences either by chemical or enzymatic synthesis. These molecules were primarily designed to increase the complexity and stability of natural biopolymers for the development of nucleic acid-based drugs and, lately, for various biotechnological purposes [7, 8]. The chemically redesigned nucleic acids (xeno-nucleic acids, XNA) [9] were successfully applied as therapeutics, including antisense and antigene oligonucleotides, noncoding RNA (siRNA, miRNA, and shRNA), aptamers, and nucleic acid enzymes [7, 10, 11]. The XNA molecules showed to be excellent building blocks of nanostructures, biosensors, and bioassays [12, 13]. Recently, the artificial nucleic acids have been utilized *in vivo* as additional types of genetic biopolymers [14–20].

Enzymatic and Chemical Synthesis of Nucleic Acid Derivatives, First Edition.
Edited by Jesús Fernández-Lucas and María-José Camarasa Rius.

Since modified nucleic acid fragments are important components for various applications, they should be easily made available in high qualities and quantities by chemical or enzymatic synthesis. The enzymatic synthesis was employed in the production of XNA molecules when the chemical approach failed. For example, the chemical synthesis of artificial sequences is limited by 20–100 nt, therefore, it is problematic to produce long genetic fragments without additional steps of ligation and polymerization. On the other hand, using the enzymatic methods, long (>100 bp) unnatural nucleic acids can be obtained. Virtually fully modified genes (~1000 bp) can be synthesized in one-pot polymerase chain reaction (PCR) [19–21].

Moreover, the enzymatic synthesis is an essential step of *in vitro* selection procedure or SELEX (Systematic Evolution of Ligands by EXponential enrichment) that includes repetitive rounds of amplification and selection of functional oligonucleotides (aptamers) [22–24]. Additionally, some functional groups of phosphoramidites (the building blocks of chemical synthesis) are not compatible with the chemistry of solid-phase synthesis and need specific protection [25].

The common methods of enzymatic synthesis of modified nucleic acids, such as primer extension reaction (PEX), PCR, and *in vitro* transcription (Tx) give the opportunity to synthesize diverse modified fragments within few hours, although the enzymatic protocols have some drawbacks as well. The template-directed enzymatic synthesis of XNA molecules needs to use the specific polymerases capable of incorporating XNA monomers and, at the same time, recognizing XNA templates. Furthermore, the modified nucleoside triphosphates (dNxTP) should be readily available to produce a variety of synthetic fragments.

Various types of XNA molecules and their monomers have been produced and comprehensively reviewed by others [26–29]. In this chapter, we will primarily discuss the enzymatic synthesis of modified nucleic acids that were able to bypass *in vitro* replication and evolution processes, and serve as genetic templates *in vivo*. We also emphasize the production of very long synthetic biopolymers composed of multiple modifications that benefit the construction of fully artificial gene and genome. We summarize the modifications according to their type, e.g. base-, sugar-, and phosphodiester backbone-modified XNA, as well as special types of the modified nucleic acids (L-DNA) that were successfully PCR amplified. Therefore, we focus on XNAs that meet several requirements:

- The XNA molecules should form duplexes with RNA, DNA, and each other. The introduction of unnatural blocks should not interfere with the double-stranded (ds) structure formation.
- The XNA should be biocompatible with either natural or mutant DNA polymerases. They should not disturb the polymerization reaction while providing the excellent accuracy of incorporation.
- The XNA biopolymers should support enzymatic replication producing many copies of itself by PCR amplification.
- The enzymatic synthesis of relatively long and highly modified XNA, with more than two types of modified letters, is preferable.

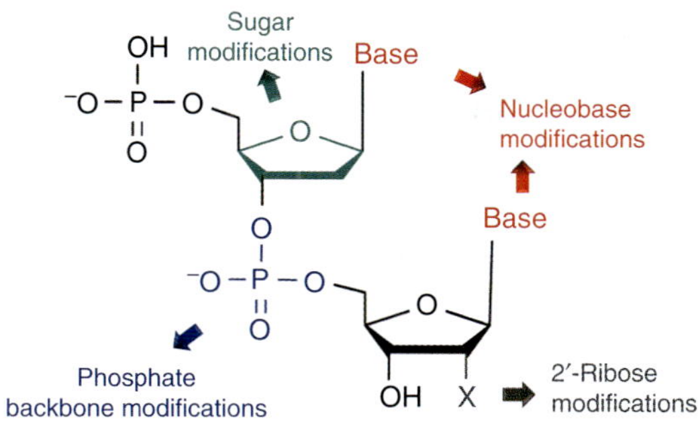

Figure 7.1 Possible chemical alteration of nucleotide structure.

- The XNA should mimic the functions of natural nucleic acids in terms of transferring the genetic information, *in vitro* or *in vivo*. Thus, they should support efficient *in vitro* evolution or *in vivo* information transfer.

7.2 Types of XNA Biomolecules

Chemical modifications of nucleic acids can be added to the nucleobases, sugar moieties, and phosphodiester backbone (Figure 7.1). Modification of the nucleobases can improve the base-paring stability and specificity. The modification of the phosphodiester backbone improves nuclease resistance and pharmacokinetic properties of synthetic nucleic acid therapeutics. The modification of the sugar moieties has an influence on various nucleic acid properties, such as duplex-forming ability, nuclease resistance, and toxicity in a cell. Ideally, the combination of these modifications could be used to obtain biomolecules that are able to form stable duplexes together with an improved half-life in serum.

Additionally, the introduction of modification into a nucleic acid structure can enhance the chemical diversity of natural biomolecules. Four natural nucleotides represent limited functionality and chemical diversity when compared to amino acids with their 21 diverse functional groups. DNA and RNA lack carboxylate, basic, and positively charged side chain groups, which are widely represented in proteins by different amino acids. These limitations can be overcome by the introduction of additional functional groups into four bases, 2′-OH position of the sugar, and phosphodiester linkage. Multilateral diversifications can, therefore, significantly enhance the chemical potential of natural biopolymers.

7.3 Enzymatic Synthesis of XNA and DNA Polymerases

The enzymatic synthesis of modified oligonucleotides by the incorporation of functionalized 2′-deoxynucleoside triphosphates (dNTPs) mainly includes PEX [30] with different mesophilic or thermophilic DNA polymerases, PCR [31] with thermostable DNA polymerases, and *in vitro* transcription [32] with DNA-dependent RNA polymerases (RNAPs).

PEX and Tx produce one or several copies of single-stranded XNA molecule using the initial unmodified DNA sequence as a template. In both cases, the DNA or RNA polymerases should recognize modified nucleoside triphosphates and efficiently incorporate them in the extending strand. These methods are suitable for the incorporation of diverse modified groups (including bulky groups) and are mainly applicable for XNA synthesis of relatively short fragments (20–120 bp bearing modifications in one strand only).

On the other hand, PCR amplification allows the production of large fragments containing modifications in both DNA strands. In PCR amplification, fully modified sequences 40–2000 bp can be synthesized using either short synthetic oligonucleotides or plasmid DNA as templates [20, 21, 33]. PCR includes the repeated cycles of denaturation and extension of original target sequences with two primers and requires highly thermostable polymerases active at 68–75 °C. Using only 10 cycles of PCR amplification, ~1000 copies of dsXNA molecules can be obtained in less than one hour reaction. Moreover, several types of dNTPs bearing different functional groups can be employed simultaneously in PCR giving a high-density modified product.

Although PCR is a powerful method to produce high-density long modified material, it is also a complicated process that requires simultaneous recognition of modified triphosphates and polymerase tolerance to a modified template. Therefore, the optimal combination of thermostable DNA polymerases and dNxTP should be carefully evaluated for the efficient XNA synthesis during PCR. The formation of the full-length modified strands is generally strongly dependent on both the nature of substrates and the type of DNA polymerase [34, 35].

Over the years various DNA polymerases have been evaluated for their PCR amplification ability of XNA fragments. Some of these DNA-dependent XNA polymerases are commercially available (Table 7.1). They mainly belong to the A- or B-families of polymerases from thermophilic or hyperthermophilic organisms and either possess or lack $3' \rightarrow 5'$ exonuclease activity (*exo*$^-$). The most commonly used thermostable DNA polymerases for XNA polymerization are *Taq* of *Thermus aquaticus*, KOD Dash of *Pyrococcus kodakaraensis*, Vent *exo*$^-$ of *Thermococcus litoralis*, Deep Vent from *Pyrococcus* species GB-D, 9°N from *Thermococcus* species 9°N-7, *Pwo* of *Pyrococcus woesei*, and Phusion that is *Pyrococcus*-like enzyme [25, 44].

The data from many studies suggest that family B DNA polymerases are superior and have an expanded substrate repertoire than family A polymerases [34, 35, 45–48]. Although with small alterations of nucleobase structures such as dUTP, 5-chloro-dUTP, 7-deaza-dATP (Figure 7.2a), DNA polymerases from family A (e.g. *Taq*) performed better in PCR compared to family B enzymes [21, 48].

Besides the commercial DNA polymerases, genetically engineered DNA polymerases with enhanced efficiency of the dNxTPs incorporation and extension during PCR have been developed [28, 49, 50]. These examples include XNA polymerases from Holliger group capable of synthesizing 1 kb cyanine dye-modified DNA [37] and 2 kb phosphorothioate substituted DNA [33]. Romesberg and coworkers generated new variants of the Stoffel fragment of *Taq* polymerase that could result in PCR amplification of 2′-modified oligonucleotides [38, 51].

Table 7.1 Thermostable DNA polymerases used for PCR amplification with modified dNTP.[a)]

Polymerase	Specification	$3' \rightarrow 5'$ *exo* activity[b)]	Family	Extension rate (kb min^{-1})	Error rate ($\times 10^{-6}$)
9°N_m	*Thermococcus* species 9°N-7 variant E143D	Reduced	B	nd	nd
AccuPrime Pfx	Hot Start KOD variant	Yes	B	nd	1.6
Bst	Large fragment of DNA polymerase I from *Bacillus stearothermophilus*	No	A	nd	15
Deep VentR™	*Pyrococcus* species GB-D.	Yes	B	1.4	12–2.7
D-Dpo4–3C [36]	An artificial mirror-image mutant of *Sulfolobus solfataricus* DNA polymerase IV	No	Y	nd	nd
E10 [37]	*Pfu exo*⁻ (V93Q, D141A, E143A) variant with V337I, E399D, N400D, R407I, Y546H	No	B	nd	400
KOD1	*Thermococcus kodakaraensis*	Yes	B	6.0–7.8	7.6–2.6
KOD Dash or KOD XL	Blend of KOD and KOD *exo*⁻	Both	B	nd	3–4 < Taq
KOD *exo*⁻	KOD *exo*⁻ variant N210D	No	B	nd	nd
KOD FX	Hot Start KOD variant	Yes	B	nd	10 < Taq
M1 [33]	*Taq* variant G84A, D144G, K314R, E520G, F598L, A608V, E742G	No	A	nd	1.6 > Taq
OneTaq	Blend of Taq and Deep Vent	Both	A & B	nd	140
Pfu	*Pyrococcus furiosus*	Yes	B	0.5–1.5	2.8–0.7
Phusion	*Pfu* variant fused with sso7d[c)]	Yes	B	nd	2.6–0.4
Pwo	*Pyrococcus woesei*	Yes	B	nd	2.4

(Continued)

Table 7.1 (Continued)

Polymerase	Specification	3′ → 5′ *exo* activity[b]	Family	Extension rate (kb min^{-1})	Error rate ($\times 10^{-6}$)
SFM4–3 [38]	Stoffel fragment of *Taq* V518A, N583S, D655N, I614E, E615G, E681K, E742Q, M747R	No	A	nd	17 000–2800
SFM4–6 [38]	Stoffel fragment of *Taq* D655N, I614E, E615G, L657M, E681K, E742N, M747R	No	A	750	nd
Taq	*Thermus aquaticus* YT1	No	A	1–4.8	560–8
Tgo	*Thermococcus gorgonarius*	Yes	B	1.5	5.6–3.5
Therminator	9°N™ *exo*⁻ variant D141A/E143A, A485L	No	B	nd	nd
Tth	*Thermus thermophilus* HB8	No	A	1.5–2.0	77
Vent	*Thermococcus litoralis*	Yes	B	1	57–2.8
Vent *exo*⁻	Vent exo⁻ variant D141A and E143A	No	B	1	190

a) Table adapted from different sources [39–43], New England Biolabs website (https://www.neb.com) and Thermo Fisher Scientific website (www.thermofisher.com); nd – not determined.

b) *exo*⁻ – DNA polymerases lacking 3′ → 5′ exonuclease activity.

c) Sso7d is a ssDNA binding protein from hyperthermophilic archaebacteria *Sulfolobus solfataricus*.

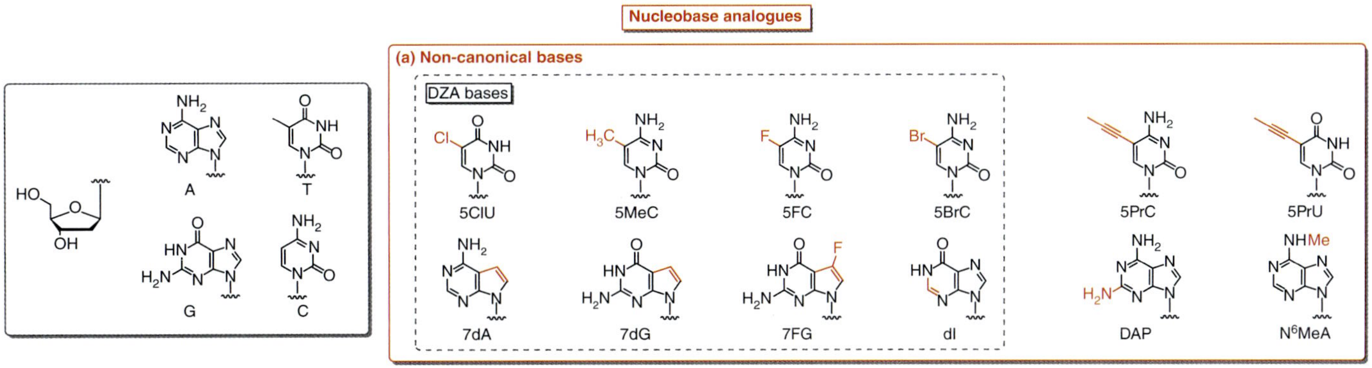

Figure 7.2 Examples of modified nucleobases with intact Watson–Crick base pairing (C5-substituted pyrimidine and N7-substituted purine analogues) that were successfully applied in PCR amplification. (a) Non-canonical bases. (b) Analogues bearing amino-acid-like groups. (c) Analogues bearing fluorescent tags.

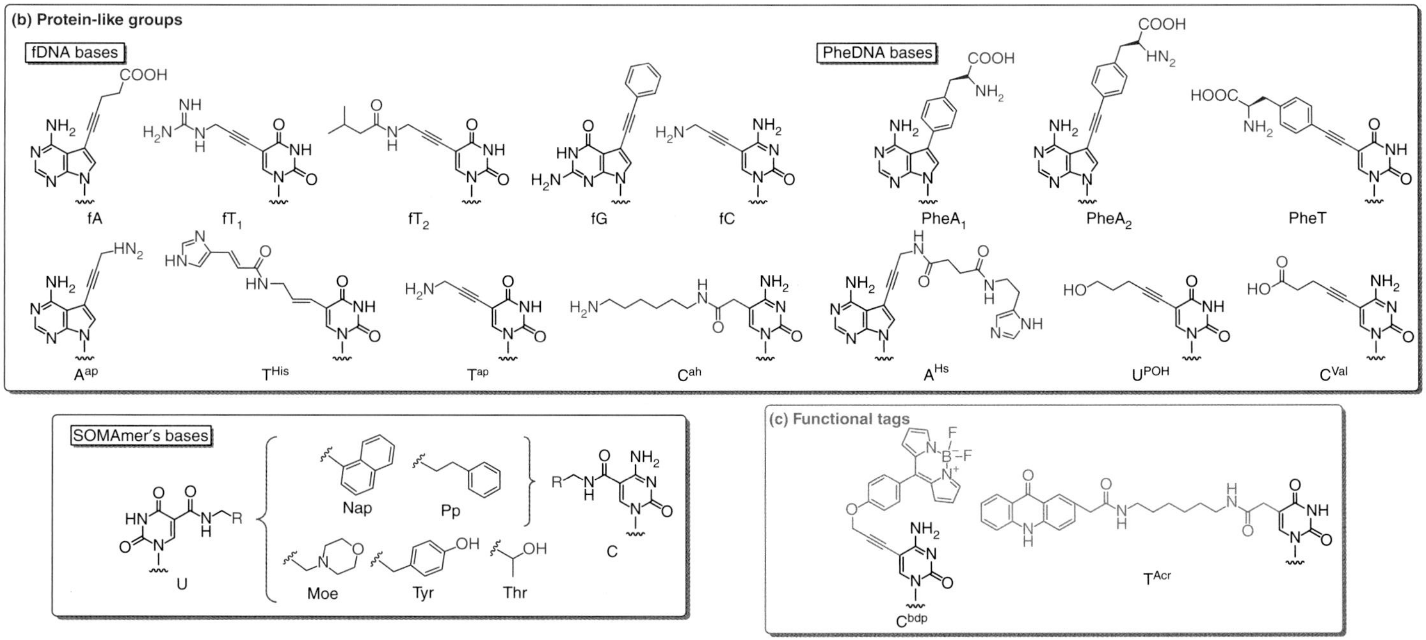

Figure 7.2 (*Continued*)

Moreover, the buffer composition can considerably influence polymerase specificity to modified dNTP substrates, and therefore, enhance the efficiency of PCR amplification of XNA molecules. It is well known that the increased concentrations of divalent ions (Mn^{2+} and Mg^{2+}, 1 mM and 3–5 mM, respectively) and polymerase (50–100 U ml^{-1}) improve the incorporation of modified triphosphates, while decreasing the fidelity of DNA polymerases [20]. PCR enhancers and different additives (e.g. DMSO – dimethylsulfoxide, formamide, SSB – single-stranded DNA-binding protein, and betaine) can surprisingly improve the yield of PCR with modified dNTP [39, 45, 52, 53] from no product formation to full-length PCR product (Table 7.2).

7.4 Base-Modified XNAs (Base-XNAs)

Base-modified XNA can be divided into those that preserve the Watson–Crick pairing of nucleic acid structure (nucleobase analogues) and those that differ from classical Watson–Crick base pairing motifs (unnatural base pairs or UBPs).

7.4.1 Nucleobase Analogues

Nucleobase analogues are derivatives of natural bases with the intact Watson–Crick base pairing motifs. The preferable positions for the alteration of nucleobase structure are the C5-position of pyrimidines and the N7-position of purines since they reside in the major groove of the DNA helix and do not interfere with duplex stability and enzymatic polymerization of DNA [25, 37, 80, 81]. Generally, diverse DNA polymerases can effectively process the corresponding nucleoside triphosphates as substrates and incorporate them in the nucleic acid structure [19–21, 45, 59, 81–83].

The studies mostly focused on the alteration of only one type of nucleobase in a dNxTP set leaving the other three bases unmodified. The functional groups were primarily introduced using 5-substituted pyrimidines (dUTP or dCTP) [34, 35, 48, 84–87] and rarely using 7-substituted-7-deaza-purines [83, 88]. This enables incorporation of only one kind of functional group resulting in the partial replacement of nucleic acids. In order to overcome the lack of functionality of natural nucleic acids, multiple chemical groups can be introduced simultaneously using different types of dNxTP during PCR [89].

In practice, such multiple incorporations of dNxTP were difficult to achieve and only a few examples of successful PCR amplification with more than one kind of substituted dNTP have been reported. The simultaneous incorporation of two, [21, 54, 56–58], three [59, 60], or all four [19, 20, 45, 46, 52] different base-modified triphosphates proceeded during PCR. A summary of PCR synthesis of highly base-modified XNA is shown in Table 7.2.

7.4.1.1 Non-Canonical Nucleotides

Recently, we have demonstrated that base-modified oligonucleotides can be efficiently amplified by PCR with natural *Taq* DNA polymerase, providing long fragments up to 2 kb (A:T-substituted) and 1.5 kb (fully modified)

Table 7.2 Different modifications of nucleic acid that were successfully PCR amplified.

Modification	Length, bp[a)]	Letters[b)]	Polymerase[c)]	Special PCR conditions	Applications	References
1. Base modifications						
1.1 Non-canonical nucleobases						
DZA_1 (7dA:5ClU:G:C)	2074	A + T	*Taq*, OneTaq	Standard ThermoPol buffer and PCR cycle	Synthesis of random libraries, genetic templates, inhibition of restriction enzymes	[19–21]
DZA_2 (7dA:5ClU:7dG:5MeC)	1541	All 4				
DZA_3 (7dA:5ClU:7dG:5FC)	1541	All 4				
DZA_4 (7dA:5ClU:7FG:5MeC)	149	All 4				
DZA_5 (7dA:5ClU:7FG:5FC)	149	All 4				
DZA_6 (7dA:5ClU:dI:5BrC)	1023	All 4				
DAP:dI	160	G + A	*Taq*	100 U ml^{-1} DNA polymerase	Study of antibiotic specificity	[54, 55]
DAP:5PrU DAP:5PrC or DAP:5MeC 5PrU:5MeC	96, 104	A + T A + C T + C	*Taq* (JumpStart, a hot start variant)	Real-Time PCR (qPCR)	Optimization of qPCR	[56]
1.2 Amino acid-modified nucleobases						
fDNA	79	All 4	*Pwo* and Vent *exo*$^-$	10% DMSO, 5% formamide, 0.75 M betaine, 50 mM TMAC, ±0.36 μg SSB	DNA functionalization	[45, 52]
PheDNA	98	A + T	*Pwo*	2% DMSO, 100 U ml^{-1} *Pwo*, denaturation at 98 °C	DNA functionalization	[57]

A^{ap} : T^{His}	62, 98	A + T	*Taq*		DNAzymes	[58]
DAP or N_6MeA:T^{ap}:C^{ah}	77	A + T + C	Vent *exo*⁻		DNAzymes	[59]
A^{Hs}:U^{POH}:C^{Val}	98	A + T + C	Vent *exo*⁻		DNAzymes	[60]
SOMAmer's bases	40	T + C	KOD *exo*⁻	PEX with 500 μM dNTP/dNxTP; extension 68 °C – 6–8 h	Aptamer to PCSK9	[61]
1.3 Tagged nucleobases						
dC^{bdp}	98	C	KOD Dash (XL)			[62]
dT^{Acr}	108	T	KOD Dash (XL)		DNA probes	[63]
Bio-dU	2500	T	M1 variant of *Taq*	50 μM dNTP or dNxTP; extension 68 °C – 20 min, 50 cycles		[33]
FITC-dA	400	A			Microarrays	
CyDNA	1000	C	E10 mutant of *Pfu*		Microarray and microfluidic	[37]
BioDNA	300	All 4	Vent *exo*⁻			[46]

(Continued)

Table 7.2 (Continued)

Modification	Length, bp[a)]	Letters[b)]	Polymerase[c)]	Special PCR conditions	Applications	References
1.4 Unnatural base pairs (UBP)						
isoG:isoC	88	2 UBP	Titanium Taq			[64]
P:Z	51–57	3–4 UBP	*Taq*, Phusion	100 μM dNTP, 50 μM dZTP, 600 μM dPTP, pH 8.0 (*Taq*), pH 7.0 (Phusion)	Cell-SELEX/ Cell-LIVE	[65]
Ds:Pa	150	1 UBP	Vent	$40\ U\ ml^{-1}$, extension 65 °C – 4 min		[66]
Ds:Px	75	1 UBP	Deep Vent	4 mM $MgSO_4$ total, extension 65 °C – 4 min		[67]
	82, 97, 98	1–3 Ds[d)]	AccuPrime Pfx	50 μM dDsTP and dPxTP, 400 μM dNTP, extension 65 °C – 2.5–3.5 min	Aptamer to $hVEGF_{165}$, IFN-γ, and vWF	[68, 69]
5SICS:NaM	102–149	1–2 UBP	OneTaq	200 μM dNTP, 100 μM dNaM, 3 mM $MgSO_4$	SSO[e)]	[70, 71]
TPT3:NaM	134	1 UBP	OneTaq, *Taq*			[70]
2. Sugar modifications						
2.1 Pentose-XNAs						
LNA	43	A	$9°N_m$	2.5 mM $MnCl_2$ and 2 mM $MgSO_4$, 0.1 M betaine, 750 μM dNTP	Potential ASO[f)], aptamer, and LNAzymes	[72]

	43, 50	A	KOD	3 mM $MgSO_4$, 0.2 mg ml^{-1} BSA, 250 μM dNTP/LNA-ATP		[73]
4′-ThioDNA	104	All 4	KOD Dash	100 U ml^{-1} KOD Dash, 2% DMSO, extension 72 °C – 10 min	*In vivo* transcription in mammalian cells	[18]
	362	1 any				[10]
2.2 2′-Ribose-XNAs						
2′-OMe-DNA	75, 100	1 any	SFM4–3, SFM4–6	Extension: 50 °C – 20 min; 40 cycles; 3.5 mM $MgCl_2$		[38]
2′-F-DNA	75, 100	2 any	SFM4–3			
2′-F-DNA	68	A + G	SFM4–3	Asymmetric elution PCR with extension 50 °C – 20 min	Aptamer to HNE	[74]
2′-F-DNA	24–68	U + C	SFM4–3	Asymmetric qPCT, extension 50 °C – 1 h; 20 cycles; 3.5 mM $MgCl_2$		[75]
2′-N_3-DNA	75	1 any	SFM4–3	0.1–1 mM $MnCl_2$, 3.5 mM $MgCl_2$, 10–20 cycles, extension 50 °C – 1 h	DNA hydrogel	[51]

(Continued)

Table 7.2 (Continued)

Modification	Length, bp[a)]	Letters[b)]	Polymerase[c)]	Special PCR conditions	Applications	References
2′-Cl-, ANA, 2′-NH_3-DNA	75	A	SFM4–3			
3. Phosphate backbone modifications						
αS-DNA	70	All 4	*Taq*	10 mM Mg^{2+}	Potential ASO and aptamers	[76]
	2000	All 4	M1 variant of *Taq*	68 °C – 20 min, 50 cycles,100 μM αS-dNTP		[33]
	74	A + C	*Taq*	4 mM $MgCl_2$, 72 °C – 3′	Aptamer to TGF-β1	[77]
	80	A + T	*Ex Taq*	4 mM $MgCl_2$	Aptamer to AGEs	[78]
	68	All 4	*Taq*	4 mM $MgCl_2$	Aptamer to EDIII	[79]
4. Special type of modified DNA						
L-DNA	118, 207	All 4	D-Dpo4–3C	200 μM L-dNTP, 1 M betaine, 1 mM DTT	L-DNA aptamers	[36]

a) The maximal length of PCR product.
b) The type(s) and number of modified dNxTP used in enzymatic reaction.
c) The polymerases that demonstrated the best amplification efficiency with the corresponding dNxTP.
d) Ds is a hydrophobic unnatural base 7-(2-thienyl)imidazo[4,5-*b*]pyridine.
e) SSO is a semi-synthetic organism.
f) ASO is an antisense oligonucleotide.

[20, 21], which we denoted as "DZA" (Figure 7.2a and Table 7.2) [19]. The DZA sequences consist of different combinations of synthetic non-canonical nucleosides, namely 5-chloro-2′-deoxyuridine (5ClU), 5-methyl-2′-deoxycytidine (5MeC), 5-fluoro-2′-deoxycytidine (5FC), 7-deaza-2′-deoxyadenosine (7dA), 7-deaza-2′-deoxyguanosine (7dG), 7-fluoro-7-deaza-2′-deoxyguanosine (7FG), and 2′-deoxyinosine (dI).

The most fruitful combinations of modified triphosphates in PCR amplification was 7dG pairing with 5FC or 5MeC together with the 5ClU:7dA base pair (DZA_2 and DZA_3, Table 7.2). This combination succeeded in amplification of 1.5 kb fully modified DZA fragment as well as short random DZA libraries (67 and 82 bp). The *in vitro* replication of a 1 kb fragment was also achieved with a nucleotide set containing dI:5BrC and 5ClU:7dA (DZA_6), although it failed in the synthesis of random DZA libraries [20].

Besides DZA_2 and DZA_3 combinations, the DZA sets containing 7FG with any of the modified cytidines (5MeC, 5FC, or 5BrC) together with 5ClU:7dA effectively supported the PCR of random libraries even with high-fidelity DNA polymerases (e.g. OneTaq and Phusion). These results demonstrate that diverse DZA libraries could be easily produced and potentially employed in the *in vitro* evolution strategy for the isolation of structurally altered aptamers and DNAzymes. Moreover, the DZA inserts can be efficiently ligated by T4 DNA ligase that allows their direct cloning without the need of a reverse transcription and further simplifies the *in vitro* evolution of DZA.

Furthermore, the DZA cassettes could efficiently protect DNA from restriction enzyme cleavage (in 24 examined cases, only $Taq_\alpha I$ cleave the DZA restriction site) [20]. This finding might facilitate the assembly of multicomponent genetic systems and molecular cloning procedures by preventing undesired cleavage of internal restriction sites.

A significant outcome of the DZA study was that artificial biomolecules composed of 7dA:5ClU:7dG:5FC (DZA_3, Table 7.2) and 7dA:5ClU:dI:5BrC (DZA_6) were able to transfer the genetic information in living organisms. By using a gene encoding for dihydrofolate reductase conferring resistance to trimethoprim (R67 DHFR, 237 bp), we demonstrated that a DZA insert could serve as a genetic template for bacterial polymerases *in vivo* (Figure 7.3). This study proved the accurate information transfer from completely modified DZA genes to DNA in bacteria (DZA → DNA → mRNA → Protein circuit). It is important to note that none of

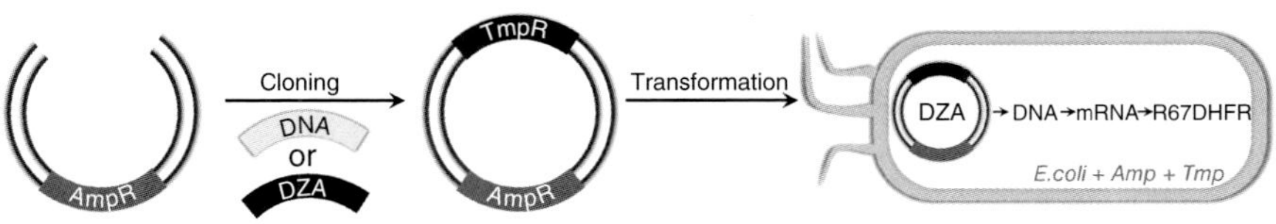

Figure 7.3 General scheme of *in vivo* studies with the R67 DHFR gene containing either natural or different DZA-modified fragments (DZA_3 or DZA_6) conferring trimethoprim resistance (*TmpR*). The samples were cloned into an ampicillin-resistant vector (*AmpR*) followed by transformation into *E. coli* cells. The cultures were grown in media containing antibiotics and the resulting viable cells were counted and analyzed.

the synthetic biopolymers possessing all four modified bases has been found to be functional in living organisms.

Overall, the four unnatural letters of DZA molecules can successfully replace the natural letters of DNA in the evolution process and in genetic information transfer. The unique properties of DZA monomers, including inhibition of restriction endonucleases and excellent substrate ability for diverse groups of DNA polymerases (*Taq*, OneTaq, Phusion, Vent *exo*$^{-}$) make them highly potential in a variety of applications.

In another study made by Kutyavin, non-canonical nucleotides were employed in Real-Time PCR amplification (qPCR) to improve the detection efficiency of AT-rich fragments [56]. It was shown that 5-substituted pyrimidines for methyl or propynyl groups (5MeC, 5PrC, and 5PrU) together with DAP (2-aminoadenosine) as adenosine alternative were very good substrates for *Taq* DNA polymerase (Figure 7.2a, Table 7.2).

Bailly and Waring also used DAP together with dI nucleoside triphosphates in PCR of 160 bp fragment [54, 55]. The DAP:T base pair has an extra hydrogen bond compared with A:T increasing the duplex stability, while dI lacks 2-amino group leading to weaker I:C base pairs than G:C (Figure 7.2a). They demonstrated that the repositioning of 2-amino group from guanosine replaced by inosine to adenosine replaced by DAP, only slightly affected PCR amplification by *Taq* DNA polymerase.

Therefore, the introduction of nucleobase analogues into nucleic acids structure can modulate the DNA duplex stability. Small modifications of nucleobases such as 5-substituted pyrimidines for halogen, methyl, or alkynyl groups, and DAP are very well-known stabilizing duplexes substitutions, while 7-deaza purines, inosine, and N^6MeA are destabilizing DNA analogues [20, 21, 55, 56, 90]. In general, they are well accepted by diverse polymerases at high multiplexity both *in vitro* and in cells. Moreover, restriction sites containing the base analogues can be efficiently protected from restriction enzyme cleavage [20].

7.4.1.2 Amino-acid-Like Groups

In order to address the functional gap of the natural biopolymers, the diverse amino-acid-like substitutions have been introduced into the nucleic acid structure through PCR with base-modified dNTP. Here we focus on the simultaneous incorporation of more than one type of modified nucleotide in DNA since it can significantly enrich the chemical repertoire of nucleic acids.

An interesting example of complete substitution of DNA for diverse chemical groups is the high-density functionalized DNA (fDNA, Figure 7.2b) [45, 52]. The double-stranded fDNA was produced by PCR using *Pwo* and Vent *exo*$^{-}$ DNA polymerases from family B, with all four base-modified dNTPs in the mixture (Table 7.2). The used dNTP carried the diverse functionality (fdNTP) including acidic (fA), basic (fT_1, fC), lipophilic (fT_2), and aromatic (fG) groups. They observed strong sequence dependence during PCR with *Pwo* polymerase where T- and A-rich fDNA sequences were preferable as templates over GC-rich fragments. This might be problematic in the reproduction of random libraries and GC-rich fragments since PCR efficiency strongly depends on the sequence context.

They also verified different buffer conditions with well-known PCR additives. The addition of DMSO (10%), formamide (5%), betaine (0.75 M), tetramethylammonium chloride (TMAC, 50 mM), and *Escherichia coli* SSB (0.36 μg) significantly enhanced PCR amplification with all four fdNTP. This study is another example of successful PCR amplification of the high-density functionalized DNA although the fDNA has never been used further in any biological and biotechnological application.

The introduction of different protein-like functionalities in nucleic acid structure can significantly increase the probability to select catalytic nucleic acid (DNAzymes). Working together multiple chemical groups can enhance the activity of DNAzymes. Encouraged by this idea, the Williams group demonstrated the simultaneous incorporation of two potentially catalytic groups (amine and imidazole) into DNA during PCR amplification by *Taq* DNA polymerase. Aminopropynyl- and imidazolyl-modified side chains were attached to A and T bases, respectively, and mimic the lysine and histidine residues that are often involved in enzymatic catalysis (A^{ap} and T^{His} in Figure 7.2b, Table 7.2).

Similar considerations inspired Kuwahara et al. to search for combinations of three base-modified triphosphates (A:T:C and natural G) that are optimal for PCR amplification [59]. During PCR, simultaneous replacement of A for non-canonical DAP or N^6MeA, T for 5-(3-aminopropynyl)-dU (T^{ap}), and C for 5-[(6-aminohexyl)-carbamoymethyl]-dC (C^{ah}) was achieved with Vent *exo*$^-$ (Figure 7.2a,b, Table 7.2).

In another study, deoxyadenosine and deoxyuridine triphosphates bearing phenylalanine-like groups ($PheA_1$, $PheA_2$, and PheT, Figure 7.2b) were examined in PCR of a 98mer template. The amplification using *Pwo* DNA polymerase was successfully performed with the corresponding $PheA_1$ or $PheA_2$ and PheT and the other two natural triphosphates with the addition of DMSO (2%), and at increased denaturing temperature (98 °C) (PheDNA, Table 7.2).

In order to select active M^{2+}-independent catalysts, Hollenstein synthesized and examined base-modified dNTP with multiple functionalities. The nucleotides were equipped with three types of chemical groups present in the active site of serine proteases, such as imidazole, hydroxyl, and carboxylic acid groups (A^{Hs}, U^{POH}, and C^{Val} are His-, Ser-, and Asp-like, respectively, Figure 7.2b) [60, 91]. The corresponding modified dNTP were excellent substrates for Vent *exo*$^-$ DNA polymerases in standard PCR conditions even when they were introduced simultaneously together with natural dGTP.

The efforts of some laboratories proved to be successful leading to the isolation of DNAzymes decorated with several amino acid functionalities that could catalyze metal-independent RNA cleavage [92–96].

Other attractive examples of oligonucleotide ligands with new protein-like functionality are SOMAmers (Slow Off-rate Modified Aptamers) [61, 97, 98]. SOMAmers are aptamers that have chemically modified pyrimidine bases bearing amino acid side chains (Figure 7.2b). This new class of ligands with picomolar to nanomolar affinity range combines the flexibility and stability of nucleic acids with the chemical diversity of proteins. The introduction of functional groups at two of the four nucleobases in DNA dramatically enhanced the probability to select ligands with higher affinities as well as improved nuclease resistance [61]. The diverse modified libraries were synthesized by KOD *exo*⁻ DNA polymerase using PEX with modified dUTP and dCTP, and were employed in the *in vitro* selection against human PCSK9 (proprotein convertase subtilisin/kexin type 9) [61]. Interestingly, although the combination of PpC and ThrU modifications was the best in library synthesis, it failed in SELEX procedure. The best inhibitor was selected from the library containing PpC and NapU with $IC_{50} = 2.8$ nM and $K_d = 14.7$ pM.

7.4.1.3 Functional Tags

DNA oligonucleotides, such as fluorophore dyes, biotin molecules, equipped with a variety of different chemical functionalities and other affinity tags for a wide range of applications can be synthesized enzymatically as well. Here we describe only tagged oligonucleotides that were synthesized by PCR amplification with high-density modification in both stands. The synthesis of redox or fluorescent labeled oligonucleotides by PEX and terminal deoxynucleotidyl transferase (TdT) was well reviewed in [25].

Using commercially available DNA polymerases, the PCR products with multiple fluorophore incorporations were achieved. As examples, KOD Dash DNA polymerase was used for the PCR amplification with F-Bodipy-dCTP ($dC^{bdp}TP$) [62] or Acridone-dTTP ($dT^{Acr}TP$) [63] analogues and the other three natural dNTP (Figure 7.2c). The corresponding fluorophore analogues completely replaced one type of nucleotide in 98 and 108 bp PCR products, respectively.

Tasara et al. systematically evaluated the various combinations of fluorophore- or biotin-labeled dNTP in PCR reactions with *Taq* or Vent *exo*⁻ DNA polymerases. The results demonstrated that Vent *exo*⁻ was able to successfully synthesize a 300 bp DNA product with the complete substitution of all natural nucleotides by their biotin-labeled counterparts (BioDNA, Table 7.2) [46].

At the same time, the Holliger group was working on the isolation of natural enzyme variants that are able to incorporate a diverse range of non-canonical substrates. These attempts resulted in an M1 mutant of *Taq* DNA polymerase that efficiently polymerized up to 400 bp fluorescein (FITC-dA) or 2.5 kb biotin (bio-dU)-labeled PCR products (Table 7.2) [33]. They also performed polymerase evolution for efficient incorporation of cyanine dye tagged dCTP (Cy3 or Cy5) [37] in PCR. The Cy3- or Cy5-containing PCR fragments (>1 kb), where all dC is replaced by a fluorophore analogue, were synthesized by E10 mutant of *Pfu* polymerase (CyDNA, Table 7.2).

The base-tagged DNA oligonucleotides with multiple fluorescent groups can significantly increase the sensitivity of DNA detection and, therefore, advance the existing biological assays, e.g. microarrays and immunoassays.

Taken together, the incorporation of 5-substituted pyrimidine and 7-substituted-7-deaza-purine nucleotides strongly depends on the structure of the functional group and linker, the type of base, and the nature of polymerase [34, 35, 48, 49, 84, 87]. Small alterations of nucleobase structure (e.g. DZA) are accepted better than bulky groups (e.g. fluorescent-tags) by diverse commercial polymerases even when all four nucleobases are substituted at high multiplexity. In general, DNA polymerases prefer to incorporate C5-modified pyrimidines than N7-substituted purines in PCR (especially 7-substituted-7-deaza-G). These clearly indicate that the choice of enzyme, template, modification, and reaction conditions might play a role in the PCR amplification efficiency.

To further chemically decorate the DNA bases, bulky groups can be introduced through copper(I)-catalyzed alkyne–azide cycloaddition (CuAAC or click chemistry) [99, 100]. These reactions involve the post-synthetic modification of 5-ethynyl-deoxyuridine-substituted PCR product. Other valuable methods of the post-synthetic derivatization of pre-modified nucleic acid fragments can be employed as well [53, 101, 102].

7.4.2 Unnatural Base Pairs

The other direction to increase the nucleobase repertoire is based on the use of UBPs. UBPs are a special class of nucleobase modifications, which have the base:base recognition motifs that do not occur in nature. They have an altered base pair structure compared to the classic Watson–Crick A:T and G:C base pairs (Figure 7.4). UBPs can be further divided into two sub-classes: those that retain hydrogen bonding and those that form hydrophobic base pairs based on steric complementary. These base pairs are often referred to as a third base pair, which is additional to the Watson–Crick base pair systems.

The most promising candidates of UBPs to serve as the third base pair in storing and coding the information are represented by three families. Hydrogen bonding P:Z developed by the Benner group [103], base pairs based on hydrophobic and packing forces Ds:Px [68], and 5SICS:NaM or TPT3:NaM [104, 105] developed by the Hirao and Romesberg laboratories, respectively (Table 7.2, Figure 7.4).

They all represent base pairs that have shown a remarkable efficiency and fidelity during PCR amplification and transcription albeit with only few UBP in the duplex structure. The base pairs developed by the Romesberg group (5SICS:NaM and TPT3:NaM) were successfully applied in the *in vivo* replication and transcription [104–106]. The other P:Z and Ds:Px succeeded as the components of aptamers with an expanded genetic code.

7.4.2.1 Hydrogen-Bonding Base Pairs

The Benner laboratory has been propagating several nucleobase analogues including isoG:isoC [107] and P:Z [108] based on the rule of rearranging donor:acceptor hydrogen bond formation (Figure 7.4a) [103]. These UBPs were designed to interact via complementary hydrogen bonding between synthetic nucleotides. DNA containing isoC:isoG was reasonably well extended by the Klenow fragment of *E. coli* DNA polymerase (KF *exo*$^-$) [107] and transcribed by T7 RNA polymerase (T7 RNAP). However, due to the tautomerization of isoG that led to mispairing with U/T bases and low PCR replication fidelity (98%), this first candidate for the third base pair was replaced by an improved P:Z base pair [103, 109]. The P:Z resulted in representing a more stable and more accurate pair with PCR amplification fidelity of 99.8% using *Taq* DNA polymerases [65]. Moreover, the successful transcription and reverse transcription of the P:Z base

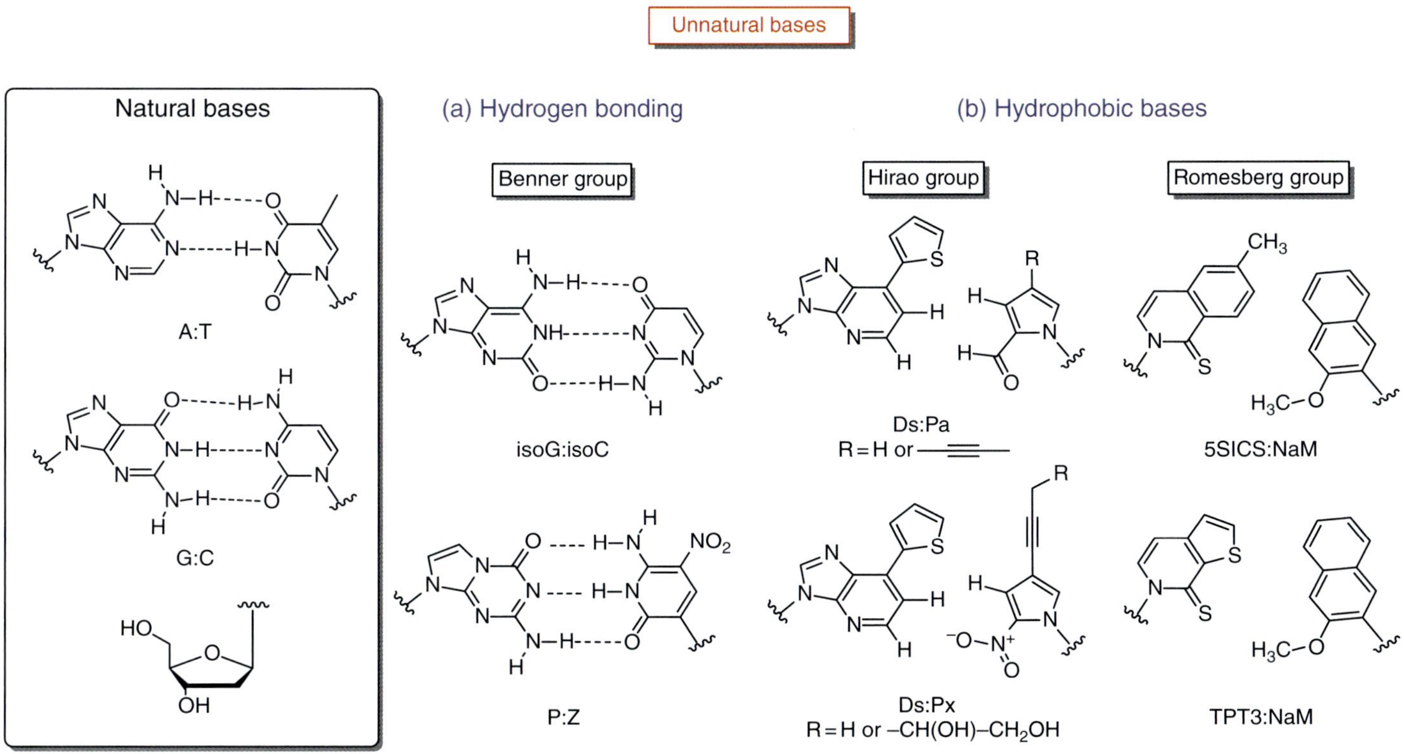

Figure 7.4 Unnatural base pairs that are not occurring in nature. (a) Hydrogen-bonding base pairs with changed Watson–Crick base pair formation. (b) Specific base pairs formed by hydrophobic interactions based on steric complementary.

pair were reported [110]. The continuation of this research resulted in the development of an Artificially Expanded Genetic Information System (AEGIS) where natural A:T, G:C, and the additional P:Z base pairs build the six-letter genetic code A:T:G:C:P:Z. This system was successfully employed in AEGIS-based Cell Laboratory *In Vitro* selection (cell-LIVE) procedure [111–114].

The natural DNA or RNA libraries result in maximum 4^n individual oligonucleotides, where n is the length of the random region. The additional unnatural base might expand the existing code increasing the probability to generate aptamers from a pool of 5^n or 6^n possible sequences with the five- or six-letter code, respectively [69, 112]. Encouraged by the idea that the additional letters can enhance the probability of selecting high-affinity aptamers, Benner and coworkers successfully identified a series of ligands recognizing breast cancer cells with K_d of 30 nM [112], liver cancer cells with K_d of 14 nM [111], Glypican 3 expressed by a surface of liver cells with K_d of 6 nM [113], and the protective antigen from *Bacillus anthracis* with K_d of 35 nM [114] using cell-LIVE procedure.

It should be noted that the best aptamer candidates contained none or a few P and Z letters [111–114]. Even when the initial library composed of the increased molar ratio of P + Z bases compare with a natural one (A:T:G:C:P:Z = 1 : 1 : 1 : 1 : 2 : 2, ~17.5 P + Z), only sequences with maximum five modifications were selected [113]. This indicates that the library depletion during *in vitro* selection procedure, either due to inefficient chemical synthesis of the initial library or low P:Z base pair retention during enzymatic steps of SELEX. Nevertheless, aptamers composed of AEGIS lost their affinity for the targets when unnatural nucleotides were exchanged with a natural counterpart.

7.4.2.2 Hydrophobic Base Pairs

Significant results in the development of alternative base pairs have been achieved toward the expansion of a genetic alphabet based on hydrophobic base pair interactions made by the groups of Hirao and Romesberg. The most recent and fruitful base pair of Hirao and coworkers was Ds:Px that exhibited high fidelity PCR replication (>99.9%, only one Ds:Px UBP) using Deep Vent DNA polymerase (Figure 7.4b, Table 7.2) [67]. Moreover, high-affinity Ds-containing aptamers with low picomolar affinity have been successfully evolved from DNA libraries containing additional Ds nucleotides in the random region (five-component genetic code) [68, 69]. The aptamers possessed the unique hydrophobic functionality of Ds and recognized the human vascular endothelial growth factor ($VEGF_{165}$), interferon-γ (IFN-γ) (1–3 Ds letters), or von Willebrand factor A-domain (vWF) (2 Ds letters) targets with 0.65, 38, or 61 pM affinity, respectively. Interesting, the replacement of Ds unnatural nucleotide for dA led to a loss affinity of the selected aptamer as in the previous examples with AEGIS-containing aptamers. This indicates that the modified nucleotides are essential for the aptamer binding to their specific target.

Romesberg's group, succeeded in the development of a unique 5SICS:NaM UBP (Figure 7.4b) that was shown to be well replicated by OneTaq DNA polymerase with high fidelity of >99.9% [71, 105]. The 5SICS:NaM was also efficiently transcribed into RNA using T7 RNAP in both ways, NaM-templated incorporation of 5SICS nucleoside triphosphate into RNA or vice versa [115].

The further optimization of the 5SICS structure yielded a TPT3 nucleotide (Figure 7.4b) [70, 116], which pairing with NaM nucleotide became the most efficiently replicated UBP identified to date [117]. These base pairs led to an important achievement in the field of synthetic biology, namely, the *in vivo* replication of an artificial base pair in bacteria using the corresponding extracellular nucleoside triphosphates [104, 105]. More recently, they also demonstrated that T7 RNAP is able to successfully transcribe mRNA and tRNA containing the NaM:TPT3 *in vivo* following by the production of the functional green fluorescein protein (GFP) [106]. Thus, the semi-synthetic organisms (SSO) based on *E. coli* strain could retrieve increased information represented by single UBP in the presence of the unnatural nucleoside triphosphates.

7.5 Sugar-Modified XNAs (Sugar-XNAs)

7.5.1 Pentose-XNA

Since the modifications of the sugar moiety result in a more profound impact on a nucleic acid structure than do base modifications, in most cases, the fully sugar-substituted XNAs are invisible to existing natural ecosystems. Thus, they are resistant to nuclease-mediated degradation and repair. They continue to remain difficult templates for the recognition by naturally available polymerases. Therefore, the PCR amplification of sugar-XNAs are challenging especially when all four sugar-modified dNTPs are presented in the mixture. This limits the possibility of enzymatic sugar-XNA production, *in vitro* selection, and XNA↔XNA information transfer.

Nevertheless, several mutant forms of DNA polymerases have been recently evolved to overcome this drawback [118]. Some of them were able to extend DNA or RNA primers using a DNA template with all four modified dNTPs, followed by reverse transcription of the sugar-XNA product back to DNA by specifically engineered XNA reverse transcriptases [118]. In this study, XNA polymerases for HNA (hexitol nucleic acids) was developed, which also seemed to work for CeNA (cyclohexenyl nucleic acids), LNA (locked nucleic acids), TNA (threose nucleic acids), ANA (arabino nucleic acids), and FANA (2′-fluoroarabino nucleic acids, Figure 7.5). The evolution of these XNA polymerases actively promoted the development of XNA aptamers, XNAzymes, and XNA nanostructures with unique structures and functions [11, 12, 118, 119].

Interestingly, the sugar-XNA molecules, HNA, CeNA, and ANA, have been successfully employed to encode the active site of the thymidylate synthase of *E. coli* [14, 15]. The binary modified systems (G and T) composed of 12 continuous HNA or CeNA nucleotides or 9 ANA nucleotides were able to serve as a template for the bacterial replication system. The TNA is another promising member of the sugar-XNA world. These molecules demonstrated efficient primer elongation up to 70 consecutive TNA monomers by Therminator or Kod-RI (a variant of KOD *exo*$^-$) polymerases, reverse transcription by SuperScript II or Bst enzymes, and succeeded in the aptamer selection [120–124]. Despite the notable characteristics of TNA, HNA, CeNA, ANA, and FANA, the direct *in vitro* replication of these XNA to XNA by any natural or synthetic polymerases is not possible yet.

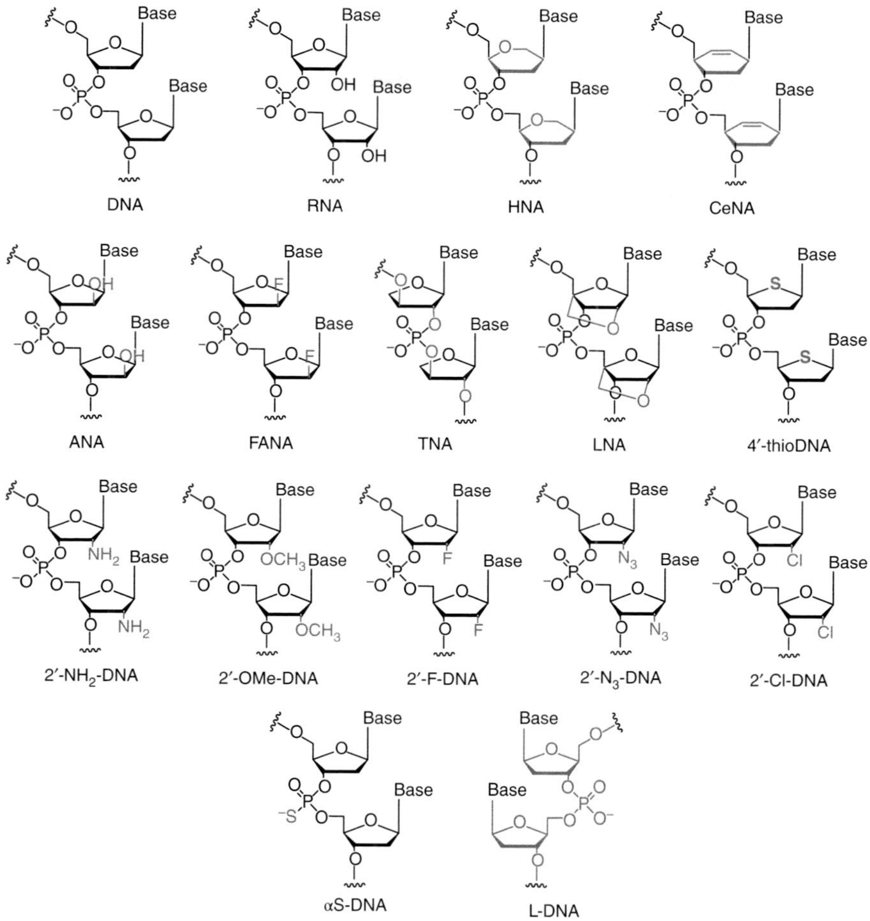

Figure 7.5 The examples of sugar, phosphodiester backbone (αS-DNA), and mirror-image (L-DNA) nucleic acid analogues.

Despite the generally poor substrate properties, some XNA molecules have been shown to be active in PCR amplification. These examples include LNA that was PCR amplified albeit with only one modified letter (LNA-A) [72] and 4′-thioDNA (2′-deoxy-4′-thioribonucleic acids) that demonstrated excellent amplification ability [47] (Figure 7.5).

PCR replication of LNA was performed with 9°N_m or KOD DNA polymerases using very short 43- or 50-mer templates where 39 nt were the complementary primer regions [72, 73]. Therefore, only a few modified LNA-A could be incorporated, 5–6 nt in one strand and 3 in another DNA strand. It is interesting to note that in the chosen PCR conditions, reactions with all four natural dNTP were either unsuccessful or led to very low product formation, while the yield of the LNA-modified product was satisfactory [72, 73].

On the other hand, samples of 4′-thioDNA with replaced 4′-oxygen atoms of the sugar with a sulfur atom were found to be excellent substrates for DNA polymerases allowing their applications as aptamers and genetic templates [47]. They possess duplex stability and nuclease resistance higher than that of the natural counterpart. A full-length PCR product as long as 104 bp was afforded with all four kinds of 4′-thio-dNTP and KOD Dash DNA polymerase. The addition of 2% DMSO and the extension of reaction time (10 minutes) improved the PCR yield more than threefold. Furthermore, partially modified 4′-thioDNAs served as templates for *in vitro* transcription by T7 RNAP and, more significantly, for *in vivo* transcription in mammalian cells [10, 18].

The same group also generated 4′-SeDNA (2′-deoxy-4′-selenoribonucleic acids) using PCR amplification with 4′-selenothymidine-5′-deoxynucleoside triphosphate (SeTTP) and the other three natural dNTPs [125]. SeTTP was readily accepted by the KOD Dash DNA polymerase and amplified leading to 104 bp products albeit with significantly reduced efficiency. Therefore, KOD Dash polymerase demonstrated an impressive tolerance to chemical modifications at the 4′-position of the sugar moiety during *in vitro* replication.

The remarkable properties of sugar-XNAs are very important for synthetic biology and therapeutic applications. Since they are invisible to natural enzymes, they can be used as a novel genetic storing system that is able to safely convey information without the integration in a host organism. Therefore, further improvements toward efficient PCR amplification of sugar-XNAs are necessary to provide direct implementation of the modified library in *in vitro* selection processes and in the generation of long sugar-XNA genes and vectors.

7.5.2 2′-Ribose-XNA

The most widely used chemical modifications to increase aptamer biostability are the alteration of 2′ position of sugar. The 2′-*O*-methyl (2′-OMe), 2′-fluoro (2′-F), 2′-amino (2′-NH_2), and 2′-azido (2′-N_3) RNA/DNAs were applied to improve the performance of nucleic acid drugs [28, 126] (Figure 7.5).

First attempts to incorporate 2′-ribomodified nucleotides into nucleic acid structures were mostly performed by T7 RNAP or their mutant variants [127–132]. The 2.5 kb partially 2′-F-substituted transcripts were produced by T7 RNAP [133] and 1.8 kb 2′-F-adenosine-substituted DNA fragments were able to act as a messenger for the expression of luciferase in *Xenopus laevis oocytes* and in a eukaryotic cell-free system [134]. Moreover, the DNA polymerases were examined to accept 2′-modified substrates. For example, a variant of Tgo DNA polymerases was genetically engineered to obtain 2′-N_3- and 2′-F-modified DNA as long as 1.7 kb [135]. Although these examples represent the possibility to introduce 2′-substitutions into a nucleic acid structure, they produced single-stranded sequences containing one or two modified nucleotides by PEX.

Only recently, Romesberg and coworkers reported successful PCR amplification of 2′-OMe- and 2′-F-containing oligonucleotides [38] by thermostable DNA polymerase SFM4–3 (a variant of the Stoffel fragment of *Taq* DNA polymerase, Table 7.1). This thermostable enzyme demonstrated the ability to efficiently catalyze PCR reactions with 75 or 100 nt DNA templates. On this basis, 2′-F-purine-modified aptamers against human neutrophil elastase (HNE) were selected with nanomolar affinity [74]. Moreover, the evolution of the thermostable SFM4–3 DNA polymerase allows developing a method for amplification of RNA and modified RNA oligonucleotides that was referred as Polymerase Chain Transcription (PCT) [75]. The method used repeated thermocycles as conventional PCR and led to the efficient production of multiple copies of short natural or 2′-modified RNA molecules.

Furthermore, the ability of the SFM4–3 mutant to tolerate 2′-azido, 2′-chloro, 2′-amino, or arabinose sugars were examined [51]. It has been shown that all four single 2′-N_3-NTP can be incorporated and amplified during PCR with SFM4–3. 2′-NH_2-, Cl-, and ANA-modified adenosine triphosphates were also incorporated by the enzyme using the 75 nt DNA template.

7.6 Phosphodiester Backbone-XNA

The modification of the phosphodiester backbone is one of the most commonly used to improve nuclease resistance and the pharmacokinetic properties of synthetic nucleic acid therapeutics. Almost all oligonucleotide-based drugs that have reached the clinical trial are phosphorothioated or boranophosphated containing α-oxygen substituted by sulfur or a borano group on the phosphate backbone [7, 136].

Since phosphorothioated oligonucleotides display attractive pharmacokinetic properties and increased lifetime *in vivo* compared to their natural counterparts, they were widely employed for the *in vitro* selection of biostable aptamers [76–79]. Partially modified aptamers with two modified αS-nucleotides together with the other natural nucleotides have been selected against transforming growth factor-β1 (TGF-β1) [77] or advanced glycation end products (AGEs) [78]. The selection of fully modified thioaptamer against dengue-2 virus envelope protein domain III (EDIII) [79] was demonstrated as well. In all cases, the commercial *Taq* DNA polymerase has been successfully used as a catalyst in PCR with the corresponding αS-dNTPs (Table 7.2).

However, PCR amplification in the presence of all four kinds of α-phosphorothioate dNTP (αS-dNTPs) with natural *Taq* polymerase can be challenging. PCR with all four αS-dNTPs performed poorly and required either increased Mg^{2+} concentration (10 mM) [76] or a mix with natural dNTP [77, 78]. Under increased concentrations of Mg^{2+}, the fully modified αS-PCR product (70 bp) was achieved albeit with multiple mutations and deletions especially when αS-DNA was used as a template [76].

To overcome the limited ability of natural *Taq* polymerase to accurately and efficiently incorporate αS-dNTPs in the course of PCR amplification, the Holliger group generated a mutant *Taq* DNA polymerase (M1, Table 7.1) [33]. The M1 mutant efficiently catalyzed the PCR amplification of 2-kb-long phosphorothioate-modified DNA with the complete substitution of natural nucleotides for αS-nucleotides (αS-DNA, Table 7.2).

7.7 A Mirror-Image L-DNA

An alternative approach to producing highly nuclease-resistant oligonucleotides for therapeutic applications is by using L-nucleotides as building blocks. L-Nucleic acids are mirror-images of the natural DNA or RNA and were successfully applied in the evolution of L-RNA aptamers or so-called Spiegelmers (NOXXON Pharma, Figure 7.5) [137]. On one side, the artificial L-nucleic acids are orthogonal that makes them completely invisible to natural nucleases; from another side, this restricts their enzymatic synthesis and PCR amplification. This was true until the moment when two groups generated mirror-image D-polymerases by solid phase peptide synthesis and chemical ligation [36, 138].

Wang et al. produced a mirror-image of the smallest known DNA polymerase (174 residues) of African swine fever virus polymerase X (ASFV pol X) [138]. The artificial D-polymerase was able to catalyze the L-DNA primer extension with L-dNTPs resulting in the functional 56-mer L-DNAzyme. Although this is a significant achievement, ASFV pol X is a thermosensitive enzyme with optimal catalytic activity at 37 °C that is not applicable to the standard PCR protocol.

In another example, Pech et al. generated a thermostable polymerase based on the enantiomer of a mutated Dpo4 (DNA polymerase IV) from *Sulfolobus solfataricus* consisting of 352 D-amino acids (Table 7.1) [36]. Remarkably, this artificial D-polymerase was able to enhance the PCR amplification of L-DNA genes as long as 207 bp using 30 PCR thermocycles with 30 minutes of extension (Table 7.2).

These mirror-image polymerases provide the possibility of direct *in vitro* evolution of Spiegelmers using repetitive rounds of selection and amplification with L-nucleic acid libraries, engineered D-polymerases, and natural L-targets.

7.8 Conclusions

Nucleic acids are among the most remarkable biomolecules occurring in nature since they store and transfer all genetic information using four-letter code (G, C, A & T, or U). Today, the biological applications of nucleic acids have gone beyond the realms of genetic information storage and transfer. The addition of a new functionality to nucleic acids enhances the capabilities of the exciting biological system. These endeavors include developing a new third base pair (UBP) capable of pairing with its specific partner and not with the natural bases, as well as evolving an additional third type of nucleic acid (XNA) capable of accurately storing and transferring genetic information in a living organism.

Notable examples of XNA synthetic biopolymers include base-modified XNAs featuring chemical modifications at the N7-position of purines and C5-position of pyrimidines (e.g. DZA and fDNA) [20, 45], as well as sugar-modified XNAs (e.g. HNA, FANA, TNA, 4′-thioDNA) [118]. The availability of natural or genetically engineered polymerases capable of synthesizing XNA sequences and transcribing them back to DNA, can further facilitate the progress in the XNA research field – Xenobiology.

Since the increasing demands in artificial nucleic acids, enzymatic methods need to be established to synthesize the high-density XNA molecules that are not able to be produced by solid-phase chemical synthesis. Today, there are only a few examples of successful PCR amplification of fully substituted sequences (with all four modified nucleotides) containing non-canonical nucleotides (DZA) [19–21], amino acid-like groups (fDNA) [45, 52], biotin residues (BioDNA) [46], 4′-thio-nucleotides (4′-thioDNA) [47], phosphorothioates (αS-DNA) [33], and unique L-nucleotides (L-DNA) [36]. However, only PCR with αS-DNA and DZA monomers succeeded in amplification of >1.5 kb fully morphed fragments, and only DZA proved to serve as a genetic messenger *in vivo*.

PCR amplification of XNA is one of the most efficient methods to produce thousands of copies of the initial template in a few hours with satisfactory yields. Virtually very long XNA molecules (>500 bp) can be PCR amplified and assembled in artificial vectors, genes, or even genome using enzymatic or chemical ligation [19, 139]. Optimization of PCR procedure of XNA sequences might open horizons for *in vitro* isolation of novel catalytic nucleic acids with a functionality that has not been accessible before. Broadening chemical diversity of aptamers might lead to novel three-dimensional structures of XNA and increased affinities to a wider range of targets. At the same time, the redesigned structure of XNA molecules protects them from digestion by different nucleases and restriction enzymes.

The combination of advanced technologies such as high-fidelity XNA-polymerase engineering, advanced chemistry of nucleic acid, direct ligation of XNA inserts, and XNA sequencing will endorse the enzymatic synthesis of artificial biopolymers, and thereby their availability for genetic manipulations and selection.

References

1 Johnson, T.B. and Coghill, R.D. (1925). Researches on pyrimidines. C111. The discovery of 5-methyl-cytosine in tuberculinic acid, the nucleic acid of the tubercle bacillus. *J. Am. Chem. Soc.* 47 (11): 2838–2844.

2 Dunn, D.B. and Smith, J.D. (1958). The occurrence of 6-methylaminopurine in deoxyribonucleic acids. *Biochem. J.* 68 (4): 627–636.

3 Carell, T., Brandmayr, C., Hienzsch, A. et al. (2012). Structure and function of noncanonical nucleobases. *Angew. Chem. Int. Ed.* 51 (29): 7110–7131.

4 Wyatt, G.R. and Cohen, S.S. (1953). The bases of the nucleic acids of some bacterial and animal viruses: the occurrence of 5-hydroxymethylcytosine. *Biochem. J.* 55 (5): 774–782.

5 Takahashi, I. and Marmur, J. (1963). Replacement of thymidylic acid by deoxyuridylic acid in the deoxyribonucleic acid of a transducing phage for *Bacillus subtilis*. *Nature* 197 (4869): 794–795.

6 Kirnos, M.D., Khudyakov, I.Y., Alexandrushkina, N.I., and Vanyushin, B.F. (1977). 2-Aminoadenine is an adenine substituting for a base in S-2L cyanophage DNA. *Nature* 270 (5635): 369–370.

7 Morihiro, K., Kasahara, Y., and Obika, S. (2017). Biological applications of xeno nucleic acids. *Mol. BioSyst.* 13 (2): 235–245.

8 Pinheiro, V.B. and Holliger, P. (2012). The XNA world: progress towards replication and evolution of synthetic genetic polymers. *Curr. Opin. Chem. Biol.* 16 (3–4): 245–252.

9 Herdewijn, P. and Marlière, P. (2009). Toward safe genetically modified organisms through the chemical diversification of nucleic acids. *Chem. Biodivers.* 6 (6): 791–808.

10 Tarashima, N., Ando, H., Kojima, T. et al. (2016). Gene silencing using 4′-thioDNA as an artificial template to synthesize short hairpin RNA without inducing a detectable innate immune response. *Mol. Ther. Nucleic Acids* 5 (1): e274.

11 Taylor, A.I., Pinheiro, V.B., Smola, M.J. et al. (2014). Catalysts from synthetic genetic polymers. *Nature* 518 (7539): 427–430.

12 Taylor, A.I., Beuron, F., Peak-Chew, S.-Y. et al. (2016). Nanostructures from synthetic genetic polymers. *ChemBioChem* 17 (12): 1107–1110.

13 Ma, Q., Lee, D., Tan, Y.Q. et al. (2016). Synthetic genetic polymers: advances and applications. *Polym. Chem.* 7 (33): 5199–5216.

14 Pezo, V., Liu, F.W., Abramov, M. et al. (2013). Binary genetic cassettes for selecting XNA-templated DNA synthesis in vivo. *Angew. Chemie Int. Ed.* 52 (31): 8139–8143.

15 Pezo, V., Schepers, G., Lambertucci, C. et al. (2014). Probing ambiguous base-pairs by genetic transformation with XNA templates. *ChemBioChem* 15 (15): 2255–2258.

16 Krueger, A.T., Peterson, L.W., Chelliserry, J. et al. (2011). Encoding phenotype in bacteria with an alternative genetic set. *J. Am. Chem. Soc.* 133 (45): 18447–18451.

17 Delaney, J.C., Gao, J., Liu, H. et al. (2009). Efficient replication bypass of size-expanded DNA base pairs in bacterial cells. *Angew. Chem. Int. Ed.* 48 (25): 4524–4527.

18 Maruyama, H., Furukawa, K., Kamiya, H. et al. (2015). Transcription of 4′-thioDNA templates to natural RNA in vitro and in mammalian cells. *Chem. Commun.* 51 (37): 7887–7890.

19 Eremeeva, E., Abramov, M., Margamuljana, L. et al. (2016). Chemical morphing of DNA containing four noncanonical bases. *Angew. Chem. Int. Ed.* 55 (26): 7515–7519.

20 Eremeeva, E., Abramov, M., Margamuljana, L., and Herdewijn, P. (2017). Base-modified nucleic acids as a powerful tool for synthetic biology and biotechnology. *Chem. Eur. J.* 23 (40): 9560–9576.

21 Eremeeva, E., Abramov, M., Marlière, P., and Herdewijn, P. (2016). The 5-chlorouracil:7-deazaadenine base pair as an alternative to the dT:dA base pair. *Org. Biomol. Chem.* 15 (1): 168–176.

22 Ellington, A.D. and Szostak, J.W. (1990). In vitro selection of RNA molecules that bind specific ligands. *Nature* 346 (6287): 818–822.

23 Tuerk, C. and Gold, L. (1990). Systematic evolution of ligands by exponential enrichment: RNA ligands to bacteriophage T4 DNA polymerase. *Science (80-.)* 249 (4968): 505–510.

24 Stoltenburg, R., Reinemann, C., and Strehlitz, B. (2007). SELEX--a (r)evolutionary method to generate high-affinity nucleic acid ligands. *Biomol. Eng.* 24 (4): 381–403.

25 Hocek, M. (2014). Synthesis of base-modified 2′-deoxyribonucleoside triphosphates and their use in enzymatic synthesis of modified DNA for applications in bioanalysis and chemical biology. *J. Org. Chem.* 79 (21): 9914–9921.

26 Hollenstein, M. (2012). Nucleoside triphosphates – building blocks for the modification of nucleic acids. *Molecules* 17 (12): 13569–13591.

27 Kuwahara, M. and Sugimoto, N. (2010). Molecular evolution of functional nucleic acids with chemical modifications. *Molecules* 15 (8): 5423–5444.

28 Lipi, F., Chen, S., Chakravarthy, M. et al. (2016). *In vitro* evolution of chemically-modified nucleic acid aptamers: pros and cons, and comprehensive selection strategies. *RNA Biol.* 13 (12): 1232–1245.

29 Chen, T., Hongdilokkul, N., Liu, Z. et al. (2016). The expanding world of DNA and RNA. *Curr. Opin. Chem. Biol.* 34: 80–87.

30 Langer, P.R., Waldrop, A.A., and Ward, D.C. (1981). Enzymatic synthesis of biotin-labeled polynucleotides: novel nucleic acid affinity probes. *Proc. Natl. Acad. Sci. U.S.A.* 78 (11): 6633–6637.

31 Saiki, R., Scharf, S., Faloona, F. et al. (1985). Enzymatic amplification of beta-globin genomic sequences and restriction site analysis for diagnosis of sickle cell anemia. *Science* 230 (4732): 1350–1354.

32 Boyer, P.D., Krebs, E.G., and Tamanoi, F. (1970). *The Enzymes*. Academic Press.

33 Ghadessy, F.J., Ramsay, N., Boudsocq, F. et al. (2004). Generic expansion of the substrate spectrum of a DNA polymerase by directed evolution. *Nat. Biotechnol.* 22 (6): 755–759.

34 Held, H.A. and Benner, S.A. (2002). Challenging artificial genetic systems: thymidine analogs with 5-position sulfur functionality. *Nucleic Acids Res.* 30 (17): 3857–3869.

35 Kuwahara, M., Nagashima, J., Hasegawa, M. et al. (2006). Systematic characterization of 2′-deoxynucleoside- 5′-triphosphate analogs as substrates for DNA polymerases by polymerase chain reaction and kinetic studies on enzymatic production of modified DNA. *Nucleic Acids Res.* 34 (19): 5383–5394.

36 Pech, A., Achenbach, J., Jahnz, M. et al. (2017). A thermostable d-polymerase for mirror-image PCR. *Nucleic Acids Res.* 45 (7): 3997–4005.

37 Ramsay, N., Jemth, A.S., Brown, A. et al. (2010). CyDNA: synthesis and replication of highly cy-dye substituted DNA by an evolved polymerase. *J. Am. Chem. Soc.* 132: 5096–5104.

38 Chen, T., Hongdilokkul, N., Liu, Z. et al. (2016). Evolution of thermophilic DNA polymerases for the recognition and amplification of C2′-modified DNA. *Nat. Chem.* 8 (6): 556–562.

39 Śpibida, M., Krawczyk, B., Olszewski, M., and Kur, J. (2017). Modified DNA polymerases for PCR troubleshooting. *J. Appl. Genet.* 58 (1): 133–142.

40 Zhang, L., Kang, M., Xu, J., and Huang, Y. (2015). Archaeal DNA polymerases in biotechnology. *Appl. Microbiol. Biotechnol.* 99 (16): 6585–6597.

41 McInerney, P., Adams, P., and Hadi, M.Z. (2014). Error rate comparison during polymerase chain reaction by DNA polymerase. *Mol. Biol. Int.* 2014: 1–8.
42 Terpe, K. (2013). Overview of thermostable DNA polymerases for classical PCR applications: from molecular and biochemical fundamentals to commercial systems. *Appl. Microbiol. Biotechnol.* 97 (24): 10243–10254.
43 Van Pelt-Verkuil, E., Van Belkum, A., and Hays, J.P. (2008). Taq and other thermostable DNA polymerases. In: *Principles and Technical Aspects of PCR Amplification*, 103–118. Dordrecht: Springer-Verlag.
44 Sawai, H., Ozaki, A.N., Satoh, F. et al. (2001). Expansion of structural and functional diversities of DNA using new 5-substituted deoxyuridine derivatives by PCR with superthermophilic KOD dash DNA polymerase. *Chem. Commun.* 0 (24): 2604–2605.
45 Jäger, S., Rasched, G., Kornreich-Leshem, H. et al. (2005). A versatile toolbox for variable DNA functionalization at high density. *J. Am. Chem. Soc.* 127 (2): 15071–15082.
46 Tasara, T., Angerer, B., Damond, M. et al. (2003). Incorporation of reporter molecule-labeled nucleotides by DNA polymerases. II. High-density labeling of natural DNA. *Nucleic Acids Res.* 31 (10): 2636–2646.
47 Kojima, T., Furukawa, K., Maruyama, H. et al. (2013). PCR amplification of 4′-ThioDNA using 2′-Deoxy-4′-thionucleoside 5′-triphosphates. *ACS Synth. Biol.* 2 (9): 529–536.
48 Kuwahara, M., Takahata, Y., Shoji, A. et al. (2003). Substrate properties of C5-substituted pyrimidine 2′-deoxynucleoside 5′-triphosphates for thermostable DNA polymerases during PCR. *Bioorg. Med. Chem. Lett.* 13 (21): 3735–3738.
49 Lapa, S.A., Chudinov, A.V., and Timofeev, E.N. (2016). The toolbox for modified aptamers. *Mol. Biotechnol.* 58 (2): 79–92.
50 Dellafiore, M.A., Montserrat, J.M., and Iribarren, A.M. (2016). Modified nucleoside triphosphates for in-vitro selection techniques. *Front. Chem.* 4: 18.
51 Chen, T. and Romesberg, F.E. (2017). Enzymatic synthesis, amplification, and application of DNA with a functionalized backbone. *Angew. Chem. Int. Ed.* 56 (45): 14046–14051.
52 Jäger, S. and Famulok, M. (2004). Generation and enzymatic amplification of high-density functionalized DNA double strands. *Angew. Chem. Int. Ed.* 43 (25): 3337–3340.
53 Raindlová, V., Pohl, R., Šanda, M., and Hocek, M. (2010). Direct polymerase synthesis of reactive aldehyde-functionalized DNA and its conjugation and staining with Hydrazines. *Angew. Chem. Int. Ed.* 49 (6): 1064–1066.
54 Bailly, C. and Waring, M.J. (1995). Transferring the purine 2-amino group from guanines to adenines in DNA changes the sequence-specific binding of antibiotics. *Nucleic Acids Res.* 23 (6): 885–892.
55 Bailly, C. and Waring, M.J. (1998). The use of diaminopurine to investigate structural properties of nucleic acids and molecular recognition between ligands and DNA. *Nucleic Acids Res.* 26 (19): 4309–4314.
56 Kutyavin, I.V. (2008). Use of base-modified duplex-stabilizing deoxynucleoside 5′-triphosphates to enhance the hybridization properties of primers

and probes in detection polymerase chain reaction. *Biochemistry* 47 (425): 13666–13673.

57 Čapek, P., Cahová, H., Pohl, R. et al. (2007). An efficient method for the construction of functionalized DNA bearing amino acid groups through cross-coupling reactions of nucleoside triphosphates followed by primer extension or PCR. *Chem. Eur. J.* 13 (21): 6196–6203.

58 Gourlain, T., Sidorov, A., Mignet, N. et al. (2001). Enhancing the catalytic repertoire of nucleic acids. II. Simultaneous incorporation of amino and imidazolyl functionalities by two modified triphosphates during PCR. *Nucleic Acids Res.* 29 (9): 1898–1905.

59 Kuwahara, M., Hososhima, S.S., Takahata, Y. et al. (2003). Simultaneous incorporation of three different modified nucleotides during PCR. *Nucleic Acids Symp. Ser.* 3 (1): 37–38.

60 Hollenstein, M. (2013). Deoxynucleoside triphosphates bearing histamine, carboxylic acid, and hydroxyl residues – synthesis and biochemical characterization. *Org. Biomol. Chem.* 11 (31): 5162–5172.

61 Gawande, B.N., Rohloff, J.C., Carter, J.D. et al. (2017). Selection of DNA aptamers with two modified bases. *Proc. Natl. Acad. Sci. U.S.A.* 114 (11): 2898–2903.

62 Dziuba, D., Pohl, R., and Hocek, M. (2014). Bodipy-labeled nucleoside triphosphates for polymerase synthesis of fluorescent DNA. *Bioconjug. Chem.* 25 (11): 1984–1995.

63 Shoji, A., Hasegawa, T., Kuwahara, M. et al. (2007). Chemico-enzymatic synthesis of a new fluorescent-labeled DNA by PCR with a thymidine nucleotide analogue bearing an acridone derivative. *Bioorg. Med. Chem. Lett.* 17 (3): 776–779.

64 Johnson, S.C., Sherrill, C.B., Marshall, D.J. et al. (2004). A third base pair for the polymerase chain reaction: inserting isoC and isoG. *Nucleic Acids Res.* 32 (6): 1937–1941.

65 Yang, Z., Chen, F., Alvarado, J.B., and Benner, S.A. (2011). Amplification, mutation, and sequencing of a six-letter synthetic genetic system. *J. Am. Chem. Soc.* 133 (38): 15105–15112.

66 Hirao, I., Kimoto, M., Mitsui, T. et al. (2006). An unnatural hydrophobic base pair system: site-specific incorporation of nucleotide analogs into DNA and RNA. *Nat. Methods* 3 (9): 729–735.

67 Yamashige, R., Kimoto, M., Takezawa, Y. et al. (2012). Highly specific unnatural base pair systems as a third base pair for PCR amplification. *Nucleic Acids Res.* 40 (6): 2793–2806.

68 Kimoto, M., Yamashige, R., Matsunaga, K. et al. (2013). Generation of high-affinity DNA aptamers using an expanded genetic alphabet. *Nat. Biotechnol.* 31 (5): 453–457.

69 Matsunaga, K., Kimoto, M., and Hirao, I. (2017). High-affinity DNA aptamer generation targeting von Willebrand factor A1-domain by genetic alphabet expansion for systematic evolution of ligands by exponential enrichment using two types of libraries composed of five different bases. *J. Am. Chem. Soc.* 139 (1): 324–334.

70 Li, L., Degardin, M., Lavergne, T. et al. (2014). Natural-like replication of an unnatural base pair for the expansion of the genetic alphabet and biotechnology applications. *J. Am. Chem. Soc.* 136 (3): 826–829.

71 Malyshev, D.A., Dhami, K., Quach, H.T. et al. (2012). Efficient and sequence-independent replication of DNA containing a third base pair establishes a functional six-letter genetic alphabet. *Proc. Natl. Acad. Sci.* 109 (30): 12005–12010.

72 Veedu, R.N., Vester, B., and Wengel, J. (2008). Polymerase chain reaction and transcription using locked nucleic acid nucleotide triphosphates. *J. Am. Chem. Soc.* 130 (26): 8124–8125.

73 Veedu, R.N., Vester, B., and Wengel, J. (2009). Efficient enzymatic synthesis of LNA-modified DNA duplexes using KOD DNA polymerase. *Org. Biomol. Chem.* 7 (7): 1404–1409.

74 Thirunavukarasu, D., Chen, T., Liu, Z. et al. (2017). Selection of 2′-fluoro-modified aptamers with optimized properties. *J. Am. Chem. Soc.* 139 (8): 2892–2895.

75 Chen, T. and Romesberg, F.E. (2017). Polymerase chain transcription: exponential synthesis of RNA and modified RNA. *J. Am. Chem. Soc.* 139 (29): 9949–9954.

76 Andreola, M., Calmels, C., Michel, J. et al. (2000). Towards the selection of phosphorothioate aptamers. *Eur. J. Biochem.* 267 (16): 5032–5040.

77 Kang, J., Lee, M.S., Copland, J.A. et al. (2008). Combinatorial selection of a single stranded DNA thioaptamer targeting TGF-β1 protein. *Bioorg. Med. Chem. Lett.* 18 (6): 1835–1839.

78 Higashimoto, Y., Matsui, T., Nishino, Y. et al. (2013). Blockade by phosphorothioate aptamers of advanced glycation end products-induced damage in cultured pericytes and endothelial cells. *Microvasc. Res.* 90: 64–70.

79 Gandham, S.H.A., Volk, D.E., Lokesh, G.L.R. et al. (2014). Thioaptamers targeting dengue virus type-2 envelope protein domain III. *Biochem. Biophys. Res. Commun.* 453 (3): 309–315.

80 Cahová, H., Panattoni, A., Kielkowski, P. et al. (2016). 5-Substituted pyrimidine and 7-substituted 7-deazapurine dNTPs as substrates for DNA polymerases in competitive primer extension in the presence of natural dNTPs. *ACS Chem. Biol.* 11 (11): 3165–3171.

81 Ingale, S.A. and Seela, F. (2016). 7-Deaza-2′-deoxyguanosine: selective nucleobase halogenation, positional impact of space-occupying substituents, and stability of DNA with parallel and antiparallel strand orientation. *J. Org. Chem.* 81 (18): 8331–8342.

82 Hottin, A. and Marx, A. (2016). Structural insights into the processing of nucleobase-modified nucleotides by DNA polymerases. *Acc. Chem. Res.* 49 (3): 418–427.

83 Seela, F. and Röling, A. (1992). 7-Deazapurine containing DNA: efficiency of c7GdTP, c7 AdTP and c7IdTP incorporation during PCR amplification and protection from endodeoxyribonuclease hydrolysis. *Nucleic Acids Res.* 20 (1): 55–61.

84 Lee, S.E., Sidorov, A., Gourlain, T. et al. (2001). Enhancing the catalytic repertoire of nucleic acids: a systematic study of linker length and rigidity. *Nucleic Acids Res.* 29 (7): 1565–1573.

85 Sakthivel, K. and Barbas, C.F. (1998). Expanding the potential of DNA for binding and catalysis: highly functionalized dUTP derivatives that are substrates for thermostable DNA polymerases. *Angew. Chem. Int. Ed.* 37 (20): 2872–2875.

86 Hollenstein, M. (2012). Synthesis of deoxynucleoside triphosphates that include proline, urea, or sulfonamide groups and their polymerase incorporation into DNA. *Chem. Eur. J.* 18 (42): 13320–13330.

87 Kuwahara, M., Hanawa, K., Ohsawa, K. et al. (2006). Direct PCR amplification of various modified DNAs having amino acids: convenient preparation of DNA libraries with high-potential activities for in vitro selection. *Bioorganic Med. Chem.* 14 (8): 2518–2526.

88 Dierick, H., Stul, M., De Kelver, W. et al. (1993). Incorporation of dITP or 7-deaza dGTP during PCR improves sequencing of the product. *Nucleic Acids Res.* 21 (18): 4427–4428.

89 Liu, E., Lam, C., and Perrin, D. (2015). Synthesis and enzymatic incorporation of modified deoxyuridine triphosphates. *Molecules* 20 (8): 13591–13602.

90 Wong, K.K. and Jolla, L. (1991). PCR with 5-methyl-dCTP replacing dCTP. *Nucleic Acids Res.* 19 (5): 1081–1085.

91 Hollenstein, M. (2015). Generation of long, fully modified, and serum-resistant oligonucleotides by rolling circle amplification. *Org. Biomol. Chem.* 13 (38): 9820–9824.

92 Sidorov, A.V., Grasby, J.A., and Williams, D.M. (2004). Sequence-specific cleavage of RNA in the absence of divalent metal ions by a DNAzyme incorporating imidazolyl and amino functionalities. *Nucleic Acids Res.* 32 (4): 1591–1601.

93 Perrin, D.M., Garestier, T., and Hélène, C. (2001). Bridging the gap between proteins and nucleic acids: a metal-independent RNAseA mimic with two protein-like functionalities. *J. Am. Chem. Soc.* 123 (8): 1556–1563.

94 Hollenstein, M., Hipolito, C.J., Lam, C.H., and Perrin, D.M. (2013). Toward the combinatorial selection of chemically modified DNAzyme RNase a mimics active against all-RNA substrates. *ACS Comb. Sci.* 15 (4): 174–182.

95 Hollenstein, M., Hipolito, C.J., Lam, C.H., and Perrin, D.M. (2009). A DNAzyme with three protein-like functional groups: enhancing catalytic efficiency of M^{2+}–independent RNA cleavage. *ChemBioChem* 10 (12): 1988–1992.

96 Hollenstein, M., Hipolito, C.J., Lam, C.H., and Perrin, D.M. (2009). A self-cleaving DNA enzyme modified with amines, guanidines and imidazoles operates independently of divalent metal cations (M^{2+}). *Nucleic Acids Res.* 37 (5): 1638–1649.

97 Rohloff, J.C., Gelinas, A.D., Jarvis, T.C. et al. (2014). Nucleic acid ligands with protein-like side chains: modified aptamers and their use as diagnostic and therapeutic agents. *Mol. Ther. Nucleic Acids* 3 (10): e201.

98 Vaught, J.D., Bock, C., Carter, J. et al. (2010). Expanding the chemistry of DNA for in vitro selection. *J. Am. Chem. Soc.* 132 (12): 4141–4151.

99 Tolle, F., Brändle, G.M., Matzner, D., and Mayer, G. (2015). A versatile approach towards nucleobase-modified aptamers. *Angew. Chem. Int. Ed.* 54 (37): 10971–10974.

100 Horiya, S., MacPherson, I.S., and Krauss, I.J. (2014). Recent strategies targeting HIV glycans in vaccine design. *Nat. Chem. Biol.* 10 (12): 990–999.

101 Wirges, C.T., Timper, J., Fischler, M. et al. (2009). Controlled nucleation of DNA metallization. *Angew. Chem. Int. Ed.* 48 (1): 219–223.

102 Mehedi Masud, M., Ozaki-Nakamura, A., Kuwahara, M. et al. (2003). Modified DNA bearing 5(methoxycarbonylmethyl)-2′-deoxyuridine: preparation by PCR with thermophilic DNA polymerase and postsynthetic derivatization. *ChemBioChem* 4 (7): 584–588.

103 Yang, Z., Hutter, D., Sheng, P. et al. (2006). Artificially expanded genetic information system: a new base pair with an alternative hydrogen bonding pattern. *Nucleic Acids Res.* 34 (21): 6095–6101.

104 Malyshev, D.A., Dhami, K., Lavergne, T. et al. (2014). A semi-synthetic organism with an expanded genetic alphabet. *Nature* 509 (7500): 385–388.

105 Zhang, Y., Lamb, B.M., Feldman, A.W. et al. (2017). A semisynthetic organism engineered for the stable expansion of the genetic alphabet. *Proc. Natl. Acad. Sci.* 114 (6): 1317–1322.

106 Zhang, Y., Ptacin, J.L., Fischer, E.C. et al. (2017). A semi-synthetic organism that stores and retrieves increased genetic information. *Nature* 551 (7682): 644–647.

107 Switzer, C., Moroney, S.E., and Benner, S.A. (1989). Enzymatic incorporation of a new base pair into DNA and RNA. *J. Am. Chem. Soc.* 111 (21): 8322–8323.

108 Piccirilli, J.A., Krauch, T., Moroney, S.E., and Benner, S.A. (1990). Enzymatic incorporation of a new base pair into DNA and RNA extends the genetic alphabet. *Nature* 343 (6253): 33–37.

109 Sismour, A.M. (2005). The use of thymidine analogs to improve the replication of an extra DNA base pair: a synthetic biological system. *Nucleic Acids Res.* 33 (17): 5640–5646.

110 Leal, N.A., Kim, H.-J., Hoshika, S. et al. (2015). Transcription, reverse transcription, and analysis of RNA containing artificial genetic components. *ACS Synth. Biol.* 4 (4): 407–413.

111 Zhang, L., Yang, Z., Sefah, K. et al. (2015). Evolution of functional six-nucleotide DNA. *J. Am. Chem. Soc.* 137 (21): 6734–6737.

112 Sefah, K., Yang, Z., Bradley, K.M. et al. (2014). In vitro selection with artificial expanded genetic information systems. *Proc. Natl. Acad. Sci.* 111 (4): 1449–1454.

113 Zhang, L., Yang, Z., Le Trinh, T. et al. (2016). Aptamers against cells overexpressing glypican 3 from expanded genetic systems combined with cell engineering and laboratory evolution. *Angew. Chem. Int. Ed.* 55 (40): 12372–12375.

114 Biondi, E., Lane, J.D., Das, D. et al. (2016). Laboratory evolution of artificially expanded DNA gives redesignable aptamers that target the toxic form of anthrax protective antigen. *Nucleic Acids Res.* 44 (20): 9565–9577.

115 Seo, Y.J., Matsuda, S., and Romesberg, F.E. (2009). Transcription of an expanded genetic alphabet. *J. Am. Chem. Soc.* 131 (14): 5046–5047.
116 Dhami, K., Malyshev, D.A., Ordoukhanian, P. et al. (2014). Systematic exploration of a class of hydrophobic unnatural base pairs yields multiple new candidates for the expansion of the genetic alphabet. *Nucleic Acids Res.* 42 (16): 10235–10244.
117 Malyshev, D.A. and Romesberg, F.E. (2015). The expanded genetic alphabet. *Angew. Chem. Int. Ed.* 54 (41): 11930–11944.
118 Pinheiro, V.B., Taylor, A.I., Cozens, C. et al. (2012). Synthetic genetic polymers capable of heredity and evolution. *Science* 336 (6079): 341–344.
119 Alves, Ferreira-Bravo, I., Cozens, C., Holliger, P., and DeStefano, J.J. (2015). Selection of 2′-deoxy-2′-fluoroarabinonucleotide (FANA) aptamers that bind HIV-1 reverse transcriptase with picomolar affinity. *Nucleic Acids Res.* 43 (20): 9587–9599.
120 Yu, H., Zhang, S., and Chaput, J.C. (2012). Darwinian evolution of an alternative genetic system provides support for TNA as an RNA progenitor. *Nat. Chem.* 4 (3): 183–187.
121 Yu, H., Zhang, S., Dunn, M.R., and Chaput, J.C. (2013). An efficient and faithful in vitro replication system for Threose nucleic acid. *J. Am. Chem. Soc.* 135 (9): 3583–3591.
122 Nikoomanzar, A., Dunn, M.R., and Chaput, J.C. (2017). Evaluating the rate and substrate specificity of laboratory evolved XNA polymerases. *Anal. Chem.* 89 (23): 12622–12625.
123 Dunn, M.R., Otto, C., Fenton, K.E., and Chaput, J.C. (2016). Improving polymerase activity with unnatural substrates by sampling mutations in homologous protein architectures. *ACS Chem. Biol.* 11 (5): 1210–1219.
124 Dunn, M.R. and Chaput, J.C. (2016). Reverse transcription of threose nucleic acid by a naturally occurring DNA polymerase. *ChemBioChem* 17 (19): 1804–1808.
125 Tarashima, N., Sumitomo, T., Ando, H. et al. (2015). Synthesis of DNA fragments containing 2′-deoxy-4′-selenonucleoside units using DNA polymerases: comparison of dNTPs with O, S and Se at the 4′-position in replication. *Org. Biomol. Chem.* 13 (25): 6949–6952.
126 Lauridsen, L.H., Rothnagel, J.A., and Veedu, R.N. (2012). Enzymatic recognition of 2′-modified ribonucleoside 5′-triphosphates: towards the evolution of versatile aptamers. *ChemBioChem* 13 (1): 19–25.
127 Conrad, F., Hanne, A., Gaur, R.K., and Krupp, G. (1995). Enzymatic synthesis of 2′-modified nucleic acids: identification of important phosphate and ribose moieties in RNase P substrates. *Nucleic Acids Res.* 23 (11): 1845–1853.
128 Padilla, R. and Sousa, R. (1999). Efficient synthesis of nucleic acids heavily modified with non-canonical ribose 2′-groups using a mutantT7 RNA polymerase (RNAP). *Nucleic Acids Res.* 27 (6): 1561–1563.
129 Padilla, R. and Sousa, R. (2002). A Y639F/H784A T7 RNA polymerase double mutant displays superior properties for synthesizing RNAs with non-canonical NTPs. *Nucleic Acids Res.* 30 (24): 138e–138e.

130 Burmeister, P.E., Lewis, S.D., Silva, R.F. et al. (2005). Direct in vitro selection of a 2′-*O*-methyl aptamer to VEGF. *Chem. Biol.* 12 (1): 25–33.

131 Chelliserrykattil, J. and Ellington, A.D. (2004). Evolution of a T7 RNA polymerase variant that transcribes 2′-*O*-methyl RNA. *Nat. Biotechnol.* 22 (9): 1155–1160.

132 Ruckman, J., Green, L.S., Beeson, J. et al. (1998). 2′-Fluoropyrimidine RNA-based aptamers to the 165-amino acid form of vascular endothelial growth factor (VEGF165). Inhibition of receptor binding and VEGF-induced vascular permeability through interactions requiring the exon 7-encoded domain. *J. Biol. Chem.* 273 (32): 20556–20567.

133 Aurup, H., Williams, D.M., and Eckstein, F. (1992). 2′-Fluoro and 2-amino-2′-deoxynucleoside 5′-triphosphates as substrates for T7 RNA polymerase. *Biochemistry* 31 (40): 9636–9641.

134 Aurup, H., Siebert, A., Benseler, F. et al. (1994). Translation of 2′-modified mRNA in vitro and in vivo. *Nucleic Acids Res.* 22 (23): 4963–4968.

135 Cozens, C., Pinheiro, V.B., Vaisman, A. et al. (2012). A short adaptive path from DNA to RNA polymerases. *Proc. Natl. Acad. Sci.* 109 (21): 8067–8072.

136 Eckstein, F. (2014). Phosphorothioates, essential components of therapeutic oligonucleotides. *Nucleic Acid Ther.* 24 (6): 374–387.

137 Vater, A. and Klussmann, S. (2015). Turning mirror-image oligonucleotides into drugs: the evolution of Spiegelmer® therapeutics. *Drug Discovery Today* 20 (1): 147–155.

138 Wang, Z., Xu, W., Liu, L., and Zhu, T.F. (2016). A synthetic molecular system capable of mirror-image genetic replication and transcription. *Nat. Chem.* 8 (7): 698–704.

139 El-Sagheer, A.H. and Brown, T. (2017). Single tube gene synthesis by phosphoramidate chemical ligation. *Chem. Commun.* 53 (77): 10700–10702.

8

Synthetic Approaches to the Fleximer Class of Nucleosides – A Historic Perspective

Therese C. Ku and Katherine Seley-Radtke

University of Maryland, Baltimore County, Department of Chemistry & Biochemistry, 1000 Hilltop Circle, Baltimore, MD 21250, USA

For more than a decade, the Seley-Radtke group has designed and synthesized various classes of flexible purine nucleos(t)ides, or "fleximers." [1–13] These novel nucleosides were designed to better understand how flexibility in the nucleobase, one of the primary recognition sites of a nucleoside, affects receptor-ligand recognition and function, as well as its potential to overcome the challenges of binding site mutations. This review explores the various types of fleximers (Figure 8.1a–d) that have been synthesized and the synthetic methodologies that have been developed and improved over the years.

Fleximers have demonstrated several key advantages over the natural rigid purine nucleosides. For example, the distal guanosine fleximer (Flex-G, Figure 8.1a) proved to be an inhibitor of *S*-adenosyl-L-homocysteine hydrolase (SAHase), an adenosine-metabolizing enzyme [3]. By rotating the hemi-aminal bond into a *syn*-like conformation (Figure 8.2a), as opposed to the thermodynamically favored *anti*-conformation (Figure 8.2b), the guanosine flex-base was able to position the amino group in such a way that it mimicked the adenosine nucleobase [3, 4]. Moreover, the flex-guanosine triphosphate (Flex-GTP) was shown to be a superior substrate of human GDP-L-fucose pyrophosphorylase compared to the natural substrate guanosine triphosphate (GTP) [5], likely due to the fleximer's ability to interact with amino acids in the active site not accessible by the natural substrate (Figure 8.2c) [6]. This also allowed Flex-GTP to retain all activity when essential catalytic residues needed for GTP binding were mutated [5, 6].

More recently, a series of doubly flexible fleximers (Figure 8.1c) based on the FDA-approved acyclic nucleoside Acyclovir were shown to inhibit hard to treat viruses such as the coronaviruses [10] and filoviruses [11], while Acyclovir itself was completely inactive. Even more notable, these compounds inhibit several flaviviruses, including Dengue and Zika. These biological results, described in a recently published review that covered the biological impact of fleximers, in addition to others obtained in the past few years [14], provide strong impetus to further explore these structurally interesting molecules.

Enzymatic and Chemical Synthesis of Nucleic Acid Derivatives, First Edition.
Edited by Jesús Fernández-Lucas and María-José Camarasa Rius.

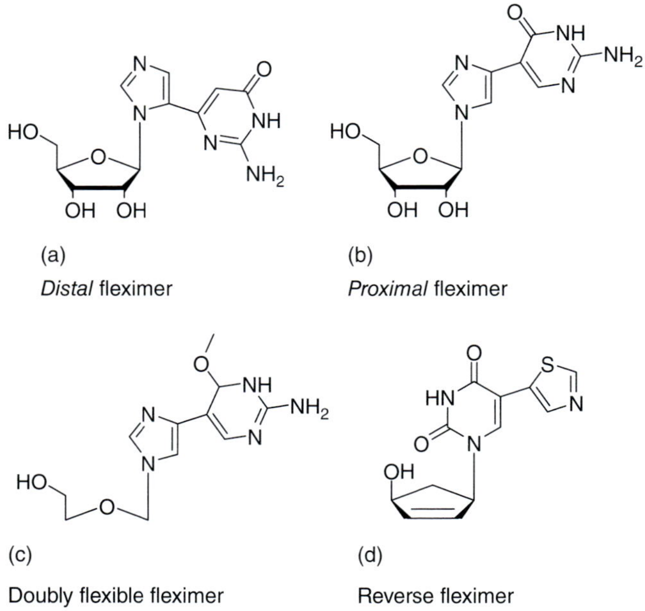

Figure 8.1 Example of Seley-Radtke's fleximers.

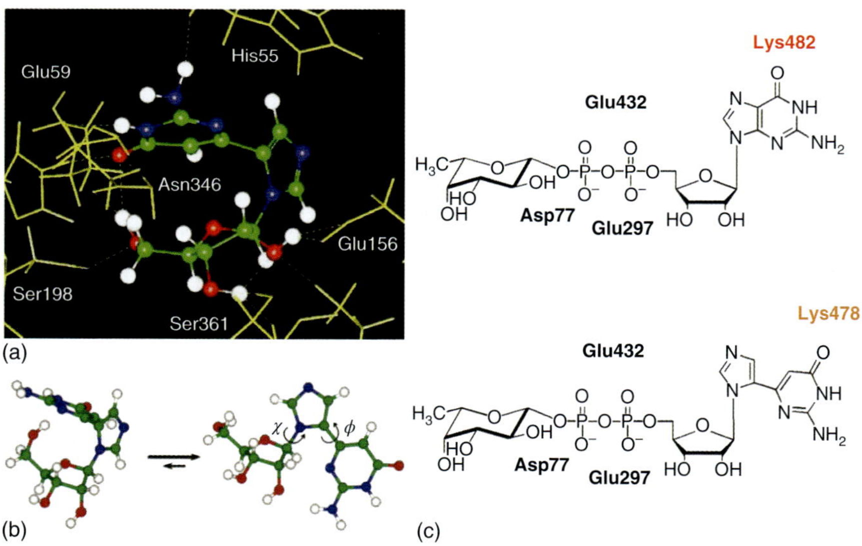

Figure 8.2 Biological advantages of fleximers [3–6].

While the biological results for these compounds have been groundbreaking in some cases, their syntheses have proven nontrivial. This review examines the methodology of fleximer synthesis in a historic fashion, presenting both published and unpublished works, as well as the most current and successful synthetic approaches used today.

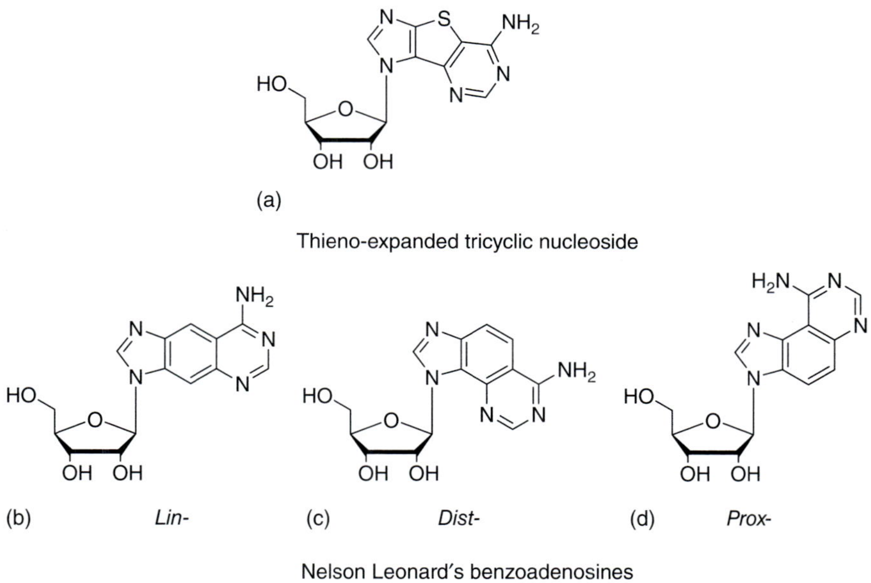

Figure 8.3 Expanded base nucleosides.

The "fleximer concept" originally arose from a separate study that focused on a series of thieno-expanded nucleoside analogues, or tricyclics (Figure 8.3a), which, in turn, was inspired by Nelson Leonard's benzene-expanded analogues (Figure 8.3b–d) [15–19]. These molecules, especially the *lin*-benzo derivatives, were among the first examples of expanded nucleobases designed to investigate the structural limitations of enzyme binding sites [15–18, 20–24]. As an extension of Leonard's work, the Seley-Radtke group designed and synthesized a series of tricyclic nucleosides [1, 2, 19, 25–27]. These were shown via *ab initio* calculations to be able to form the same hydrogen-bonding patterns as the natural nucleoside base pairs, but exhibited increased polarizability [28].

During the efforts to synthesize the tricyclic nucleosides, it was recognized that treating the tricyclics with Raney nickel would remove the bridging sulfur (Figure 8.4) [1, 2]. This would leave a carbon–carbon bond between the imidazole and pyrimidine moieties of the purine base, thereby producing a "flexible"

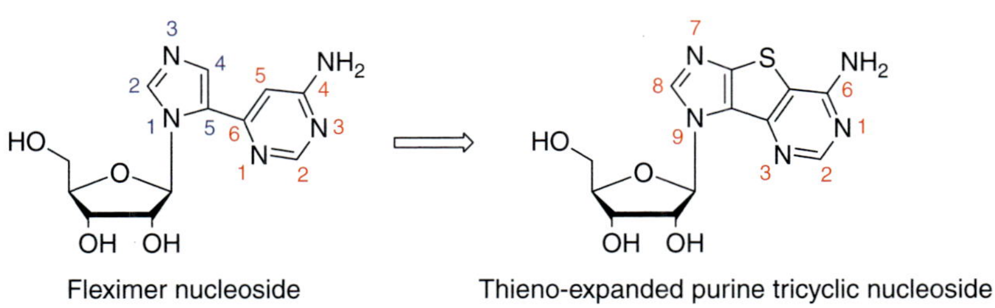

Figure 8.4 Retrosynthetic method for fleximer adenosine (numbering included for clarity).

purine nucleoside. Introducing flexibility into the nucleobase was a groundbreaking concept, especially in the field of nucleoside drug design, thus the fleximers have been of interest to many.

8.1 Distal Fleximers

8.1.1 Ribose Distal Fleximers

As mentioned, the first approach to realizing the fleximers began with the construction of the tricyclic nucleoside (Scheme 8.1) [1, 2, 19]. This involved synthesis of the tricyclic bases and then coupling the base to the sugar in the hope of avoiding the tedious sugar protection and deprotection procedures that are typical in nucleoside syntheses [19]. Although a mixture of the N7 and N9 coupled products was anticipated (following purine numbering rules), only the unwanted N7 isomer was isolated [2, 19]. Toward achieving a new approach, based on the observation that 4,5-dibromoimidazole is symmetric, Vorbrüggen coupling between **1.2** and β-D-ribofuranose-1,2,3,5-tetraacetate should, and ultimately did, yield only one product, the intermediate **1.3** [2]. The acetate protecting groups on the sugar were then replaced with more robust benzyl protecting groups to withstand the harsh reaction conditions for tricyclic base synthesis. Treatment of **1.3** with ethylmagnesium bromide (EtMgBr), followed by *N,N*-dimethylformamide (DMF) afforded the carbaldehyde **1.6** (*N*-formylpiperidine was initially used instead of DMF; however, the yields were consistently very low). Carbaldehyde **1.6** was converted to the aldoxime **1.7** using hydroxylamine hydrochloride and sodium bicarbonate ($NaHCO_3$), and subsequent dehydration in acetic anhydride (Ac_2O) yielded the nitrile **1.8**. The thieno-bicyclic intermediate **1.10** was achieved by treating the nitrile with excess potassium carbonate (K_2CO_3) and freshly made mercaptoacetamide, followed by cyclization under basic conditions. Finally, the thieno-bicycle was refluxed with triethyl orthoformate in Ac_2O to close the third ring and form the hypoxanthosine intermediate **1.11**. Since chlorination of the exocyclic enol was not possible using phosphorus oxychloride [29], the oxo moiety was converted to a thioketone intermediate that was subsequently methylated using methyliodide (CH_3I) and finally subjected to high-pressure aminolysis to produce the benzyl protected tricyclic adenosine **1.12** [1, 2].

In addition, the original method to achieve the tricyclic guanine base, using chloroformamidine hydrochloride in dimethyl sulfone [19], was too acidic and readily cleaved the glycosidic bond. Instead, the desired tricyclic guanosine **2.1** was constructed in basic conditions starting from the bicyclic intermediate **1.10** (Scheme 8.2) [1, 2]. Deprotection of the tricyclic nucleosides also proved challenging as employing the common palladium-catalyzed deprotection methods failed [30], since sulfur is a well-known palladium catalyst poison [31]. Fortunately, the benzyl groups were ultimately able to be deblocked using boron trifluoride diethyl etherate ($BF_3{\cdot}OEt_2$), although not in good yields [1, 2]. Deprotection of the fleximers, however, was much more facile since removing the sulfur first

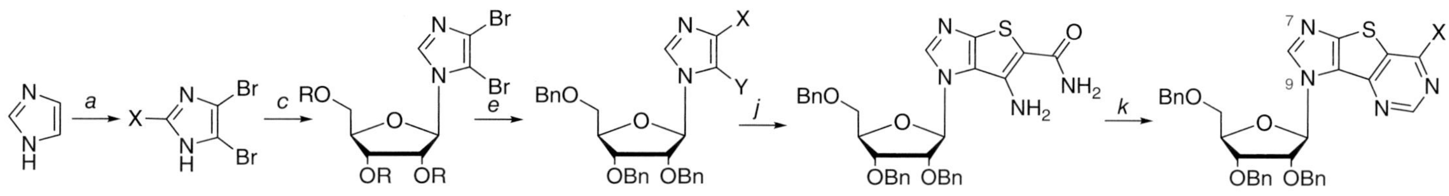

1.1 X = Br → (b) **1.2** X = H

1.3 R = Ac → (d) **1.4** R = H

1.5 X = Y = Br → (f) **1.6** X = Br, Y = CHO → (g) **1.7** X = Br, Y = CHNOH → (h) **1.8** X = Br, Y = CN → (i) **1.9** X = $SCH_2C(O)NH_2$, Y = CN

1.10

1.11 X = OH → (l) **1.12** X = NH_2

Scheme 8.1 Synthesis of thieno-expanded tricyclic adenosine and hypoxanthosine [1, 2]. Reagents and conditions: *a*. Br_2, AcOH, NaOAc, rt, three hours (71%); *b*. EtMgBr, Et_2O, rt, three hours (90%); *c*. (i) β-D-ribofuranose-1,2,3,5-tetraacetate, BSA, ACN, rt, six hours and (ii) TMSOTf, 60 °C, 18 hours (66%); *d*. NH_3, MeOH, 100 °C, three hours (quant.); *e*. (i) NaH, THF, rt, three hours and (ii) BnBr, TBAI, rt, six hours (87%); *f*. (i) EtMgBr, Et_2O, rt, four hours and (ii) DMF, rt, three hours; *g*. hydroxylamine HCl, $NaHCO_3$, EtOH, H_2O, rt, 18 hours; *h*. Ac_2O, reflux, two hours; *i*. $NH_2C(O)CH_2SH$, K_2CO_3, DMF, 55 °C, 12 hours; *j*. NaOEt, EtOH, reflux, 1.5 hours (26%, five steps); *k*. triethyl orthoformate, Ac_2O, reflux, three hours (82%); *l*. (i) P_2S_5, pyridine, reflux, 24 hours, (ii) K_2CO_3, CH_3I, CH_3OH, rt, 10 minutes, and (iii) NH_3, *n*-BuOH, 160 °C, 90 hours (60%, three steps).

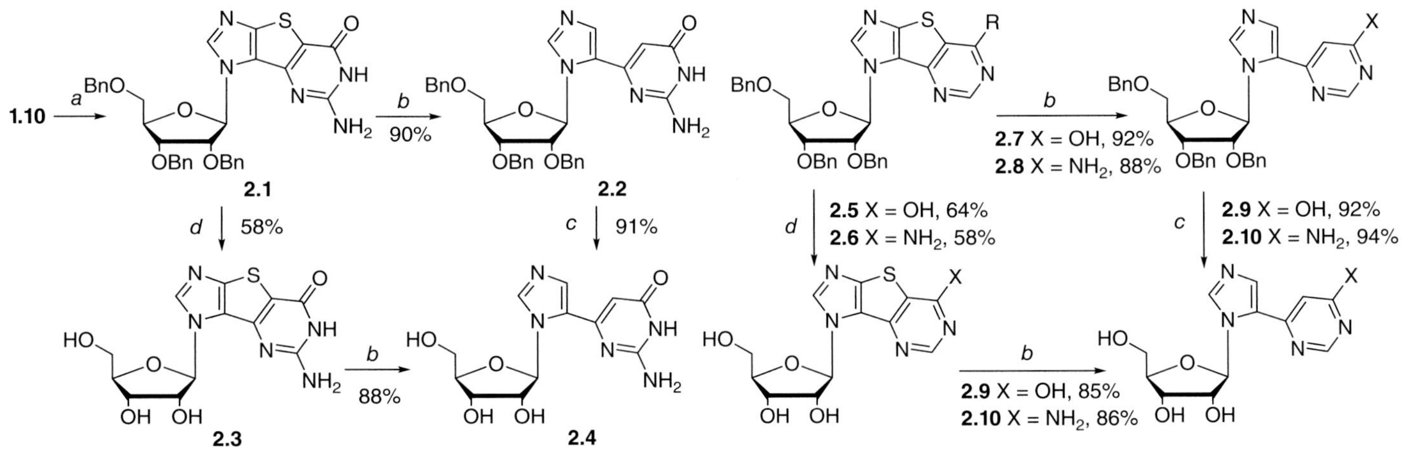

Scheme 8.2 Synthesis of thieno-expanded and fleximers guanosine, adenosine, and hypoxanthosine [1, 2]. Reagents and conditions: *a*. (i) NaOH, CS_2, MeOH, 150 °C, 18 hours, (ii) H_2O_2, MeOH, 0 °C, two hours, and (iii) NH_3, MeOH, 120 °C, 12 hours (57%); *b*. Raney Nickel, MeOH, H_2O, reflux, 18 hours; *c*. Pd/C, ammonium formate, MeOH, reflux, 18 hours; *d*. $BF_3{\cdot}OEt_2$, EtSH, CH_2Cl_2.

using Raney nickel allowed for using employing standard palladium deprotection methods, which subsequently went in good yields [2].

Owing to the tedious and low yielding nature of the methodologies that were employed for achieving these complex modified nucleosides, additional optimizations were necessary to improve the yields and reduce the reaction times. A simple, yet significant, replacement of 4,5-dibromodoimidazole with 4,5-diiodoimidazole improved yields dramatically (using the same reaction conditions, Scheme 8.3) [8, 25].

As discussed earlier, the approach to the third ring closure toward realizing the tricyclic guanosine (**2.1**) had to be modified to avoid cleavage of the glycosidic bond. Additional studies of this step proved that in addition to the formation of tricyclic guanosine (**2.1**), tricyclic xanthosine (**3.7**) as well as 2-*O*-methylxanthosine (**3.8**) were also produced by tricyclic guanosine synthesis; however, the yield ratios of the three products differed greatly depending on the reaction conditions [25]. As shown in Table 8.1, a mechanistic study to determine the conditions that could control specific outcomes was performed [25]. In this study, the ratio of sodium hydroxide (NaOH) to the starting material (**1.10**), as well as the reaction temperature, reaction time, and volume of methanol varied. Ultimately, a 1 : 1 ratio of **2.1**/**3.7** was achieved (entry 7, Table 8.1) and the undesired product **3.8** was not observed [25]. Since the tricyclic xanthosine **3.7** was also useful, these proved to be the optimal conditions for this reaction.

8.1.2 2′-Deoxyribose Distal Fleximers

Following completion of the ribose fleximer series, the 2′-deoxyribose series of tricyclics and distal fleximers were then pursued [9, 27]. The initial approach sought to modify the base modified ribose analogues by chemically removing the 2′-OH (Scheme 8.4) [25, 27]. To accomplish this, the functional groups on the bases were first protected using an *in situ* trimethylsilyl (TMS) protection, followed by the simultaneous protection of the 3′- and 5′-OH using the Markiewicz tetraisopropyldisiloxane (TIPDS) bis-protecting group [25, 27, 32]. This TIPDS protection leaves the 2′-OH unblocked, allowing for modifications [27]. Subsequent removal of the 2′-OH through Barton deoxygenation procedures [25, 26], and complete deprotection of the nucleoside provided the final 2′-deoxy tricyclic analogues **4.5**–**4.7**. These 2′-deoxy tricyclics were then treated with Raney nickel to yield the distal fleximers **4.8** and **4.9**. However, this route proved tedious as the overall yield starting from 4,5-diiodoimidazole was unacceptably low (0.5% overall yield for the 2′-deoxy tricyclic guanosine) and, therefore, a different route was designed starting with a 2′-deoxyribose sugar (Schemes 8.5 and 8.6) [9, 26, 27].

The well-known Hoffer's chlorosugar **5.3** [33, 34] was chosen due to the presence of the stereospecific anomeric chlorine. However, Hoffer's methodology (Pathway A, Scheme 8.5) proved to be quite sensitive and resulted in low yields overall, thus a more efficient method was pursued (Pathway B, Scheme 8.5) [35, 36].

After coupling of Hoffer's chlorosugar to 4,5-diiodoimidazole, the toluoyl groups had to be replaced with more robust groups once again to withstand the harsh conditions used throughout tricyclic synthesis (Scheme 8.6). As a

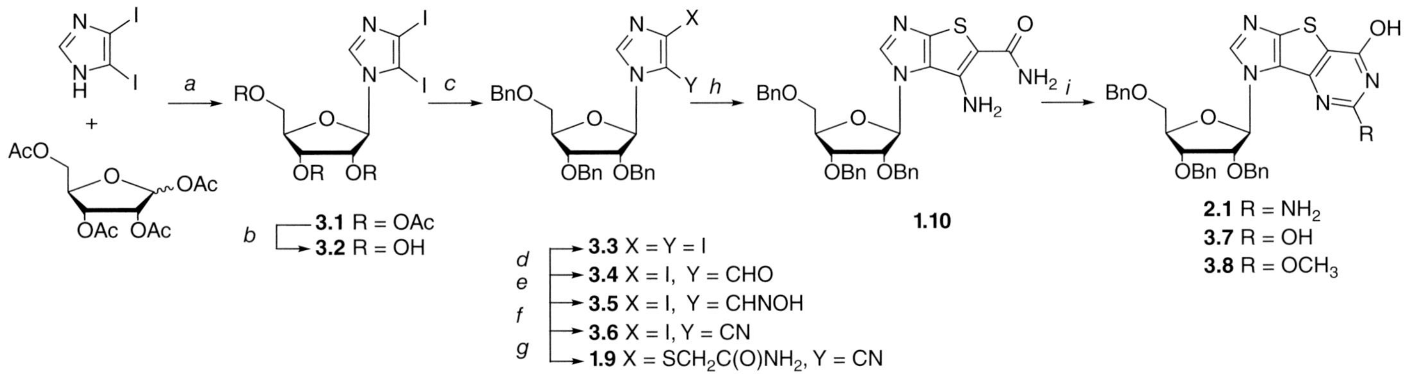

Scheme 8.3 Optimized synthesis of thieno-expanded purine nucleosides [8, 25]. Reagents and conditions: *a*. (i) BSA, ACN, four hours and (ii) TMSOTf, 60 °C, 18 hours; *b*. NH_4OH, EtOH, rt, 18 hours (75%, two steps); *c*. (i) NaH, THF, rt, three hours and (ii) BnBr, TBAI, rt, 18 hours (80%); *d*. (i) EtMgBr, THF, five hours and (ii) DMF, 18 hours (60%); *e*. $NH_2OH{\cdot}HCl$, $NaHCO_3$ or pyridine, EtOH, reflux, 2.5 hours (quant.); *f*. Ac_2O, reflux, three hours (85%); *g*. $NH_2C(O)CH_2SH$, K_2CO_3, DMF, 60 °C, 40 hours (50%); *h*. EtONa, EtOH, 80 °C, one hour (80%); *i*. (i) NaOH, CS_2, MeOH, 145 °C, 18 hours, (ii) H_2O_2, MeOH, 0 °C, two hours, and (iii) NH_3, MeOH, 120 °C, 18 hours.

Table 8.1 Mechanistic study of the terminal cyclization step of tricyclic guanosine/xanthosine.

Entry	1.10/NaOH ratio	Temperature (°C)	Reactions time (h)	Volume of MeOH (ml)	2.1/3.7/3.8 ratio
1	1 : 5	145	18	80	3 : 1 : 0
2	1 : 6	145	18	80	3 : 1 : 1
3	1 : 10	145	18	80	2 : 3 : 4
4	1 : 40	145	18	80	2 : 3 : 7
5	1 : 5	180	6	80	3 : 1 : 2
6	1 : 5	145	18	160	2 : 2 : 1
7	**1 : 5**	**145**	**18**	**250**	**1 : 1 : 0**
8	1 : 6	145	18	250	1 : 1 : trace
9	1 : 8	145	18	250	2 : 2 : 1

Bold denotes optimum ratio.
Source: Zhang et al. 2008 [25]. Reproduced with permission of Elsevier.

result, the *p*-methoxybenzyl (PMB) group was chosen as it is typically more facile to remove compared to the previously used benzyl groups. Therefore, $BF_3{\cdot}OEt_2$ was once again determined to be the best choice for final deprotection because, ceric ammonium nitrate, the deprotection reagent initially chosen (and commonly used for this deprotection), resulted in the cleavage of the glycosidic bond [9, 27]. Construction of the tricyclic base remained similar to that of the ribose analogues with a few changes: instead of using Ac_2O to obtain the nitrile **6.6**, the aldoxime **6.5** was refluxed with carbonyl diimidazole (CDI) in THF to prevent glycosidic bond cleavage as acetic acid (AcOH) was formed as by-product in the previous conditions [1, 2, 19].

Unfortunately, the previous methodology for the final cyclization step of the tricyclic guanosine proved ineffective in the presence of the 2′-deoxy sugar and, therefore, another route was developed (Scheme 8.9). The bicyclic intermediate **6.8** was treated with potassium ethyl xanthate instead of carbon disulfide for the initial cyclization. *t*-Butylhydroperoxide was used instead of hydrogen peroxide to decrease the water content. This change resulted in the guanosine product **7.1** with only a trace amount of the xanthosine analogue [9, 26, 27]. The route to the hypoxanthosine derivative utilized molecular sieves instead of Ac_2O to prevent production of AcOH while refluxing in triethyl orthoformate. Also, instead of the three-step modification of the exocyclic enol to obtain the adenosine tricyclic from the hypoxanthosine (Scheme 8.1), a two-step reaction starting by treating **7.4** with 2,4,6-triisopropylbenzenesulfonyl chloride (TPSCl) followed by aminolysis, shortened the overall reaction time by two days [25, 27]. Finally, treatment of the thieno-expanded guanosine, adenosine, and hypoxanthosine analogues with Raney nickel followed by palladium-assisted deprotection gave the desired fleximers (**9.3**, **9.8**, and **9.9**) [9, 27] (Scheme 8.7).

As described earlier, the guanosine, adenosine, and hypoxanthosine fleximer analogues were readily achieved using Raney nickel; however, surprisingly, the xanthosine fleximer was recalcitrant to removal with Raney nickel despite many

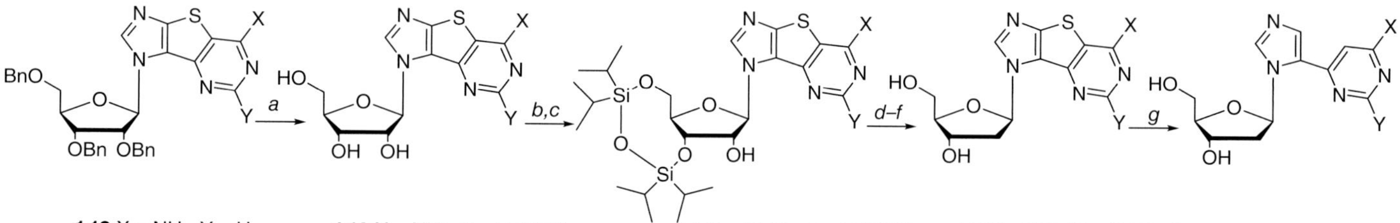

1.12 X = NH_2, Y = H
2.1 X = OH, Y = NH_2
3.7 X = Y = OH

1.12 X = NH_2, Y = H (58%)
2.6 X = OH, Y = NH_2 (58%)
4.1 X = Y = OH (88%)

4.2 X = NHBz, Y = H (52%)
4.3 X = OH, Y = NH*i*Bu (52%)
4.4 X = Y = OH (60%)

4.5 X = NH_2, Y = H (12%)
4.6 X = OH, Y = NH_2 (12%)
4.7 X = Y = OH (26%)

4.8 X = NH_2, Y = H (60%)
4.9 X = OH, Y = NH_2 (65%)

Scheme 8.4 Synthesis of 2′-deoxy fleximers starting with the ribose analogue [25, 27]. Reagents and conditions: *a*. $BF_3 \cdot OEt_2$, EtSH, CH_2Cl_2, 24 hours; *b*. (i) TMSCl, pyridine, 0 °C to rt, 2.5 hours and (ii) *for adenosine,* benzoyl chloride, 0 °C to rt, 24 hours, *for guanosine,* isobytyryl chloride, 0 °C to rt, 24 hours; *c*. 1,3-dichloro-1,1,3,3-tetraisopropyldisiloxane, pyridine, rt, 24 hours; *d*. phenyl chlorothionoformate, DMAP, rt, 24 hours; *e*. AIBN, Bu_3SnH, toluene, reflux, six hours; *f*. (i) 1 M TBAF THF, rt, four hours and (ii) NH_3, MeOH, 40 °C, 18 hours; *g*. Raney nickel, MeOH, reflux, 18 hours.

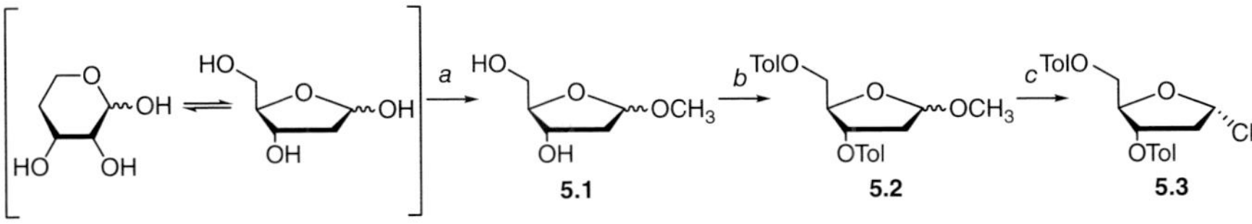

Scheme 8.5 Synthesis of Hoffer's chloro sugar [33–36]. Reagents and conditions: Pathway A (Hoffer way): *a.* HCl(g) in MeOH; *b.* *p*-TolCl, pyridine, 0 °C; *c.* HCl(g), MeOH. (65%, three steps) Reagents and conditions: Pathway B: *a.* AcCl, MeOH; *b.* *p*-TolCl, pyridine, 0 °C; *c.* AcCl, AcOH. (73%, three steps).

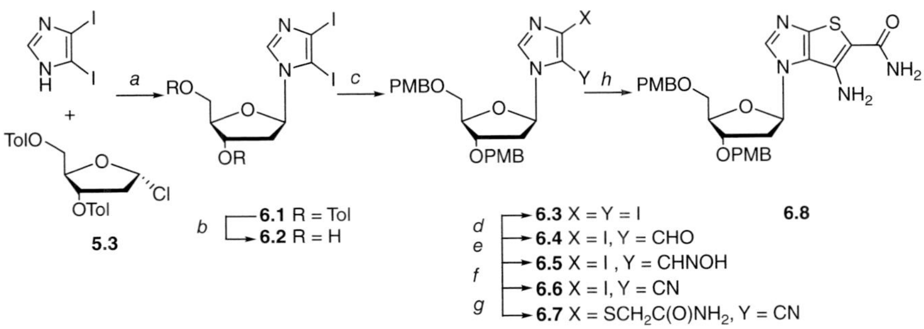

Scheme 8.6 Synthesis of 2′-deoxy thieno bicyclic nucleoside [9, 27]. Reagents and conditions: *a.* (i) NaH, CH_3CN, rt, 30 minutes and (ii) 4,5-diiodoimidazole, rt, 18 hours (84%); *b.* NaOMe, MeOH, rt, four hours (87%); *c.* (i) NaH, THF, rt, three hours and (ii) TBAI, *p*-methoxybenzylchloride, rt, 18 hours (64%); *d.* (i) EtMgBr, THF, 0 °C, 15 minutes and (ii) DMF, 0 °C, 20 minutes (90%); *e.* $NH_2OH{\cdot}HCl$, $NaHCO_3$, H_2O, EtOH, rt, 18 hours (79%); *f.* CDI, THF, 60 °C, 18 hours (60%); *g.* $NH_2C(O)CH_2SH$, K_2CO_3, DMF, 60 °C, 19 hours; *h.* EtONa, EtOH, 80 °C, six to nine hours (62% in two steps).

different attempts using various reaction times and temperatures [9]. Ultimately, the xanthosine fleximer was achieved through a different approach, using palladium-catalyzed coupling of the two heterocyclic moieties (Figure 8.5) [9].

To realize the distal xanthosine fleximer, a 5-iodoimidazole nucleoside needed to be constructed to be used in subsequent palladium-catalyzed coupling reactions (Figure 8.5). As the reactivities of the different positions of the imidazole ring are well known [37], strategic manipulation of the halogens on the 4,5-diiodoimidazole analogue **3.3** could be carried out (Scheme 8.8). The benzyl protected diiodoimidazole nucleoside **3.3** was first deiodinated using EtMgBr followed by lithiation and electrophilic trapping using hexachloroethane, to produce the 2-chloro intermediate **8.1** [38]. Chlorine was employed to occupy the most reactive C-2 position so that treatment of **8.1** with *n*-butyllithium (*n*-BuLi)/tetramethylethylenediamine could produce the transient 5-lithioimidazole, which could then be slowly quenched with bis(pinacolato)diboron to generate the boronic ester **8.2**. Unfortunately, the reaction mixture proved too complex for efficient purification, thus **8.2** could never be isolated. Alternatively, once removal of the C-4 and C-5 iodines occurred, reiodination of the imidazole would theoretically provide the

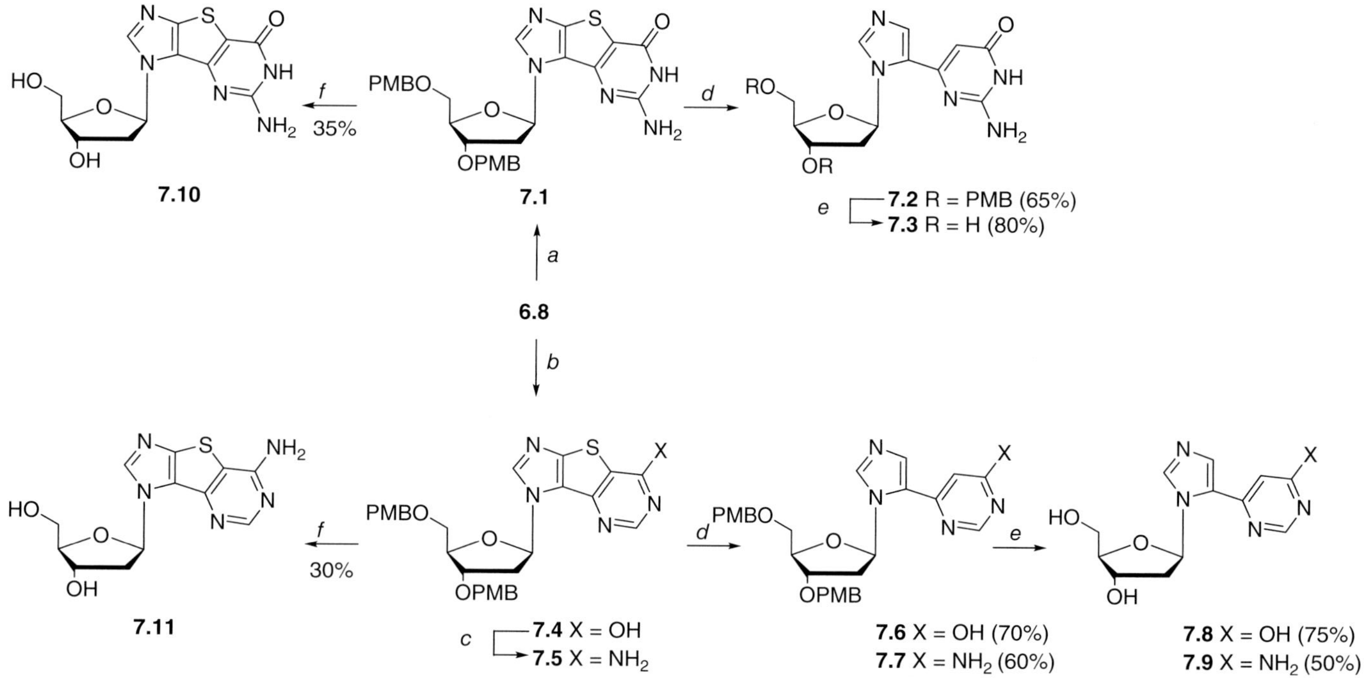

Scheme 8.7 Optimized synthesis of the thieno-expanded tricyclic and fleximer nucleosides [9, 27]. Reagents and conditions: *a*. (i) KSC(S)OEt, DMF, reflux, four hours, (ii) *t*-butylhydroperoxide, MeOH, 0 °C two hours, and (iii) NH_3, MeOH, 125 °C, 18 hours. (53%); *b*. $CH(OEt)_3$, 4 Å molecular sieves, 150 °C, six hours (88%); *c*. (i) TPSCl, DMAP, NEt_3, MeCN, three hours and (ii) NH_3, MeOH, 24 hours (65%); *d*. Raney Ni, MeOH, 65 °C, 72 hours; *e*. Pd/C, ammonium formate, EtOH, reflux, 18 hours; *f*. $BF_3 \cdot OEt_2$, EtSH, CH_2Cl_2.

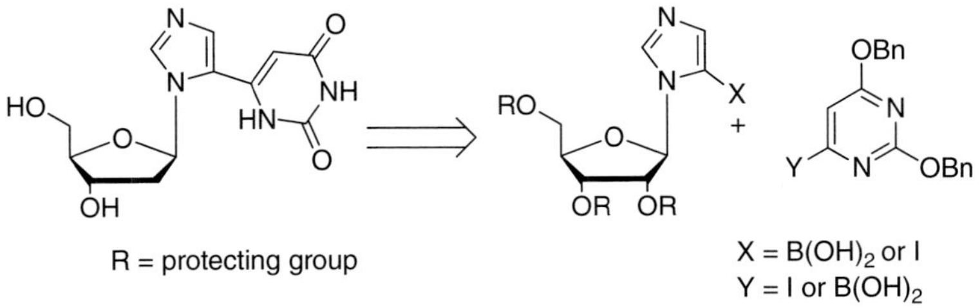

Figure 8.5 Retrosynthetic outline for constructing the xanthosine fleximer.

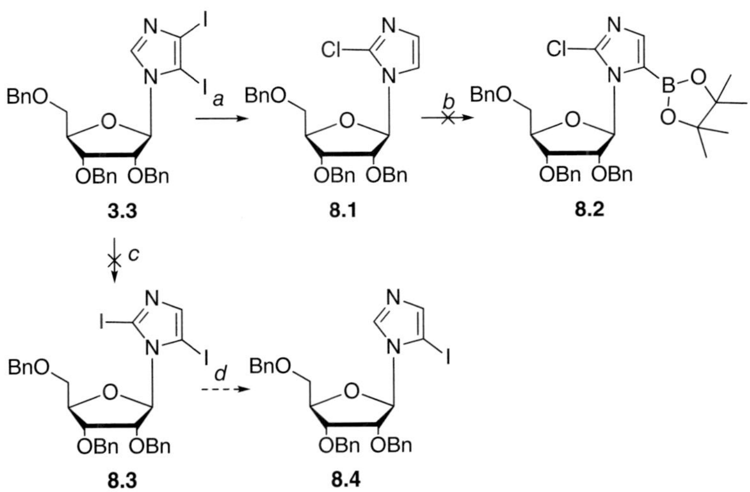

Scheme 8.8 Failed attempts to achieve the xanthosine fleximer. Reagents and conditions: *a*. (i) EtMgBr, THF, rt, three hours and (ii) *n*-BuLi, C_2Cl_6, −78 °C to rt, 2.5 hours (50%); *b*. TMEDA, *n*-BuLi, bis(pinacolato) diboron, −78 °C to rt, 18 hours; *c*. (i) EtMgBr, THF, rt, three hours and (ii) TMEDA, *n*-BuLi, I_2 −78 °C to rt, 18 hours; *d*. *n*-BuLi or EtMgBr, THF, rt, three hours.

2,5-diiodoimidazole intermediate **8.3**. This intermediate could then be selectively deiodinated at the C-2 position to give the requisite C-5-substituted imidazole **8.4** [39]. However, this methodology also failed and this strategy was also abandoned.

Next, selective deiodination of the diiodoimidazole nucleoside **3.3** was tried using sodium sulfite (Na_2SO_3) in an attempt to obtain both C-4 and C-5 mono-substituted imidazole nucleosides (Scheme 8.6); however, only the unwanted C-4 iodo isomer **6.1** was obtained as confirmed by NOESY NMR.

Since attempting to deiodinate to the correct isomer failed while diiodoimidazole was coupled to the protected ribose ring (Scheme 8.9), Hoffer's chlorosugar **7.3** was coupled to 4(5)-iodoimidazole to yield the two isomeric compounds **10.1** and **10.2** [9]. The structure of each isomer was confirmed via COSY and NOESY experiments (Scheme 8.10) [9]. Since the distal xanthosine was the desired compound, the toluoyl groups of the 5-iodo imidazole nucleoside **10.1** were replaced with *t*-butyldimethylsilane (TBDMS) [9]. Deprotection and reprotection strategies were necessary since the subsequent coupling was unsuccessful with the

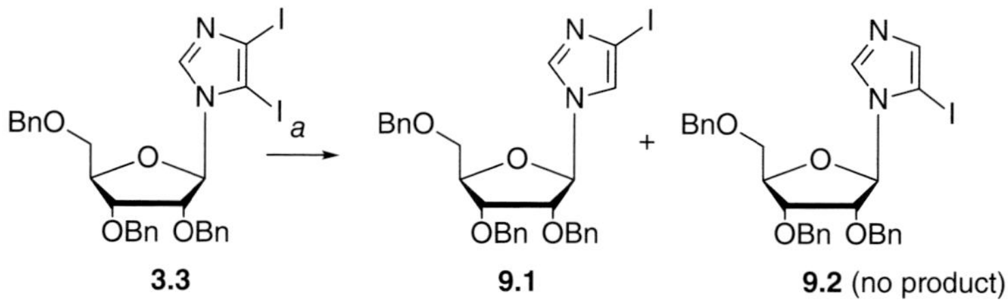

Scheme 8.9 Deiodination of **3.3** using sodium sulfite. Reagents and conditions: *a*. Na_2SO_3, DMF, H_2O, reflux, 18 hours.

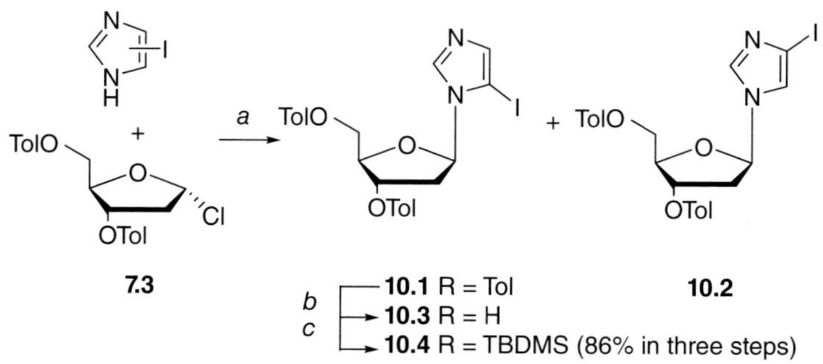

Scheme 8.10 Coupling of 4(5)-iodoimidazole to Hoffer's chlorosugar [9]. Reagents and conditions: *a*. (i) NaH, MeCN, 30 minutes and, (ii) 4(5)-iodoimidazole, 24 hours, rt; *b*. NaOMe, MeOh, four hours, rt; *c*. TBDMSCl, imidazole, DMF, 18 hours.

toluoyl groups present [9]. Through previous attempts, it was already known that the imidazole moiety would not form a stable organometallic complex [7, 8]; therefore, the pyrimidine was chosen as the most suitable organometallic coupling partner.

The xanthosine fleximer **11.5** was finally achieved via Suzuki–Miyaura coupling of **10.4** and **11.3** (Scheme 8.11), the latter of which was constructed in three steps from barbituric acid [9].

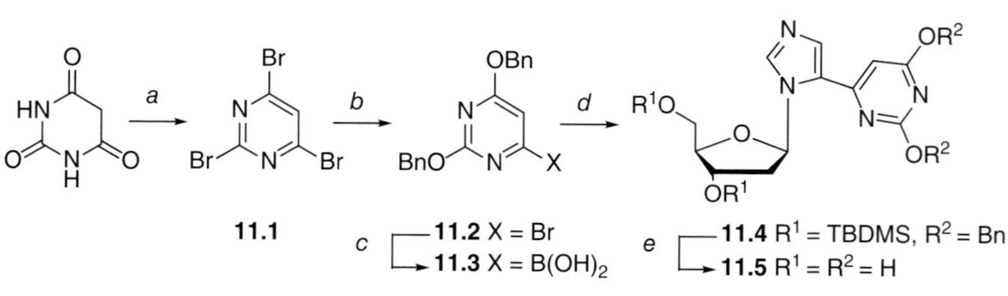

Scheme 8.11 Synthesis of 2′-deoxyxanthosine distal fleximer [9]. Reagents and conditions: *a*. $POBr_3$, *N,N*-dimethylaniline, toluene, reflux, three hours (97%); *b*. BnOH, NaOBn, benzene, rt, 18 hours (79%); *c*. (i) triisopropylborate, THF/toluene (1 : 4), −78 °C, (ii) *n*-BuLi, 18 hours, and (iii) HCl; *d*. (i) **10.4**, $Pd(PPh_3)_4$, DME, 10 minutes and (ii) $NaHCO_3$, reflux, four hours (75%); *e*. (i) ammonium formate, Pd/C, EtOH, reflux, 18 hours and (ii) TBAF, THF (53%).

8.1.3 2′-Modified Distal Fleximers

In addition to the 2′-deoxy series, substitutions at the 2′-position have become a popular modification method given the success of FDA-approved therapeutics such as Sofosbuvir [26, 40–43]. As such, the guanosine fleximers of Sofosbuvir (CH_3 "up," F "down") as well as its precursor (OH "up," CH_3 "down") were pursued. The initial attempt took the route previously tried with the 2′-deoxy Barton deoxygenation reaction presented in Scheme 8.4 [26, 42]. The TIPDS protecting group was installed onto the diiodoimidazole nucleoside intermediate **3.2** such that the 2′-OH was left unprotected and could then be oxidized using pyridinium chlorochromate (PCC). Methylation using trimethylaluminum yielded the arabinose analogue **12.3** (confirmed through NOESY). Construction of the tricyclic base utilized the same reagents and conditions found in Schemes 8.8 and 8.9 to generate the tricyclic guanosine (**12.6**) and xanthosine (**12.7**) derivatives. Unfortunately, both compounds were subsequently determined to be inactive against HCV, and due to the extremely low yields, the project was temporarily abandoned [26].

An attempt to construct a fleximer with the same modified sugar as Sofosbuvir (CH_3 "up," F "down") was also pursued [40, 41, 44]. As the previous procedure was not facile or useful (Scheme 8.12), a route starting with D-mannitose was tried (Scheme 8.13) [44]. Commercially available D-mannitol was converted to the ribolactone **13.1** in nine steps, which was then benzoyl protected at the 3′ and 5′ positions prior to reduction of the anomeric oxygen [40, 44]. Unfortunately, stereospecific orientation of the anomeric OH could not be controlled via various reducing reagents, and subsequent Mitsunobu coupling produced mainly the α-isomer instead of the desired β-isomer (**13.4**), leading to significantly low yields [40]. The guanosine fleximer **13.7** was finally obtained using reagents and conditions used in Schemes 8.3 and 8.8 [40]. The McGuigan ProTide of **13.7** was also pursued (**13.8**) in an effort to carry out a comparison to Sofosbuvir, which is a McGuigan prodrug. Unfortunately, like their predecessors, these analogues showed only minor activity against HCV (EC_{50} = 63.1 and 50.6 μM for the nucleoside and prodrug, respectively, unpublished) and the yields were once again extremely low (0.09% overall yield).

8.2 Proximal Fleximers

8.2.1 Ribose Proximal Fleximers

The ribose proximal fleximer derivatives were obtained by a much more facile route using organometallic coupling methodologies since the desired substitution patterns on the heterocycles were either commercially available or easily achievable. The desired C—C bond between the two aromatic moieties of the fleximer base could be constructed through palladium-catalyzed cross-coupling methodologies (Stille, Suzuki–Miyaura, Kumada, or Negishi), avoiding the multistep linear synthesis observed in the distal fleximer section. However, initial attempts were surprisingly unproductive, so extensive studies were undertaken

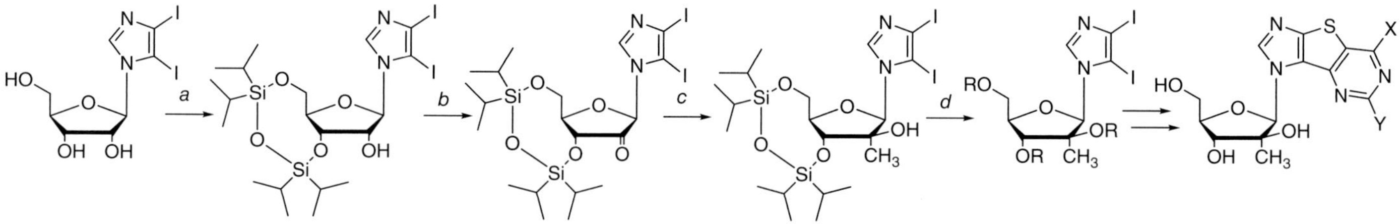

3.2 12.1 12.2 12.3 *e* 12.4 R = H → 12.5 R = Bn **12.6** X = OH, Y = NH_2 (10%, seven steps) **12.7** X = OH, Y = OH (9%, seven steps)

Scheme 8.12 Synthesis of 2′-OH/CH_3 tricyclic nucleosides [26]. Reagents and conditions: *a*. 1,3-dichloro-1,1,3,3-tetraisopropyldisiloxane, pyridine, rt, 14 hours (81%); *b*. PCC, CH_2Cl_2, 4 Å molecular sieves, rt, 12 hours (67%); *c*. $(CH_3)_3Al$, CH_2Cl_2, 0 °C to rt, two hours (73%); *d*. TBAF, THF, rt, five hours (76%); *e*. (i) NaH, THF, rt, three hours and (ii) TBAI, BnBr, rt, 18 hours (65%).

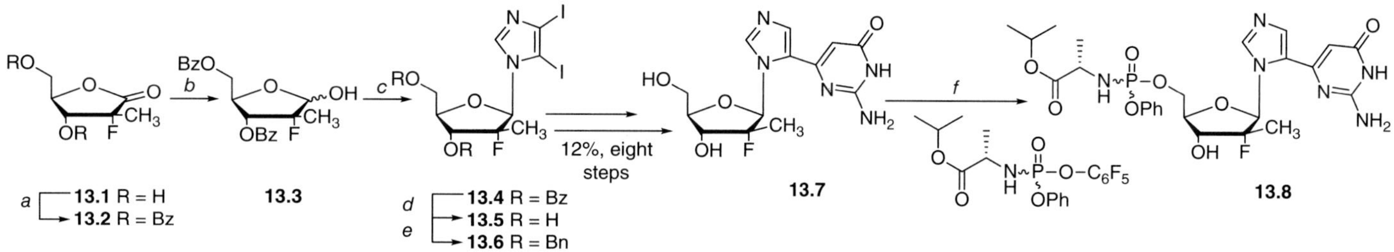

Scheme 8.13 Synthesis of 2′-CH_3/F Flex-G and its McGuigan Protide [40]. Reagents and conditions: *a*. BzCl, pyridine, 0 °C, 10 minutes, rt, 30 minutes (80%); *b*. $LiAl(OtBu)_3H$, THF, −20 °C, six hours (85%); *c*. 4,5-diiodoimidazole, Ph_3P, DIAD, THF, 0 °C to rt, 24 hours (20%); *d*. MeONa, MeOH, 0 °C, one hour (85%); *e*. (i) NaH, DMF, 0 °C, one hour and (ii) TBAI, BnBr, rt, four hours (81%); *f*. *t*-BuMgCl, DMF, 0 °C to rt, 18 hours.

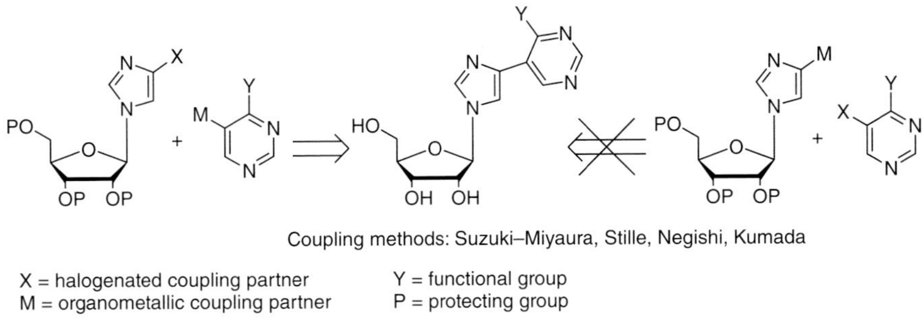

Figure 8.6 Retrosynthetic outline for synthesizing proximal analogues.

using a number of those approaches before suitable coupling conditions were ultimately found.

Since the 4-iodoimidazole nucleoside could be obtained easily through various dehalogenation techniques mentioned previously, that approach was straightforward. From there, however, a decision had to be made as to which aromatic moiety would be the organometallic partner and which would be the halogenated coupling partner (Figure 8.6). Initially, attempts focused on installing the organometallic moieties onto the imidazole ring, which proved highly ineffective. The best yields occurred after forming the organostannane complex with the imidazole (<5%) using *n*-BuLi in the presence of tributyltin chloride; however, the major product was the completely deiodinated imidazole nucleoside. For forming the organoborate, Grignard and organozinc reagents were also attempted; however, the former was never generated and while the latter was shown to have formed in situ, it did not yield the correct product during the coupling step with the halopyrimidine. Because of these challenges, the pyrimidine was selected as the organometallic coupling partner.

The first attempt to synthesize the guanosine proximal fleximer started with bis-1,3-methyl protected 5-iodouracil (**14.1**, Scheme 8.14). The 5-tributylstannane was then installed using palladium catalysis to give 5-tri-*n*-butylstannyl-1,3-dimethylpyrimidine **14.2**. It is important to note that the synthesis of the corresponding metalated pyrimidines applying Kumada and Negishi methodogies were also attempted, although only the stannyl pyrimidine was successfully made [8].

Unfortunately, all attempts at employing palladium-catalyzed coupling proved ineffective despite using various palladium catalysts and conditions [8]. Instead, adding excess cuprous thiophene-2-carboxylate (CuTC) ultimately produced **14.3** in a 20% yield [8]. Synthesis of the desired protected guanosine proximal fleximer **14.4** was then attempted by refluxing **14.3** with guanidine and sodium ethoxide in ethanol, although only a trace amount of the product was found (identified through mass spectrometry) [8].

The guanosine proximal fleximer was finally achieved via Suzuki–Miyaura coupling with 5-(dihydroxyboryl)-2,4-bis(benzyloxy)pyrimidine **15.3** and **13.1** (Scheme 8.15) [8]; a procedure described by Schinazi and Prusoff showed that slow, dropwise addition of *n*-BuLi to an aryl halide and alkyl borate provided excellent yield of the boronic acid [45]. The di-OBn intermediate **15.4** was

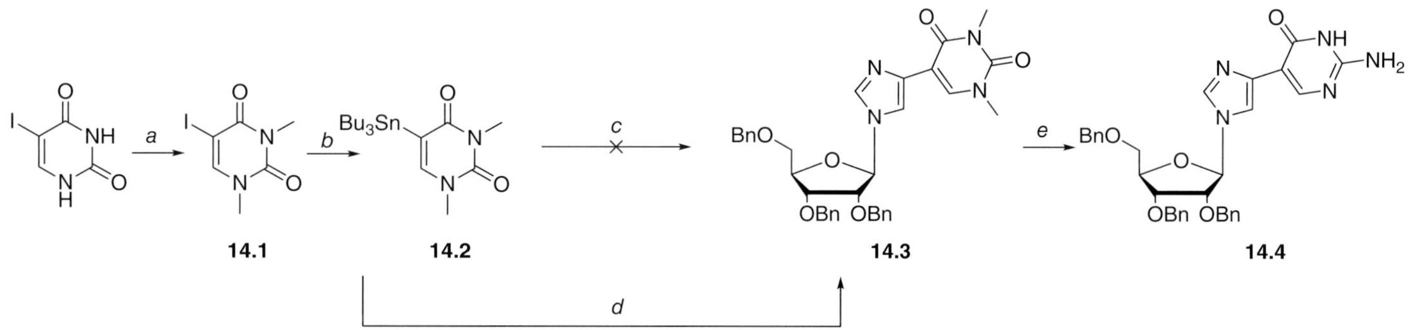

Scheme 8.14 First synthesis of the guanosine proximal fleximer [8]. Reagents and conditions: *a*. $(CH_3)_2SO_4$, H_2O, reflux, two hours (80%); *b*. $(SnBu_3)_2$, $PdCl_2(PPh_3)_2$, toluene, 90 °C, four hours (62%); *c*. **7.1**, Pd catalysts; *d*. **7.1**, CuTC, NMP, rt, 18 hours (20%); *e*. guanidine, NaOEt, EtOH, reflux, 24 hours (trace).

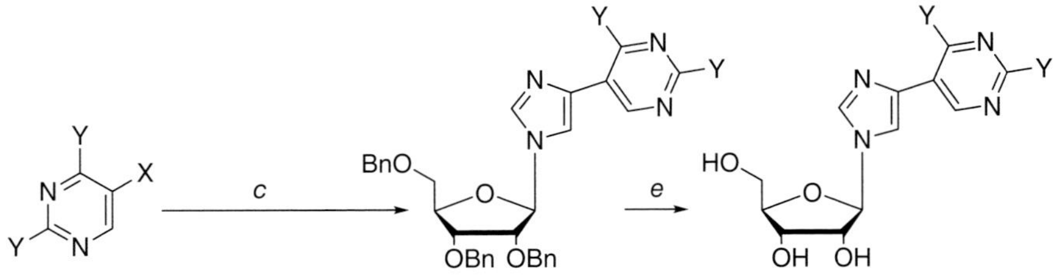

a: 15.1 X = Br, Y = Cl
b: 15.2 X = Br, Y = OBn
15.3 X = B(OH)$_2$, Y = OBn

d: 15.4 Y = OBn
15.5 Y = NH$_2$

15.6 Y = OH (88%)
15.7 Y = NH$_2$ (90%)

Scheme 8.15 Synthesis of xanthosine and 2,6-diaminopurine proximal fleximers [8]. Reagents and conditions: *a*. NaH, BnOH, toluene, rt, 18 hours (91%); *b*. (i) triisopropylborate, THF/toluene (1 : 4), −78 °C, (ii) *n*-BuLi, rt, 18 hours, and (iii) HCl (95%); *c*. (i) **13.1**, Pd(PPh$_3$)$_4$, DME, 10 minutes and (ii) NaHCO$_3$, reflux, four hours (67%); *d*. NH$_3$, CH$_3$OH, 170 °C, 96 hours (40%); *e*. Pd/C, ammonium formate, EtOH, reflux, two hours.

converted to the diamino intermediate **15.5**, and deprotection of **15.4** and **15.5** yielded the xanthosine and 2,6-diaminopurine proximal fleximers **15.6** and **15.7**, respectively [8].

As seen in Scheme 8.16, the isoguanosine fleximer **16.1** was obtained by treating **15.5** with sodium nitrite, which selectively converted the C-2 pyrimidine amine to a ketone [8]. The guanosine fleximer **16.2** was then obtained by refluxing aqueous sodium bisulfite with **15.7** [8].

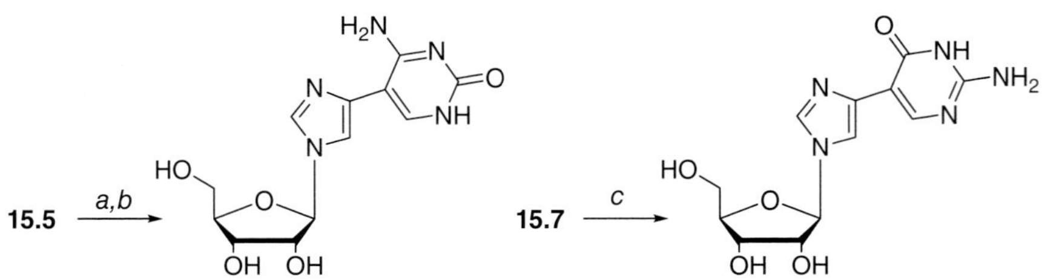

16.1 16.2

Scheme 8.16 Synthesis of isoguanosine and guanosine proximal fleximers [8]. Reagents and conditions: *a*. NaNO$_2$, AcOH, H$_2$O/THF (1 : 1), 60 °C, two hours; *b*. Pd/C, ammonium formate, EtOH, reflux, two hours (41%, two steps); *c*. NaHSO$_3$, H$_2$O, 60 °C, 10 hours (88%).

Since the proximal adenosine and hypoxanthine fleximer was unable to be realized through palladium-catalyzed coupling, the pyrimidine moiety was generated through a cyclization methodology (Scheme 8.17) [7]. Commercially available histidine monohydrate was treated with sodium hypochlorite to give 4(5)-cyanomethylimidazole **17.1** [7]. Interestingly, only Clorox® bleach consistently produced the desired product whereas solutions of sodium hypochlorite obtained from commercially available chemical suppliers did not, nor did various "house brand" bleaches. These reactions led to poor yields or no reaction at all. Speculation was that the house brands might be more dilute thereby leading to either poor yields or lack of reaction, although this does not explain the

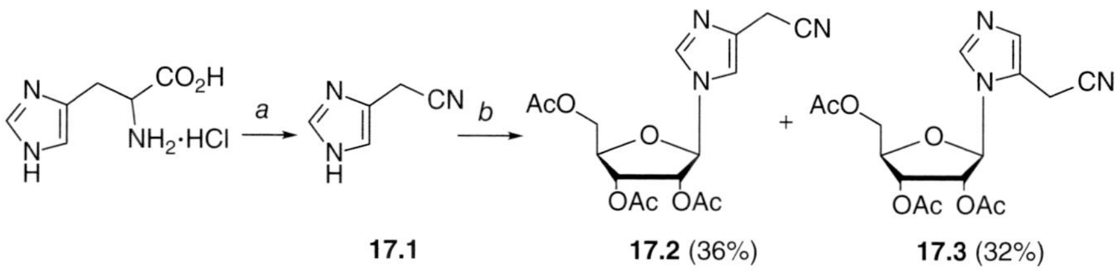

Scheme 8.17 Synthesis of 4- or 5-cyanomethylimidazole nucleoside from histidine [7]. Reagents and conditions: *a*. NaOCl, H_2O, 10–20 °C, three hours, then rt, 18 hours (70%); *b*. (i) β-d-ribofuranose-1,2,3,5-*O*-tetraacetate, BSA, ACN, three hours and (ii) TMSOTf, 60 °C, 18 hours.

lack of reaction with the commercially available solutions. Finally, Vorbrüggen coupling with β-D-ribofuranose-1,2,3,5-*O*-tetraacetate gave a mixture of the 4- and 5-cyanomethylimidazole nucleosides (**17.2** and **17.3**, respectively) with a nearly 1 : 1 ratio [7].

In situ formation of the proximal adenosine fleximer from **17.2** (Scheme 8.18) was accomplished via a [4+2] Diels–Alders cycloaddition using sodium methoxide and 1,3,5-triazine, which was followed by a retro-Diels–Alders fragmentation of the intermediate **18.1** that then formed the desired product **18.2** [7]. The acetate protecting groups were also removed during this reaction; however, this particular reaction also resulted in poor yield (15%, 3.8% overall from histidine) [7]. The hypoxanthine fleximer **18.3** was then synthesized from the adenosine analogues using standard diazotization conditions, followed by hydrolysis [7, 46].

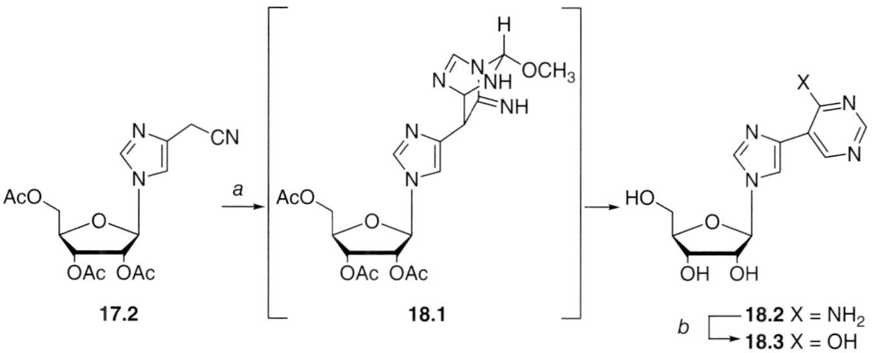

Scheme 8.18 Synthesis of the proximal adenosine fleximer through a Diels–Alders cycloaddition [7]. Reagents and conditions: *a*. 1,3,5-triazine, NaOMe, MeOH, 40 °C, 18 hours (15%); *b*. $NaNO_2$, AcOH, H_2O/THF (1 : 1), 80 °C, five hours (68%).

8.2.2 2′-Deoxyribose Proximal Fleximers

Schemes 8.20 and 8.21 outline the attempts made at synthesizing different pyrimidine organometallic complexes for use in either Suzuki–Miyaura or Stille coupling [9]. Prior to constructing the organoborate or stannyl pyrimidine, halogens were installed at the C-5 position of the pyrimidines if starting materials could not be purchased commercially. The halogenation reactions proved to be straightforward as the C-5 position is the least electron deficient

[47], which was also synergistically more reactive due to by the presence of electron-donating substituents at the C-2 and C-4 [46]. Organoborate synthesis similar to Scheme 8.16 was then attempted on several 2,4-subsitituted 5-halopyrimidnes (Scheme 8.19); however, unlike **15.3**, all efforts to isolate and characterize these boronic acids failed [9]. The purification step may have removed the -$B(OH)_2$ substituent as Suzuki–Miyaura coupling using unpurified boronic acids did yield the correct products, although resulting in low yields and leading to difficulties with purification.

Synthesis of the 5-stannyl pyrimidines proved to be more facile and with reasonable yields (Scheme 8.20) [9]. Once the halogens were installed, the tributylstannyl group was successfully substituted via palladium catalysis [9, 48].

Palladium-catalyzed cross-coupling was then possible once the pyrimidine organometal complexes were achieved [9]. In that regard, 4,5-diiodoimidazole was coupled to the Hoffer's chlorosugar **7.3** in the same fashion as seen in Scheme 8.8 (Scheme 8.21) [9, 27]. Once again, the toluoyl protecting groups of the 2′-deoxyribose imidazole nucleoside **21.1** were replaced with benzyl protecting groups prior to removal of the C-5 iodine using Grignard conditions (**21.4**) such that palladium-catalyzed coupling could then occur to achieve the proximal analogues [9].

Through conventional Suzuki and Stille heteroaryl cross-coupling methods (Scheme 8.22), followed by the deprotection steps, the guanosine (**22.3**), adenosine (**22.5**), hypoxanthosine (**22.7**), xanthosine (**22.9**), and 2,6-diaminopurine (**22.12**) proximal fleximers were then obtained [9].

8.2.3 Carbocyclic Proximal Fleximers

Carbocyclic nucleosides are a naturally occurring type of sugar- modified nucleoside wherein the furanose oxygen is replaced with a carbon [49]. For example, Neplanocin A (NpcA) and Aristeromycin (Ari) are both naturally occurring carbocyclic adenosine analogues that have exhibited potent antiviral, antiparasitic, and anticancer properties, although both are associated with significant cytotoxicity due to their ability to be converted to their corresponding triphosphate nucleotides by adenosine kinase [49]. The carbocyclic triphosphates closely mimic ATP in ATP-recognizing enzymes, and are also metabolized by adenosine deaminase, leading to toxic metabolites [50–56]. Because of the potential of this class of nucleosides, particularly given their activity against viruses such as Ebola, the Seley-Radtke laboratory was interested in exploring carbocyclic fleximers. Thus, a series of proximal carbocyclic 5′-truncated deazaneplanocin fleximers were constructed (Scheme 8.23) [57]. The truncated forms of both compounds, wherein the 4′-CH_2OH is removed, drastically lowers the cytotoxicity because these analogues cannot be phosphorylated [58]. In addition, it was found that the truncated 3-deaza analogues of NpcA and Ari were more effective inhibitors than their parent compounds [56, 59–61].

To construct the Npc sugar, cyclopentanone **23.1** was synthesized from D-ribose [62], and using a stereospecific reduction of the ketone via Luche conditions provided the "α" oriented allyl alcohol **23.2** [57]. Mitsunobu coupling of 4,5-diiodoimidazole with the sugar gave the desired β-product as inversion at

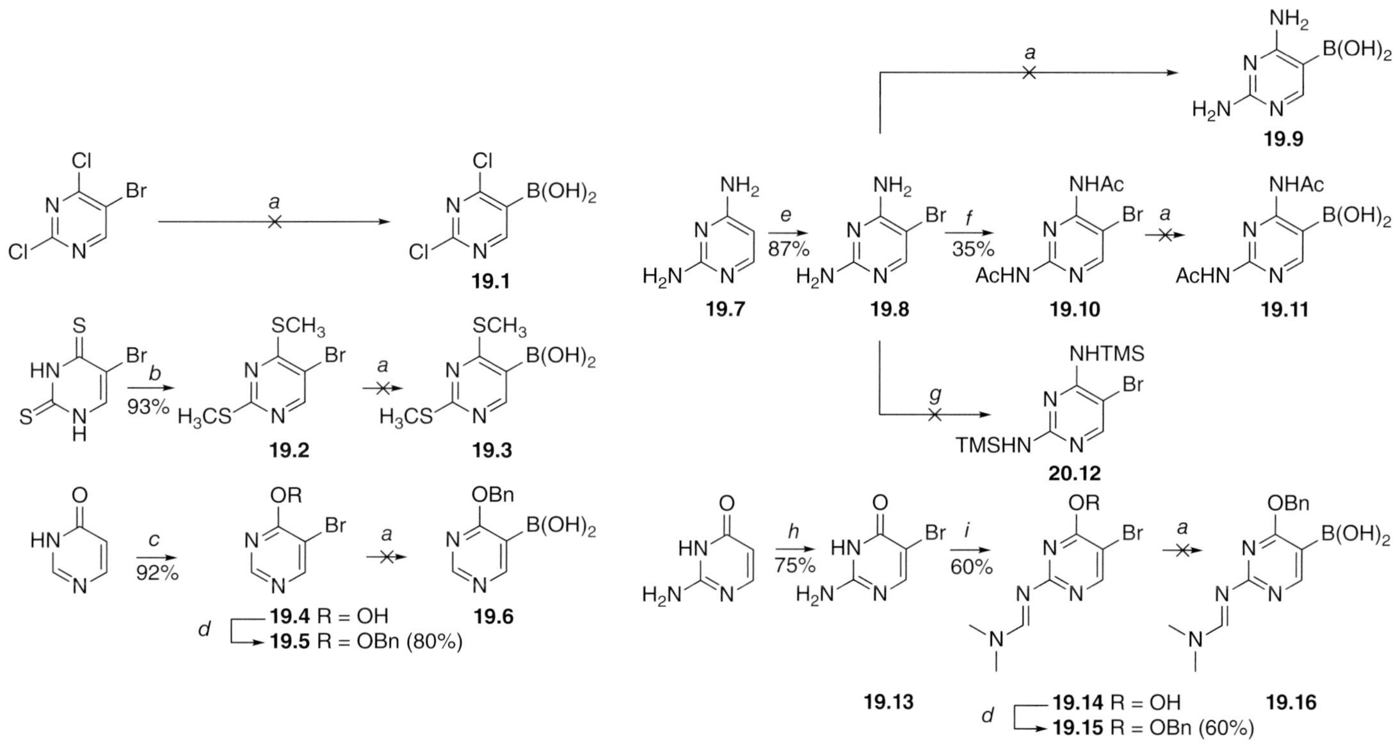

Scheme 8.19 Synthesis of boronic acid pyrimidine precursors [9]. Reagents and conditions: *a*. (i) triisopropylborate, THF/toluene (1 : 4), −78 °C, (ii) *n*-BuLi, rt, 18 hours, and (iii) HCl; *b*. NaOH, CH_3I, H_2O, 0 °C, four hours; *c*. Br_2, $CHCl_3$, MeOH, 12 hours; *d*. DIAD, PPh_3, BnOH, DMF, 1.5 hours; *e*. (i) NBS, $CHCl_3$, rt, 12 hours and (ii) CH_2Cl_2, 1 M NaOH; *f*. AcCl, pyridine, 0 °C, two hours; *g*. TMSCl, pyridine; *h*. (i) Br_2, H_2O, rt, two hours and (ii) 0 °C, 18 hours; *i*. DMF–DMAc, DMF, rt, 18 hours.

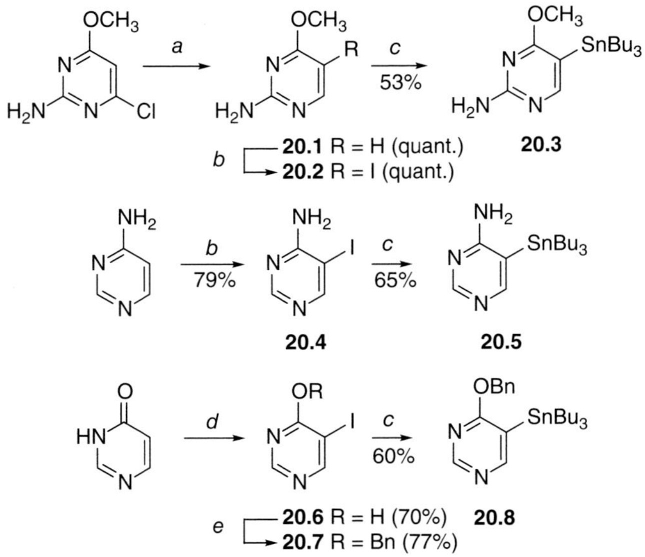

Scheme 8.20 Synthesis of tributylstannyl pyrimidine precursors [9]. Reagents and conditions: *a*. H_2, Pd/C, DIPEA, EtOAc, EtOH, rt, three hours; *b*. NIS, AcOH, 80 °C, two hours; *c*. $(SnBu_3)_2$, $Pd_2dba_3 \cdot CHCl_3$, 65 °C, three hours; *d*. NaOH, I_2, 80 °C, 18 hours; *e*. DIAD, PPh_3, BnOH, DMF, 1.5 hours.

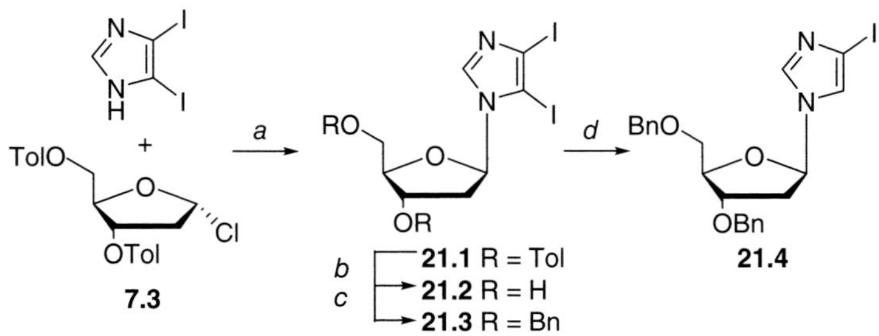

Scheme 8.21 Synthesis of 4-iodoimidazole 2′-deoxynucleoside precursor [9, 27]. Reagents and conditions: *a*. (i) NaH, CH_3CN, rt, 30 minutes and (ii) 4,5-diiodo-imidazole, rt, 24 hours, rt.; *b*. NaOMe, MeOH, four hours, rt; *c*. (i) NaH, THF, rt, three hours and (ii) TBAI, BnBr, rt, 18 hours (85%); *d*. EtMgBr, THF, 0 °C, three hours (80%).

the anomeric site takes place [57]. The C-5 iodine was removed through Grignard conditions, and subsequent Stille coupling using 2-chloro-3-stannylpyridine produced the 6-chloro-3-deazapurine analogue **23.5**, which was then converted to the 3-deazaadenosine and 3-deazahypoxanthosine derivatives [57]. Products **23.8–23.10** were achieved following isopropylidene deprotection [30, 57].

8.2.4 Proximal Fleximers from Other Groups

In addition to our efforts, other groups have now begun to explore the fleximer scaffold given their biological potential. For example, Sylvie Pochet's group in France has studied whether the fleximer bases could be coupled to either ribose

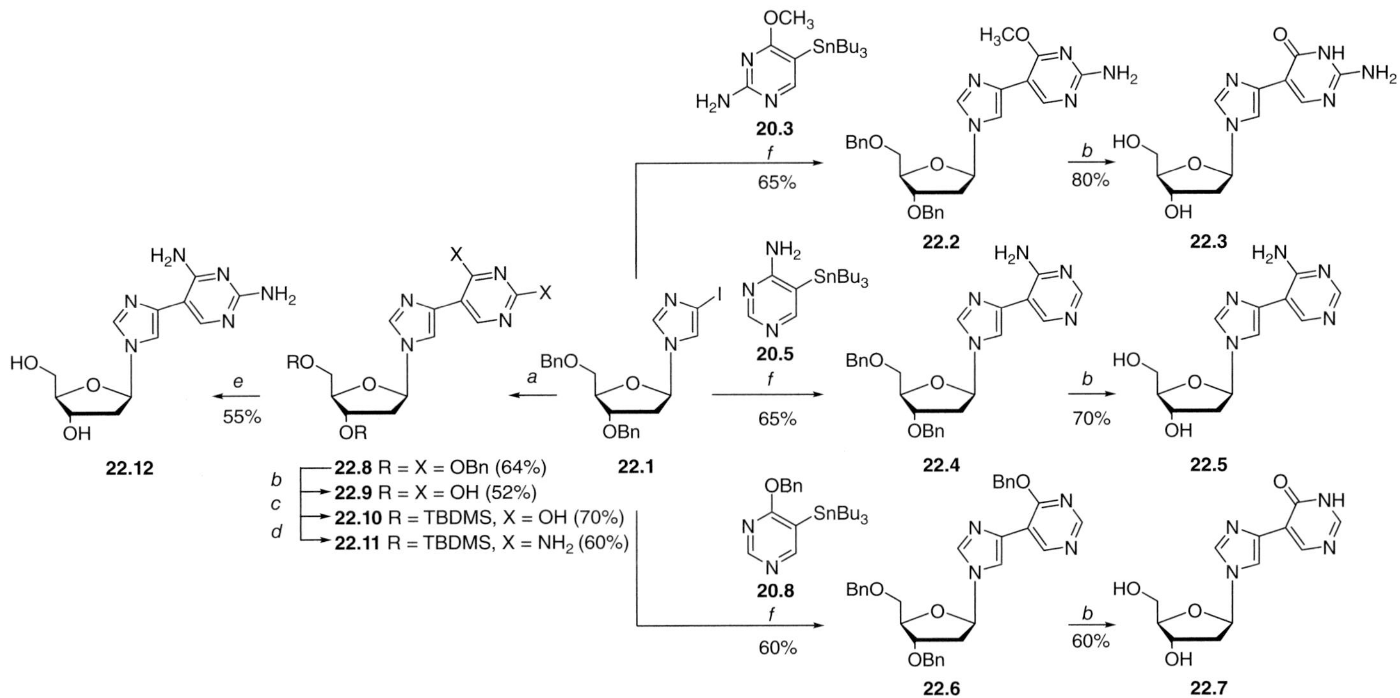

Scheme 8.22 Suzuki–Miyaura and Stille couplings to achieve proximal fleximers guanosine, adenosine, hypoxanthosin, and xanthosine [9]. Reagents and conditions: *a*. **15.3**, $Pd(PPh_3)_4$, DME, $NaHCO_3$, reflux, four hours; *b*. ammonium formate, Pd/C, reflux, 18 hours; *c*. TBSMSCl, imidazole, DMF, rt, 18 hours; *d*. (i) TBSCl, DMAP, TEA, CH_3CN, rt, 18 hours and (ii) NH_3, THF, 110 °C, 72 hours; *e*. TBAF, THF, rt, 18 hours; *f*. $Pd_2dba_3{\cdot}CHCl_3$, DMF, 100 °C, 18 hours.

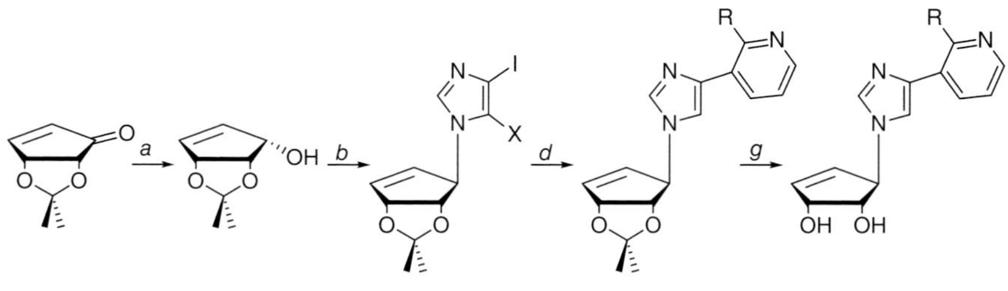

23.1 23.2 c 23.3 X = I → 23.4 X = H e 23.5 R = Cl → 23.6 R = NH2, 23.7 R = OH f 23.8 R = Cl (86%), 23.9 R = NH2 (46%, two steps), 23.10 R = OH (43%)

Scheme 8.23 Synthesis of 5′-truncated 1-deazaneplanocin fleximer derivatives [57]. Reagents and conditions: *a*. $NaBH_4$, MeOH, $CeCl_3{\cdot}7H_2O$, rt, one hour (quant.); *b*. 4,5-diiodoimidazole, DPPE, DIAD, THF, rt, 18 hours (41%); *c*. EtMgBr, THF, 0 °C, one hour (85%); *d*. 2-chloro-3-stannylpyridine, $Pd(PPh_3)_4$, CuBr, 1,4-dioxane, 120 °C, 12 hours (64%); *e*. (i) hydrazine, 80 °C, one hour and (ii) $TiCl_3$, THF, 70 °C, four hours; *f*. AcOH, 120 °C (56%); *g*. TFA/H_2O/THF (1 : 1 : 5), 18 hours.

or 2′-deoxyribose through enzymatic transglycosylation rather than synthetic coupling methodologies in an effort to shorten the synthetic approaches [63, 64]. Through a microwave-assisted Suzuki–Miyaura cross-coupling procedure [65], various 4-(hetero)aryl-1*H*-imidazoles were first constructed, four of which resemble the Seley-Radtke proximal fleximers (Scheme 8.24).

Enzymatic transglycosylation was then accomplished using *N*-deoxyribosyltransferase II of *Lactobacillus leichmannii* (*Ll*-NDT) and *Escherichia coli* purine nucleoside phosphorylase (PNP). They determined the relative substrate selectivity as well as transglycosylation efficiency and regioselectivity of N1 versus N3 substitution (imidazole numbering) [63, 64]. Results in Table 8.2 show that both enzymes could utilize all four fleximer bases as substrates; however, *Ll*-NDT proved to be the more efficient enzyme with better yields. Surprisingly, products from both N1 (**24.7**, the thermodynamic product) and N3 (**24.11**, the kinetic product) substitution were observed for *Ll*-NDT with the dimethoxy analogue **24.3** at two hours, although only the proximal fleximer **24.7** was observed at the conclusion of the reaction. Also, the *t*-butyl group of compound **24.8** after 16 hours and a double-glycosylated product **24.16** were observed [63–65].

Hudson's group has also pursued several fleximers (Scheme 8.25) [67]. At that time, the Hudson's group was attempting to derivatize 5-ethynyl-2′-deoxycytidines with various azides via a copper-catalyzed azide-alkyne Huisgen cycloaddition (CuAAC) to form 5-triazoylcytidine analogues [68]. A similar approach using β-azidoribose [66] and 5-ethynyl-pyrimidines yielded "click" fleximers (**25.6–25.9**) where the imidazole was replaced with a triazole. Thus, the desired 5-(alkyn-1-yl)pyrimidine analogues were obtained in reasonable yields via Sonogashira coupling of 5-iodopyrimidines and TMS-acetylene, followed by removal of the TMS group [69–72]. The β-azidoribose was achieved through a Vorbrüggen coupling between azidotrimethylsilane and 1-*O*-acetyl-2,3,5-tri-*O*-benzoyl-β-D-ribofuranose [66]. CuAAC between the

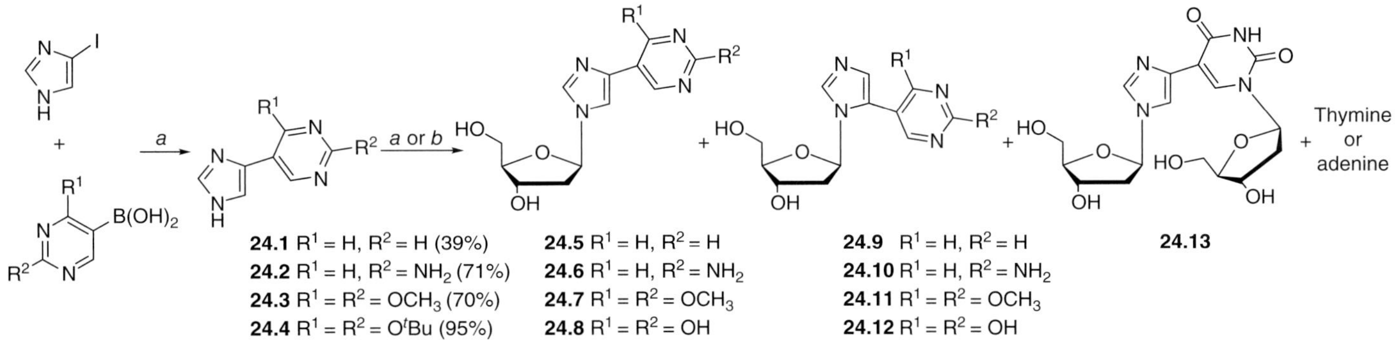

24.1 $R^1 = H, R^2 = H$ (39%)
24.2 $R^1 = H, R^2 = NH_2$ (71%)
24.3 $R^1 = R^2 = OCH_3$ (70%)
24.4 $R^1 = R^2 = O^tBu$ (95%)

24.5 $R^1 = H, R^2 = H$
24.6 $R^1 = H, R^2 = NH_2$
24.7 $R^1 = R^2 = OCH_3$
24.8 $R^1 = R^2 = OH$

24.9 $R^1 = H, R^2 = H$
24.10 $R^1 = H, R^2 = NH_2$
24.11 $R^1 = R^2 = OCH_3$
24.12 $R^1 = R^2 = OH$

24.13

Scheme 8.24 Transglycosylation products [63, 65]. Reagents and conditions: *a*. $Pd(PPh_3)_4$, Na_2CO_3, H_2O, DMF, 110–120 °C microwave, one to two hours; *b*. thymidine, L/-NDT, citrate buffer pH 6.5, 50 °C; *c*. adenosine, *E. coli* PNP, phosphate buffer pH 7.4, 50 °C.

Table 8.2 Fleximer products via transglycosylation through Ll-NDT or PNP and their yield percent [63, 65].

Entry	Base	Enzyme	Product	% Conversion after 2 h[a)]	% Conversion after 18 h[a)]	% Final conversion[a)]	Yield (%)[b)]
1	**24.1**	*Ll*-NDT	**24.5**	31	nd	95	60
2	**24.2**	*Ll*-NDT	**24.6**	30	nd	91	50
3	**24.3**	*Ll*-NDT	**24.7/24.11**	20/24	nd	0/90	0/82
4	**24.4**	*Ll*-NDT	**24.8/24.16**	7	nd	70	17/37
5	**24.1**	PNP	**24.5**	nd	8	76	43
6	**24.2**	PNP	**24.6**	nd	nd	nd	51
7	**24.3**	PNP	**24.7**	nd	40	67	51
8	**24.4**	PNP	**24.8**	nd	nd	nd	12

nd, Not determined because of the partial solubility of acceptors in the media

a) % Conversions were determined by reverse-phase HPLC analysis of an aliquot of the incubation mixtures monitored at 254 nm with a diode array detector.

b) Values refer to isolated yields after reverse phase HPLC purification of the incubation mixtures for specified times and amounts of enzyme.

azide and alkyne produced the desired benzoyl protected click-fleximers; however, the conditions required to produce these analogues were somewhat harsh and lengthy reflux was required whereas normal CuAAC conditions typically require only room temperature stirring [73].

8.3 "Reverse" Fleximers

Another series of fleximers that the Seley-Radtke group pursued was the "reverse" fleximer series. These 5-aryl-substituted uracil analogues can also be viewed as N-3- substituted purine analogues. While Herdewijn and coworkers focused on the modified ribose derivatives, and many of them showed interesting anti-HSV and VZV activities [74–77], the Seley-Radtke group pursued the corresponding carbocyclic uracil-substituted, or fleximer isoxanthosine, analogues (Figure 8.7) [12, 13]. Previous studies on these isoanalogues determined that substituting a thiazole for the imidazole moiety of the purine scaffold imparted excellent SAHase inhibition [78], and since interesting activity was observed for Herdewijn's ribose derivatives, it was decided to couple the corresponding reverse fleximer bases to several carbocyclic sugars that had been previously studied by the Seley-Radtke group and others (Scheme 8.23) [12, 13, 57, 78, 79].

In contrast to imidazole, other 5-membered heteroaryl groups were able to act as the organometallic partner. As shown in Scheme 8.26, facile construction of 2-(tributylstannyl)thiophene **26.1** and furan **26.2** was accomplished by treating either thiophene or furan with *n*-BuLi and quenching with tributyltin chloride [80]. The thiazole required more manipulation, thus the C-2 was first protected with TMS (**26.3**) so the stannyl could be installed at the C-5 position (**26.4**) [81].

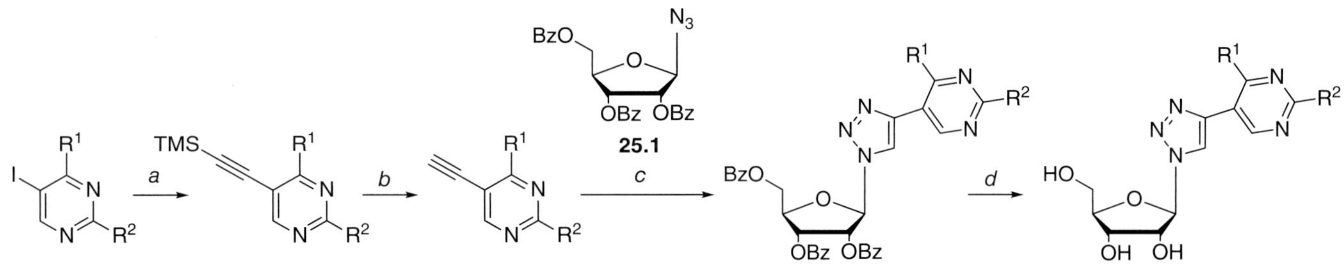

25.2 $R^1 = R^2 = OH$ (90%)
25.3 $R^1 = NH_2$, $R^2 = OH$ (80%)
25.4 $R^1 = NH_2$, $R^2 = H$ (77%)
25.5 $R^1 = R^2 = NH_2$ (quant.)

25.6 $R^1 = R^2 = OH$ (91%)
25.7 $R^1 = NH_2$, $R^2 = OH$ (47%)
25.8 $R^1 = NH_2$, $R^2 = H$ (58%)
25.9 $R^1 = R^2 = NH_2$ (90%)

Scheme 8.25 Robert Hudson's "click" fleximers [66]. Reagents and conditions: *a*. TMS-acetylene, $Pd(PPh_3)_4$, CuI, DMF or THF, NEt_3; *b*. (i) 1 M NaOH and (ii) HCl; *c*. CuI, DMEDA, THF, NEt_3, reflux; *d*. NH_3, MeOH, 0 °C to rt.

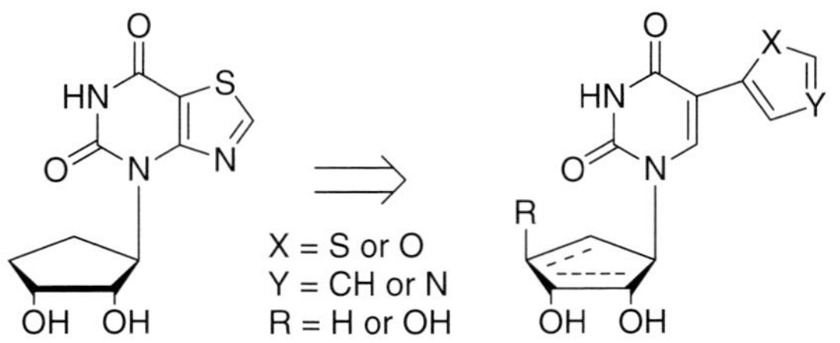

Isoxanthosine analogue

Figure 8.7 Retrosynthetic outline for synthesizing reverse fleximer analogues.

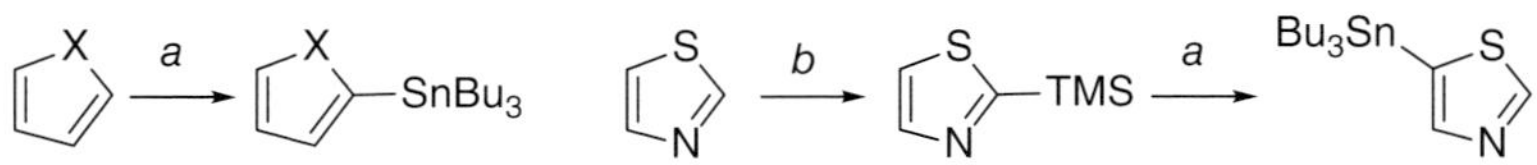

26.1 X = S (95%)
26.2 X = O (93%)
26.3
26.4 (87%), two steps

Scheme 8.26 Synthesis of thophene, furan, and thiazole organostannanes [12, 13]. Reagents and conditions: *a*. *n*-BuLi, THF, Bu_3SnCl, −78 °C; *b*. *n*-BuLi, THF, TMSCl, −78 °C.

The truncated neplanocin analogues were achieved through a Mitsunobu reaction (Scheme 8.27) using **23.2** and N^3-benzoyl-5-bromouracil in the presence of diisopropyl azodicarboxylate (DIAD) and 1,2-bis(diphenylphosphino)ethane (DPPE), instead of the traditional triphenylphosphine [13, 82]. Owing to the orientation of the alcohol, the pyrimidine was able to couple in the desired β-orientation (**27.1**) [62]. Stille coupling with the thiophene, furan, and thiazole (**26.1**, **26.2**, **26.4**), followed by deprotection in acidic conditions gave the carbocyclic reverse fleximers **27.3–27.5**. The Ari analogues were achieved in an analogous manner; however, it was observed that using N^3-benzoyluracil because the subsequent hydrogenation reduction of the double bond using palladium on carbon on the carbocycle would also remove the halogens. Therefore, the C-5 bromine of the pyrimidine was installed through *in situ* formation of azido bromide (**27.9**), and Stille coupling occurred in the same fashion as the truncated neplanocin analogues. Deprotection of the isopropylidene group occurred through standard aqueous acid hydrolysis [30].

Another potent group of carbocyclic nucleosides, which have an OH group directly attached at the 4′-position of the carbocyclic ring, are known as 5′-nor carbocyclic nucleosides [12, 82–93]. This modification also overcomes the toxicity seen with Npc and Ari as the OH group cannot reach the appropriate residue in the kinase binding site to be phosphorylated [49, 94]. Thus, the 5′-nor carbocyclic sugar **28.3** [95] was synthesized by first cracking dicyclopentadiene to obtain cyclopentadiene using previously reported methods [12]. Treatment with peracetic acid in the presence of sodium acetate and sodium bicarbonate yielded epoxide **28.1** [12, 95]. Reaction of the epoxide with acetic acid and triphenylphosphine yielded *meso*-diacetate **28.2** and subsequent resolution with *Pseudomonas cepacia* lipase stereospecifically yielded **28.3** [12]. Tsuji–Trost coupling with the uracil derivative [96] yielded the protected halogenated carbocyclic uridine **28.4** [12]. Deprotection of the N^3-benzoyl followed by Stille coupling in a manner identical to Scheme 8.28 yielded the desired products **28.6–28.8** [12, 30].

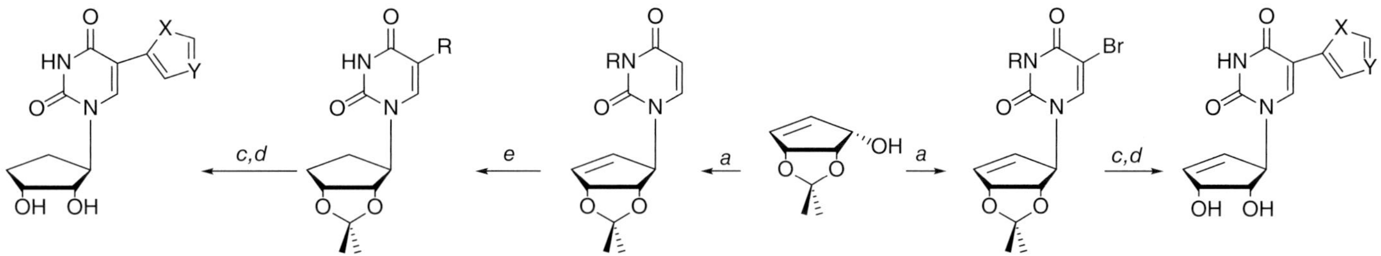

27.10 X = S, Y = CH (90%, two steps)
27.11 X = O, Y = CH (90%, two steps)
27.12 X = S, Y = N (89%, two steps)

f: **27.8** R = H (quant.) → **27.9** R = Br (48%)

b: **27.6** R = Bz (70%) → **27.7** R = H (99%)

23.2

b: **27.1** R = Bz (78%) → **27.2** R = H (62%)

27.3 X = S, Y = CH (55%, two steps)
27.4 X = O, Y = CH (58%, two steps)
27.5 X = S, Y = N (42%, two steps)

Scheme 8.27 Synthesis of neplanocin and aristeromycin reverse fleximers [13]. Reagents and conditions: *a*. N^3-benzoyluracil or N^3-benzoyl-5-bromouracil, DPPE, DIAD, CH_3CN, rt, 18 hours; *b*. NH_3, MeOH, rt, 12 hours; *c*. for thiophene, **26.1**, 1,4-dioxane, $PdCl_2(PPh_3)_2$, 90 °C, 18 hours, for furan, **26.2**, 1,4-dioxane, $PdCl_2(PPh_3)2$, 90 °C, 18 hours; for thiazole, **26.4**, THF, $Pd(PPh_3)_4$, reflux, 72 hours; *d*. TFA/H_2O (2 : 1), rt, six hours; *e*. Pd/C, MeOH, H_2, 25 psi; *f*. NBS, NaN_3, DME, H_2O, rt, 24 hours.

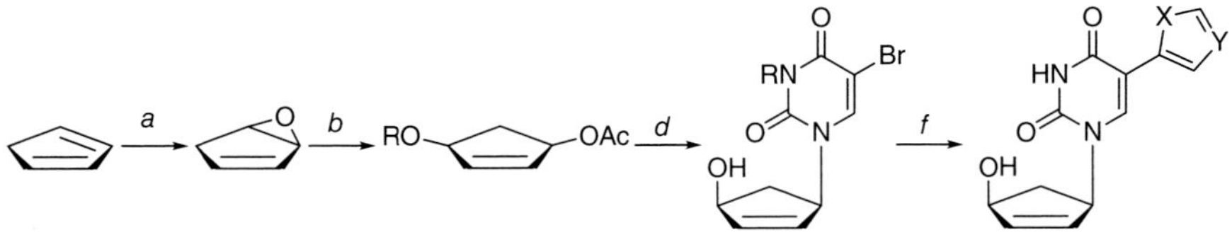

28.1

c: **28.2** R = Ac → **28.3** R = H

e: **28.4** R = Bz → **28.5** R = H

28.6 X = S, Y = CH (89%)
28.7 X = O, Y = CH (89%)
28.8 X = S, Y = N(68%)

Scheme 8.28 Carbocyclic 5′-nor "reverse" fleximers [12]. Reagents and conditions: *a*. peracetic acid, NaOAc, $NaHCO_3$, CH_2Cl_2, four hours; *b*. AcOH, $Pd(PPh_3)_4$, THF, 0 °C to rt, 18 hours (39%, two steps); *c*. *Pseudomonas cepacia* lipase, potassium phosphate buffer pH 7.4, acetone, 1 N NaOH, four hours, (49%); *d*. NaH, DMF, Pd_2dba_3, DPPP, N^3-benzoylated-5-bromouracil, 55 °C, 72 hours (31%); *e*. NH_3, MeOH, rt, three hours (77%); *f*. for thiophene, **26.1**, 1,4-dioxane, $PdCl_2(PPh_3)_2$, 120 °C, 18 hours; for furan, **26.2**, 1,4-dioxane, $PdCl_2(PPh_3)_2$, 120 °C, 18 hours; for thiazole, **26.4**, THF, $Pd(PPh_3)_4$, reflux, 72 hours.

Although some antiviral activities against human cytomegalovirus and varicella zoster virus were seen for these analogues, they were also associated with cytoxicity against all tested cell lines and thus were not pursued further [12].

8.4 Acyclic Fleximers

The most recent series of fleximers the Seley-Radtke group has pursued involves a doubly flexible scaffold obtained by combining the acyclic sugar of Acyclovir with the flex-base approach [10, 11]. As base flexibility had already demonstrated interesting results when coupled to ribose, it was perceived that the effect would be even more pronounced when coupled to an acyclic sugar. The proximal acyclic fleximers were chosen as the first series of doubly flexible targets (Schemes 8.29 and 8.30). In that regard, 2-aceteoxyethyl acetoxymethylether (**29.1**) [97] was

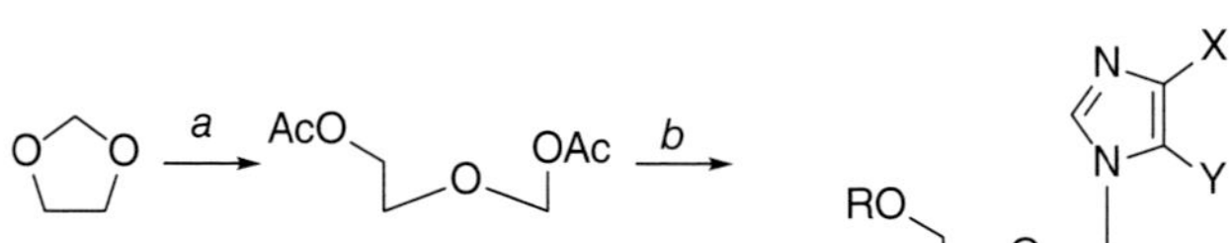

29.1

Pathway A

c, d, e: **29.2** R = Ac, X = Y = I → **29.3** R = H, X = Y = I → **29.4** R = Bn, X = Y = I → **29.5** R = Bn, X = I, Y = H

Pathway B

f, g: **29.2** R = Ac, X = Y = I → **29.6** R = H, X = I, Y = H → **29.7** R = Ac, X = I, Y = H

Scheme 8.29 Synthesis of acyclic halogenated imidazole nucleosides [10, 11]. Reagents and conditions: *a*. Ac_2O, H_2SO_4, −15 °C to rt, 18 hours (52%); *b*. 4,5-diiodoimidazole, BSA, TMSOTf, CH_3CN, rt to 90 °C, 18 hours (47%); *c*. NH_4OH, EtOH, rt, 16 hours (74%); *d*. (i) NaH, THF, rt, three hours and (ii) TBAI, BnBr, rt, 12 hours (70%); *e*. EtMgBr, THF, −15 °C to rt, four hours (78%); *f*. Na_2SO_3, 30% EtOH/H_2O, 120 °C, 18 hours (84%); *g*. Ac_2O, TEA, DMAP, CH_2Cl_2, rt, two hours (97%).

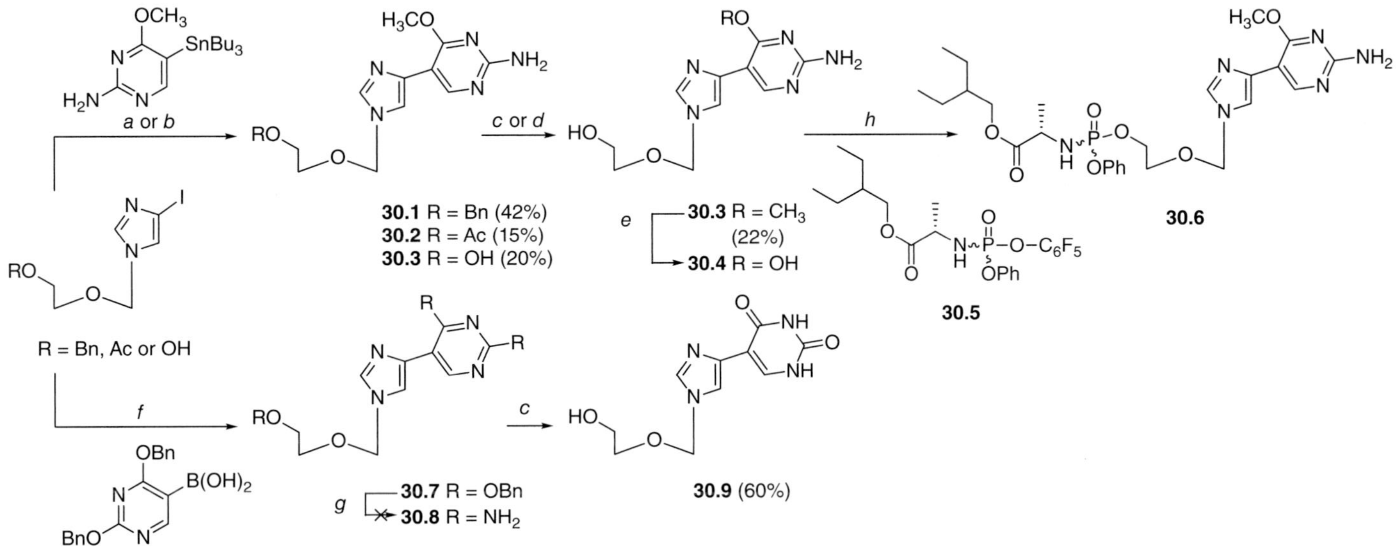

Scheme 8.30 Acyclovir fleximers' construction through palladium- catalyzed cross-coupling reactions [10, 11]. Reagents and conditions: *a*. for Bn, $Pd(PPh_3)_4$, CuI, TBAF, DMF, 45 °C, 18 hours (42%); for Ac and OH, *b*. Pd_2dba_3·$CHCl_3$, PPh_3, CuI, TBAF, DMF, 50 °C, 18 hours (20%); *c*. for Bn, Pd/C, ammonium formate, EtOH, 120 °C, 18 hours; *d*. for Ac, NH_4OH, EtOH, rt, 18 hours; *e*. BBr_3, CH_2Cl_2, −15 °C to rt, 18 hours (49%); *f*. $Pd(PPh_3)_4$, $NaHCO_3$, 1,2-dimethoxyethane, reflux, four hours (21%); *g*. NH_3, MeOH, 210 °C; *h*. **30.5**, *t*-BuMgCl, THF, rt, 2.5 hours.

Table 8.3 Antiviral activites of doubly flexible fleximers (EC_{50} in μM) [10, 11].

Compound	MERS	SARS	SUDV	MARV	EBOV	DENV	ZIKV	YFV
30.2	5.3 ± 0.7	11.9 ± 0.2	20 ± 10	70 ± 27	nd[a)]	>100	>100	68
30.3	10.1 ± 1.2	28.1 ± 0.2	>100	>100	2.2 ± 0.3	32	>100	32
30.6	nd[a)]	nd[a)]	7 ± 2	62 ± 13	27.2 ± 2.2	27	31	28

a) nd, Not determined

coupled to 4,5-diiodoimidazole through Vorbrüggen methods. From here, two paths were taken. Initially, the 5′-acetate protecting group was removed and substituted with a benzyl group such that the 4-iodo acyclic nucleoside (**29.5**) could be obtained using Grignard conditions [10]. Fortuitously, however, it was found that removal of the iodine from the acyclic sugar coupled intermediate **29.2** using aqueous sodium sulfite also removed the acetate group, and subsequent reacetylation drastically improved the overall yield as the benzyl protection step could be bypassed [11].

Stille and Suzuki heteroaryl coupling reactions (Scheme 8.30) were used analogous to those shown in Scheme 8.22 [10, 11]. Conventional deprotections yielded the acyclic proximal guanosine and xanthosine fleximers (**30.4** and **30.9**, respectively) [10, 11, 30]. As interesting biological activities against coronaviruses were observed for **30.3**, the McGuigan ProTide **30.6** was also constructed using conditions similar to those in Scheme 8.13.

As mentioned previously, these doubly flexible nucleos(t)ides have shown low micromolar activities (Table 8.3) against various families of viruses coronaviruses (Middle East Respiratory Syndrome, MERS; Severe Acute Respiratory Syndrome, SARS) [10], filoviruses (Ebola, EBOV; Sudan, SUDV; Marburg, MARV) [11], and flaviviruses (Dengue, DENV; Zika, ZIKV; yellow fever, YFV) (manuscript in preparation). The broad-spectrum activities of these compounds are especially impressive considering that they are also nontoxic; toxicity is a common pitfall of nucleos(t)ide therapeutics due to their lack of selectivity and inhibition of human polymerases. Because of these findings, additional structure–activity relationship studies on the acyclic fleximers are currently being carried out toward enhancing their therapeutic profiles against these and other hard to treat viruses.

8.5 Conclusion

A structurally diverse group of fleximers have been synthesized and explored for their biological impact not only by the Seley-Radkte group, but by other groups as well, thus highlighting the importance of this unique class of nucleoside analogues. While the synthetic routes to some of the fleximers have been optimized and/or simplified over the years, in many cases their syntheses remain tedious. Furthermore, it is clear that modifications to the sugar moiety have a significant influence on the approach used for constructing the base, thus there is no "one size fits all" approach that can be used for the synthetic strategies. The significant biological effects exhibited by the fleximers, particularly the most recent results for the doubly flexible fleximers, make the synthetic challenges worth pursuing. Additional studies currently underway are focused on exploring new classes of fleximers based on other FDA-approved nucleos(t)ide drugs to determine the effect of flexibility on activity compared to those on the parent compounds. In parallel, the ongoing mechanistic studies being carried out in our collaborators' laboratories with nucleoside recognizing enzymes involved in viral replication such as methyltransferases and viral polymerases will serve to better elucidate their mechanism(s) of action, particularly for the acyclic fleximers, given their apparent broad spectrum activity. In summary, while the biological results to date have been, in some cases, groundbreaking, much remains to be done to fully understand the implications of flexibility on antiviral drug design and, therefore, we look forward to exploring new challenges with these interesting (but often times difficult to obtain) nucleoside analogues.

References

1 Seley, K.L., Zhang, L., and Hagos, A. (2001). "Fleximers". Design and synthesis of two novel split nucleosides. *Org. Lett.* 3: 3209–3210.

2 Seley, K.L., Zhang, L., Hagos, A., and Quirk, S. (2002). "Fleximers". Design and synthesis of a new class of novel shape-modified nucleosides. *J. Org. Chem.* 67: 3365–3373.

3 Seley, K.L., Quirk, S., Salim, S. et al. (2003). Unexpected inhibition of *S*-adenosyl-L-homocysteine hydrolase by a guanosine nucleoside. *Bioorg. Med. Chem. Lett.* 13: 1985–1988.

4 Polak, M., Seley, K.L., and Plavec, J. (2004). Conformational properties of shape modified nucleosides – fleximers. *J. Am. Chem. Soc.* 126: 8159–8166.

5 Quirk, S. and Seley, K.L. (2005). Substrate discrimination by the human GTP fucose pyrophosphorylase. *Biochemistry* 44: 10854–10863.

6 Quirk, S. and Seley, K.L. (2005). Identification of catalytic amino acids in the human GTP fucose pyrophosphorylase active site. *Biochemistry* 44: 13172–13178.

7 Seley, K.L., Salim, S., and Zhang, L. (2005). "Molecular chameleons". Design and synthesis of C-4-substituted imidazole fleximers. *Org. Lett.* 7: 63–66.

8 Seley, K.L., Salim, S., Zhang, L., and O'Daniel, P.I. (2005). "Molecular chameleons". Design and synthesis of a second series of flexible nucleosides. *J. Org. Chem.* 70: 1612–1619.

9 Wauchope, O.R., Velasquez, M., and Seley-Radtke, K. (2012). Synthetic routes to a series of proximal and distal 2′-deoxy fleximers. *Synthesis* 44: 3496–3504.

10 Peters, H.L., Jochmans, D., de Wilde, A.H. et al. (2015). Design, synthesis and evaluation of a series of acyclic fleximer nucleoside analogues with anti-coronavirus activity. *Bioorg. Med. Chem. Lett.* 25: 2923–2926.

11 Yates, M.K., Raje, M.R., Chatterjee, P. et al. (2017). Flex-nucleoside analogues – novel therapeutics against filoviruses. *Bioorg. Med. Chem. Lett.* 27: 2800–2802.

12 Zimmermann, S.C., Sadler, J.M., Andrei, G. et al. (2011). Carbocyclic 5′-nor "reverse" fleximers. Design, synthesis, and preliminary biological activity. *MedChemComm* 2: 650–654.

13 Zimmermann, S.C., Sadler, J.M., O'Daniel, P.I. et al. (2013). "Reverse" carbocyclic fleximers: synthesis of a new class of adenosine deaminase inhibitors. *Nucleosides Nucleotides Nucleic Acids* 32: 137–154.

14 Seley-Radtke, K. Flexibility – not just for yoga anymore! *Antiviral Chem. Chemother.* 2017: 25.

15 Leonard, N.J., Morrice, A.G., and Sprecker, M.A. (1975). Linear Benzoadenine – stretched-out analog of adenine. *J. Org. Chem.* 40: 356–363.

16 Morrice, A.G., Sprecker, M.A., and Leonard, N.J. (1975). Angular benzoadenines – 9-aminoimidazo[4,5-F]quinazoline and 6-aminoimidazo[4,5-H]quinazoline. *J. Org. Chem.* 40: 363–366.

17 Leonard, N.J. and Hiremath, S.P. (1986). Dimensional probes of binding and activity. *Tetrahedron* 42: 1917–1961.

18 Leonard, N.J. (1982). Dimensional probes of enzyme-coenzyme binding-sites. *Acc. Chem. Res.* 15: 128–135.

19 Seley, K.L., Januszczyk, P., Hagos, A. et al. (2000). Synthesis and antitumor activity of thieno-separated tricyclic purines. *J. Med. Chem.* 43: 4877–4883.

20 Leonard, N.J. (1985). Adenylates – bound and unbound. *Biopolymers* 24: 9–28.

21 Leonard, N.J., Petric, A., and Rykowski, A. (1988). Defined dimensional alterations in enzyme substrates – birch reduction of Lin-Benzopurines – a contribution to information concerning the binding-sites of adenosine-deaminase and xanthine-oxidase. *J. Org. Chem.* 53: 3873–3875.

22 Kauffman, R.F., Lardy, H.A., Barrio, J.R. et al. (1978). Dimensional probes of enzyme binding-sites of adenine-nucleotides – interaction of Lin-Benzoadenosine 5′-diphosphate and triphosphate with mitochondrial Atp-Synthetase, purified Atpase, and adenine-nucleotide carrier. *Biochemistry* 17: 3686–3692.

23 Leonard, N.J., Scopes, D.I.C., Vanderlijn, P., and Barrio, J.R. (1978). Dimensional probes of enzyme binding-sites of adenine-nucleotides – biological effects of widening adenine ring by 2.4 a. *Biochemistry* 17: 3677–3685.

24 Schmidt, M.J., Truex, L.L., Leonard, N.J. et al. (1978). Effect of lin-benzoadenosine and lin-benzoadenosine 3′:5′-monophosphate on cyclic AMP-dependent protein kinase activity in vitro. *J. Cyclic Nucleotide Res.* 4: 201–207.

25 Zhang, Z.B., Wauchope, O.R., and Seley-Radtke, K.L. (2008). Mechanistic studies in the synthesis of a series of thieno-expanded xanthosine and guanosine nucleosides. *Tetrahedron* 64: 10791–10797.

26 Wauchope, O.R., Tomney, M.J., Pepper, J.L. et al. (2010). Tricyclic 2′-C-modified nucleosides as potential anti-HCV therapeutics. *Org. Lett.* 12: 4466–4469.

27 Wauchope, O.R., Johnson, C., Krishnamoorthy, P. et al. (2012). Synthesis and biological evaluation of a series of thieno-expanded tricyclic purine 2′-deoxy nucleoside analogues. *Bioorg. Med. Chem.* 20: 3009–3015.

28 O'Daniel, P.I., Jefferson, M., Wiest, O., and Seley-Radtke, K.L. (2008). A computational study of expanded heterocyclic nucleosides in DNA. *J. Biomol. Struct. Dyn.* 26: 283–292.

29 Gerster, J.F., Jones, J.W., and Robins, R.K. (1963). Purine nucleosides. IV. The synthesis of 6-halogenated 9-β-D-Ribofuranosylpurines from inosine and guanosine1. *J. Org. Chem.* 28: 945–948.

30 Wuts, P.G.M. and Greene, T.W. (2007). *Greene's Protective Groups in Organic Synthesis*, 4e. Hoboken, NJ: Wiley-Interscience, xxviii, 1082 p.

31 Escandón, L.S., Ordóñez, S., Vega, A., and Díez, F.V. (2008). Sulphur poisoning of palladium catalysts used for methane combustion: effect of the support. *J. Hazard. Mater.* 153: 742–750.

32 Markiewicz, W.T. (1979). Tetraisopropyldisiloxane-1,3-diyl, a group for simultaneous protection of 3′-hydroxy and 5′-hydroxy functions of nucleosides. *J. Chem. Res., Synop.* 24–25.

33 Hoffer, M. (1960). α-Thymidin. *Chem. Ber.* 93: 2777–2781.

34 Hoffer, M., Duschinsky, R., Fox, J.J., and Yung, N. (1959). Simple synthesis of pyrimidine-2′-deoxy-ribonucleosides. *J. Am. Chem. Soc.* 81: 4112–4113.

35 Tawarada, R., Seio, K., and Sekine, M. (2008). Synthesis and properties of oligonucleotides with iodo-substituted aromatic aglycons: investigation of possible halogen bonding base pairs. *J. Org. Chem.* 73: 383–390.

36 Rolland, V., Kotera, M., and Lhomme, J. (1997). Convenient preparation of 2-deoxy-3,5-di-*O*-*p*-toluoyl-α-D-erythro-pentofuranosyl chloride. *Synth. Commun.* 27: 3505–3511.

37 Joule, J.A. and Mills, K. (2009). *Heterocyclic Chemistry*, 5e. Hoboken, NJ: Wiley-VCH, xxviii, 689 p.

38 Primas, N., Mahatsekake, C., Bouillon, A. et al. (2008). A new boronic-acid based strategy to synthesize 4(5)-(het)aryl-1*H*-imidazoles. *Tetrahedron* 64: 4596–4601.

39 Holden, K.G., Mattson, M.N., Cha, K.H., and Rapoport, H. (2002). Synthesis of chiral pilocarpine analogues via a C-8 ketone intermediate. *J. Org. Chem.* 67: 5913–5918.

40 Chen, Z., Jochmans, D., Ku, T. et al. (2015). Bicyclic and tricyclic "expanded" nucleobase analogues of Sofosbuvir: new scaffolds for hepatitis C therapies. *ACS Infect. Dis.* 1: 357–366.

41 Chen, Z., Ku, T.C., and Seley-Radtke, K.L. (2015). Thiophene-expanded guanosine analogues of gemcitabine. *Bioorg. Med. Chem. Lett.* 25: 4274–4276.

42 Tomney, M., Korba, B., Zimmermann, S., and Seley-Radtke, K. (2010). Synthesis of a series of 2′-modified tricyclic nucleosides as potential HCV agents. *Antiviral Res.* 86: A68–A68.

43 De Clercq, E. and Li, G. (2016). Approved antiviral drugs over the past 50 years. *Clin. Microbiol. Rev.* 29: 695–747.

44 Wang, P.Y., Chun, B.K., Rachakonda, S. et al. (2009). An efficient and diastereoselective synthesis of PSI-6130: a clinically efficacious inhibitor of HCV NS5B polymerase. *J. Org. Chem.* 74: 6819–6824.

45 Schinazi, R.F. and Prusoff, W.H. (1985). Synthesis of 5-(dihydroxyboryl)-2′-deoxyuridine and related boron-containing pyrimidines. *J. Org. Chem.* 50: 841–847.
46 Smith, J.G. (2008). *Organic Chemistry*, 2e. Boston: McGraw-Hill, xxxvi, 1175 p., A-18, C-2, G-18, I-29.
47 Brown, D.J. (1962). *The Pyrimidines*, 774. New York: Interscience Publishers.
48 Minakawa, N., Kojima, N., Hikishima, S. et al. (2003). New base pairing motifs. The synthesis and thermal stability of oligodeoxynucleotides containing imidazopyridopyrimidine nucleosides with the ability to form four hydrogen bonds. *J. Am. Chem. Soc.* 125: 9970–9982.
49 Marquez, V.E. (1996). Carbocyclic nucleosides. In: *Advances in Antiviral Drug Design*, vol. 2 (ed. E. De Clercq), 89–146. Elsevier.
50 Borchardt, R.T., Keller, B.T., and Patel-Thombre, U. (1984). Neplanocin A. A potent inhibitor of *S*-adenosylhomocysteine hydrolase and of vaccinia virus multiplication in mouse L929 cells. *J. Biol. Chem.* 259: 4353–4358.
51 Borchardt, R.T. and Wu, Y. (1976). *S*-Aristeromycinyl-L-homocysteine, a potent inhibitor of *S*-adenosylmethionine-dependent transmethylations. *J. Med. Chem.* 19: 197–198.
52 Guranowski, A., Montgomery, J.A., Cantoni, G.L., and Chiang, P.K. (1981). Adenosine analogues as substrates and inhibitors of *S*-adenosylhomocysteine hydrolase. *Biochemistry* 20: 110–115.
53 De Clercq, E. (2005). John Montgomery's legacy: carbocyclic adenosine analogues as SAH hydrolase inhibitors with broad-spectrum antiviral activity. *Nucleosides Nucleotides Nucleic Acids* 24: 1395–1415.
54 van Brummelen, A.C., Olszewski, K.L., Wilinski, D. et al. (2009). Co-inhibition of plasmodium falciparum *S*-adenosylmethionine decarboxylase/ornithine decarboxylase reveals perturbation-specific compensatory mechanisms by transcriptome, proteome, and metabolome analyses. *J. Biol. Chem.* 284: 4635–4646.
55 Wolfe, M.S. and Borchardt, R.T. (1991). *S*-adenosyl-L-homocysteine hydrolase as a target for antiviral chemotherapy. *J. Med. Chem.* 34: 1521–1530.
56 Hasobe, M., Liang, H., Ault-Riche, D.B. et al. (1993). (1′*R*, 2′*S*, 3′*R*)-9-(2′, 3′-Dihydroxycyclopentan-1′-yl)-adenine and −3-deaza-adenine: analogues of aristeromycin which exhibit potent antiviral activity with reduced cytotoxicity. *Antiviral Chem. Chemother.* 4: 245–248.
57 Zimmermann, S.C., O'Neill, E., Ebiloma, G.U. et al. (2014). Design and synthesis of a series of truncated neplanocin fleximers. *Molecules* 19: 21200–21214.
58 Wolfe, M.S., Lee, Y., Bartlett, W.J. et al. (1992). 4′-Modified analogues of aristeromycin and neplanocin A: synthesis and inhibitory activity toward *S*-adenosyl-L-homocysteine hydrolase. *J. Med. Chem.* 35: 1782–1791.
59 Hasobe, M., McKee, J.G., Borcherding, D.R., and Borchardt, R.T. (1987). 9-(Trans-2′,trans-3′-dihydroxycyclopent-4′-enyl)-adenine and −3-deazaadenine: analogs of neplanocin A which retain potent antiviral activity but exhibit reduced cytotoxicity. *Antimicrob. Agents Chemother.* 31: 1849–1851.

60 Bujnicki, J.M., Prigge, S.T., Caridha, D., and Chiang, P.K. (2003). Structure, evolution, and inhibitor interaction of *S*-adenosyl-L-homocysteine hydrolase from plasmodium falciparum. *Proteins* 52: 624–632.

61 Chiang, P.K. (1998). Biological effects of inhibitors of *S*-adenosylhomocysteine hydrolase. *Pharmacol. Ther.* 77: 115–134.

62 Yang, M., Ye, W., and Schneller, S.W. (2004). Preparation of carbocyclic *S*-adenosylazamethionine accompanied by a practical synthesis of (−)-aristeromycin. *J. Org. Chem.* 69: 3993–3996.

63 Vichier-Guerre, S., Dugue, L., Bonhomme, F., and Pochet, S. (2016). Expedient and generic synthesis of imidazole nucleosides by enzymatic transglycosylation. *Org. Biomol. Chem.* 14: 3638–3653.

64 Vichier-Guerre, S., Dugue, L., Bonhomme, F., and Pochet, S. (2017). An expedient synthesis of flexible nucleosides via a regiocontrolled enzymatic glycosylation of functionalized imidazoles. *Org. Biomol. Chem.* 15: 8193–8203.

65 Vichier-Guerre, S., Dugué, L., and Pochet, S. (2014). A convenient synthesis of 4(5)-(hetero)aryl-1*H*-imidazoles via microwave-assisted Suzuki–Miyaura cross-coupling reaction. *Tetrahedron Lett.* 55: 6347–6350.

66 Štimac, A. and Kobe, J. (2000). Studies on the origin of stereoselectivity in the synthesis of 1,2-trans glycofuranosyl azides. *Carbohydr. Res.* 324: 149–160.

67 St Amant, A.H., Bean, L.A., Guthrie, J.P., and Hudson, R.H.E. (2012). Click fleximers: a modular approach to purine base-expanded ribonucleoside analogues. *Org. Biomol. Chem.* 10: 6521–6525.

68 Dodd, D.W., Swanick, K.N., Price, J.T. et al. (2010). Blue fluorescent deoxycytidine analogues: convergent synthesis, solid-state and electronic structure, and solvatochromism. *Org. Biomol. Chem.* 8: 663–666.

69 Janeba, Z., Balzarini, J., Andrei, G. et al. (2006). Synthesis and biological evaluation of 5-(alkyn-1-yl)-1-(*p*-toluenesulfonyl)uracil derivatives. *Can. J. Chem.* 84: 580–586.

70 Hudson, R.H.E. and Dambenieks, A.K. (2006). Synthesis of N1-unsubstitued 5-alkynylcytosine and derivatives thereof. *Heterocycles* 68: 1325–1328.

71 Kielkowski, P., Pohl, R., and Hocek, M. (2011). Synthesis of acetylene linked double-nucleobase nucleos(t)ide building blocks and polymerase construction of DNA containing cytosines in the major groove. *J. Org. Chem.* 76: 3457–3462.

72 Hudson, P.H.E. and Moszynski, J.M. (2006). A facile synthesis of fluorophores based on 5-phenylethynyluracils. *Synlett* 2997–3000.

73 Rostovtsev, V.V., Green, L.G., Fokin, V.V., and Sharpless, K.B. (2002). A stepwise Huisgen cycloaddition process: copper(I)-catalyzed regioselective "ligation" of azides and terminal alkynes. *Angew. Chem. Int. Ed.* 114: 2708–2711.

74 Wigerinck, P., Pannecouque, C., Snoeck, R. et al. (1991). 5-(5-Bromothien-2-yl)-2′-deoxyuridine and 5-(5-chlorothien-2-yl)-2′-deoxyuridine are equipotent to (E)-5-(2-bromovinyl)-2′-deoxyuridine in the inhibition of herpes simplex virus type I replication. *J. Med. Chem.* 34: 2383–2389.

75 De Winter, H. and Herdewijn, P. (1996). Understanding the binding of 5-substituted 2′-deoxyuridine substrates to thymidine kinase of herpes simplex virus type-1. *J. Med. Chem.* 39: 4727–4737.
76 Herdewijn, P.A.M.M. (1994). 5-Substituted-2′-deoxyuridines as anti-Hsv-1 agents – synthesis and structure-activity relationship. *Antiviral Chem. Chemother.* 5: 131–146.
77 Wigerinck, P., Snoeck, R., Claes, P. et al. (1991). Synthesis and antiviral activity of 5-heteroaryl-substituted 2′-deoxyuridines. *J. Med. Chem.* 34: 1767–1772.
78 Sadler, J.M., Mosley, S.L., Dorgan, K.M. et al. (2009). Synthetic strategies toward carbocyclic purine-pyrimidine hybrid nucleosides. *Bioorg. Med. Chem.* 17: 5520–5525.
79 Mosley, S.L., Bakke, B.A., Sadler, J.M. et al. (2006). Carbocyclic pyrimidine nucleosides as inhibitors of *S*-adenosylhomocysteine hydrolase. *Bioorg. Med. Chem.* 14: 7967–7971.
80 Pinhey, J.T. and Roche, E.G. (1988). The chemistry of organolead(IV) tricarboxylates. Synthesis and electrophilic heteroarylation reactions of 2- and 3-thienyl-, and 2- and 3-furyl-lead tricarboxylates. *J. Chem. Soc., Perkin Trans. 1* 2415–2421.
81 Zambon, A., Borsato, G., Brussolo, S. et al. (2008). Efficient access to 5-substituted thiazoles by a novel metallotropic rearrangement. *Tetrahedron Lett.* 49: 66–69.
82 O'Neil, I.A., Thompson, S., Murray, C.L., and Kalindjian, S.B. (1998). DPPE: a convenient replacement for triphynylphosphine in the staudinger and Mitsunobu reactions. *Tetrahedron Lett.* 39: 7787–7790.
83 Barnard, D.L., Stowell, V.D., Seley, K.L. et al. (2001). Inhibition of measles virus replication by 5′-nor carbocyclic adenosine analogues. *Antiviral Chem. Chemother.* 12: 241–250.
84 Hegde, V.R., Seley, K.L., and Schneller, S.W. (2000). Carbocyclic 5′-norcytidine (5′-noroarbodine). *J. Heterocycl. Chem.* 37: 1361–1362.
85 Barnard, D.L., Stowell, V., Seley, K.L. et al. (2000). Inhibition of measles virus replication by 5′-nor carbocyclic nucleoside analogs. *Antiviral Res.* 46: A65.
86 Hegde, V.R., Seley, K.L., and Schneller, S.W. (2000). Carbocyclic 5′-noruridine. *Nucleosides Nucleotides Nucleic Acids* 19: 269–273.
87 Seley, K.L. and Schneller, S.W. (1999). 7-Deaza-5′-noraristeromycin derivatives resembling L-toyocamycin and L-sangivamycin. *J. Heterocycl. Chem.* 36: 287–288.
88 Seley, K.L., Schneller, S.W., and Korba, B. (1998). Does the anti-hepatitis B virus activity of (+)-5′-noraristeromycin exist in its 4′-epimer and 4′-deoxygenated derivatives? *J. Med. Chem.* 41: 2168–2170.
89 Seley, K.L., Schneller, S.W., De Clercq, E. et al. (1998). The importance of the 4′-hydroxyl hydrogen for the anti-trypanosomal and antiviral properties of (+)-5′-noraristeromycin and two 7-deaza analogues. *Bioorg. Med. Chem.* 6: 797–801.
90 Seley, K.L., Schneller, S.W., Rattendi, D. et al. (1997). Synthesis and antitrypanosomal activities of a series of 7-deaza-5′-noraristeromycin derivatives

with variations in the cyclopentyl ring substituents. *Antimicrob. Agents Chemother.* 41: 1658–1661.

91 Seley, K.L., Schneller, S.W., Rattendi, D. et al. (1997). Synthesis and anti-trypanosomal activity of various 8-aza-7-deaza-5′-noraristeromycin derivatives. *J. Med. Chem.* 40: 625–629.

92 Seley, K.L., Schneller, S.W., Rattendi, D., and Bacchi, C.J. (1997). (+)-7-Deaza-5′-noraristeromycin as an anti-trypanosomal agent. *J. Med. Chem.* 40: 622–624.

93 Seley, K.L., Schneller, S.W., and Korba, B. (1997). A 5′-noraristeromycin enantiomer with activity towards hepatitis B virus. *Nucleosides Nucleotides* 16: 2095–2099.

94 Daelemans, D., Este, J.A., Witvrouw, M. et al. (1997). *S*-Adenosylhomocysteine hydrolase inhibitors interfere with the replication of human immunodeficiency virus type 1 through inhibition of the LTR transactivation. *Mol. Pharmacol.* 52: 1157–1163.

95 Siddiqi, S.M., Chen, X., and Schneller, S.W. (1993). Enantiospecific synthesis of 5′-noraristeromycin and its 7-deaza derivative and a formal synthesis of (−)-5′-homoaristeromycin. *Nucleosides Nucleotides* 12: 267–278.

96 Russ, P., Schelling, P., Scapozza, L. et al. (2003). Synthesis and biological evaluation of 5-substituted derivatives of the potent antiherpes agent (north)-methanocarbathymine. *J. Med. Chem.* 46: 5045–5054.

97 Rosowsky, A., Kim, S.H., and Wick, M. (1981). Synthesis and antitumor activity of an acyclonucleoside derivative of 5-fluorouracil. *J. Med. Chem.* 24: 1177–1181.

9

Synthesis of Oligonucleotides Carrying Nucleic Acid Derivatives of Biomedical and Structural Interest

Ramon Eritja, Anna Aviñó, Carme Fàbrega, Adele Alagia, Andreia F. Jorge, and Santiago Grijalvo

CIBER-BBN, Instituto de Química Avanzada de Cataluña (IQAC), Consejo Superior de Investigaciones Científicas (CSIC), Jordi Girona 18-26, 08034 Barcelona, Spain

9.1 Introduction

The double helix model for the structure of DNA was proposed by Watson and Crick in 1953 [1]. However, it was not until the late 1970s that the crystal structure was resolved from a defined DNA sequence prepared in the laboratory [2]. Since then, advances in the chemical synthesis of oligonucleotides have been crucial for the development of new fields. For example, synthetic DNA was important for the development of DNA recombinant techniques and structural analysis of nucleic acids as well as the characterization of nucleic acid–protein interactions. In addition, optimized strategies for obtaining modified nucleic acids provided good opportunities to use synthetic oligonucleotides in the biomedical field. Starting from the isolation, detection, and analysis of nucleic acid sequences and progressing to the development of the nucleic acid-based drugs, the methodologies used in the preparation of modified oligonucleotides are in constant evolution providing a wide range of versatile tools with application in therapeutics, material science, and molecular computing. The chemical modification of nucleic acids has never been so extensively applied by so many researchers.

This chapter is focused on presenting an overview of the nucleic acid derivatives developed by our group in the last decade as well as highlighting the important progresses made recently in this field. All emerging applications currently exploiting these valuable and versatile compounds are also referred. Firstly, a complete description of the research findings on oligonucleotides carrying O^6-alkylguanine derivatives involved in DNA repair will be provided. These modified nucleosides may be present as a result of DNA damage, which often occurs transiently but leads to important biological consequences. Secondly, the effects caused by the introduction of synthetic nucleotide derivatives such as conformationally restricted nucleotides, acyclic nucleotides, L-nucleotides, and C-nucleotides in the structural properties of DNA will be presented. Finally, the impact of 3′ and 5′-modifications on the gene silencing properties of short interfering RNAs (siRNAs) will be discussed. Excellent reviews have helped

Enzymatic and Chemical Synthesis of Nucleic Acid Derivatives, First Edition.
Edited by Jesús Fernández-Lucas and María-José Camarasa Rius.

compile remarkable advances in the synthesis of nucleic acids derivatives for application in gene silencing therapy [3]: antisense [4], or RNA interference [5], or aptamers [6] and the reader is referred to them. Recent findings in oligonucleotide labeling [7], DNA structure [8], and DNA nanotechnology [9], have also been duly addressed in literature.

9.2 Oligonucleotides Carrying the DNA Lesion *O*^{6}-Alkylguanine

Nucleobases are responsible for the base-paring properties of nucleic acids. Any modification in the nucleobases may lead to incorrect base pairing, which may induce potential mutations. The reactivity of DNA toward radiation and chemicals changes its molecular structure leading to the formation of DNA lesions. These lesions will be corrected by DNA repair mechanisms to preserve genome integrity [10].

Alkylating agents are a class of chemotherapeutic anticancer drugs that act at multiple sites of the DNA by generating adducts [11]. Of the four DNA bases, the guanine (G) base is the most susceptible to damage, leading to many subproducts [12, 13]. A subset of alkylating agents, including nitrosoureas and temozolomide, have a preference for alkylating guanine at the *O*^{6}-position, generating the most important lesion in terms of mutagenesis and carcinogenesis [14].

In addition to the other well-known side effects and limitations of chemotherapeutic agents, cyclophosphamide, mechlorethamide, chlorambucil, and nitrosoureas, also present problems of acquired tumor resistance, in part due to the action of DNA repair mechanism. Particularly, the DNA-repair protein human *O*^{6}-alkyl guanine DNA alkyltransferase (hAGT or *O*^{6}-methylguanine DNA methyl transferase, MGMT) is responsible for removing alkyl adducts from the *O*^{6} position of guanines (Figure 9.1a). The cytotoxic effects induced by alkyl adducts are thereby blocked and therefore the removal of the alkyl adducts plays an important role as a resistance mechanism [15]. It is well established that tumor cells frequently express high levels of hAGT, which is predictive of poor response to chemotherapeutic drugs. This effect has been observed in a large number of cancers [15, 16]. Therefore, pharmacological inhibition of hAGT can enhance the cytotoxicity of a diverse range of anticancer agents [16].

Several methods are available to characterize the activity of hAGT and its inhibition through interaction with small molecules. However, most of the methods commonly used as, for instance, radioactive assays or multiple-step enzymatic reactions [17–20] or *in vivo* hAGT labeling [21–34], require handling hazardous substances or following multiple steps, which discourage their use. For this reason, the development of novel methods based on oligonucleotides sequences for the detection of hAGT activity has been considered an important issue for the measurement of protein activity and for the development of new hAGT inhibitors.

One of the methods developed in our group for the evaluation of hAGT activity was based on the use of a 15-mer oligonucleotide sequence known as thrombin binding aptamer (TBA) [25]. This oligonucleotide forms an intramolecular

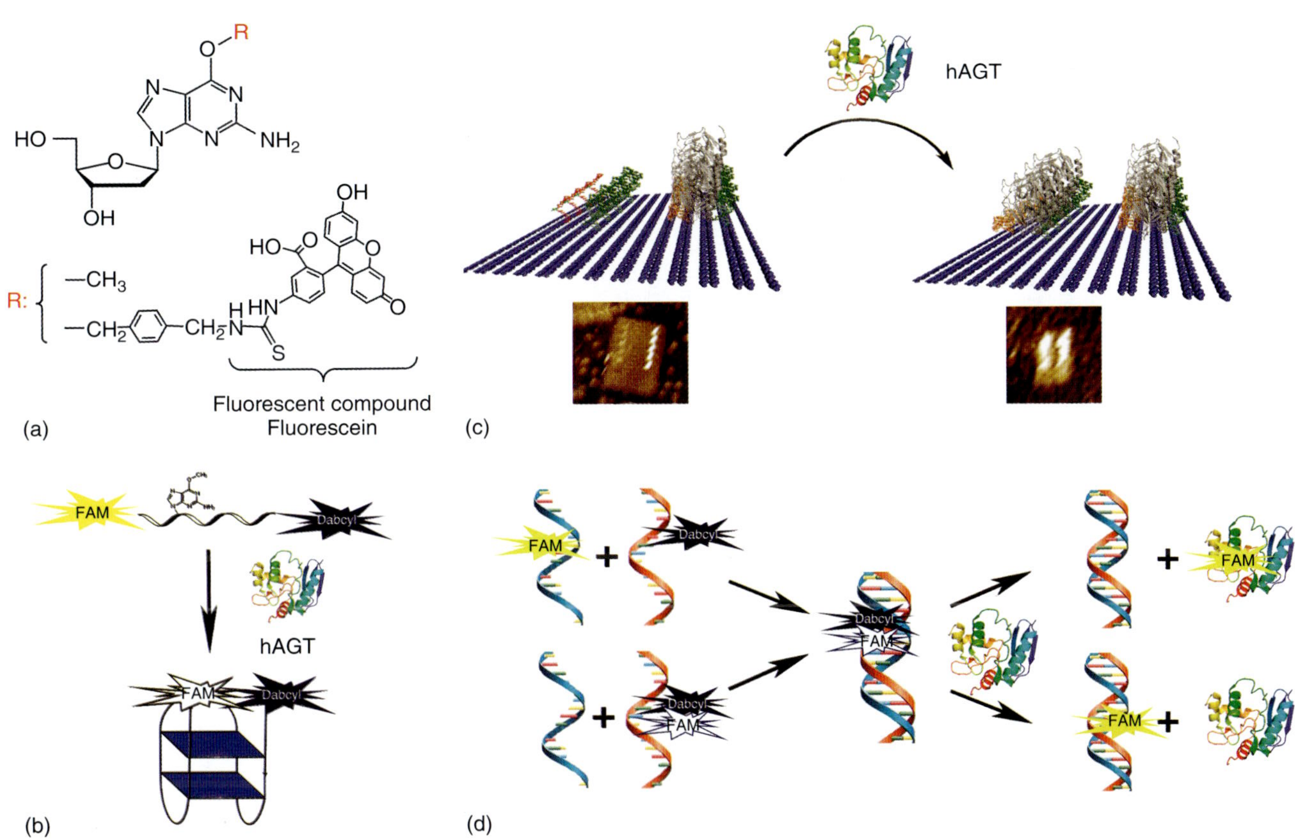

Figure 9.1 O^6-alkyl guanine derivatives and methods for the evaluation of hAGT activity. (a) Chemical structure of O^6-alkyl 2′-deoxyguaninosine derivatives. (b) Fluorescence assay based on the conformational changes of the thrombin-binding aptamer. (c) DNA origami as sensor for hAGT activity. (d) dsDNA FRET method for the quantification of hAGT activity.

G-quadruplex with two tetrads and has a strong affinity for α-thrombin. The introduction of one O^6-methylguanine residue in one of the guanine tetrads prevented quadruplex formation. In this way, the introduction of a fluorescent and a quencher molecule at the 5′ and 3′-ends of the oligonucleotide carrying one O^6-methylguanine residue allows to monitor the hAGT activity by following the decrease of the fluorescence induced by the formation of the quadruplex [26]. As described in Figure 9.1b, the process starts with the methylated TBA that is unfolded. In this conformation the quencher and the fluorophore are beyond the Foster distance. Then, the addition of hAGT removes the methyl group from the mutated guanine residue allowing the TBA to fold back into a chair-like structure. This even produces the approximation of the quencher and the fluorophore and blocks the emission of fluorescence. The hAGT activity can be quantified by the directed decrease in fluorescence.

A more elaborated system was based on the introduction of several TBAs carrying O^6-methylguanine residues in a rectangular DNA origami. The presence of methylated guanine in the TBA prevented its interaction with α-thrombin. In this way, the removal of the methyl group by hAGT could be monitored by the recruitment of α-thrombin at the O^6-methylguanine repaired sites followed by atomic force microscopy (AFM) [27] (Figure 9.1c). Another version of the use of TBAs carrying O^6-methylguanine residues and α-thrombin for the measurement of hAGT activity was from the study of the aggregation of gold nanoparticles induced by thrombin [28]. The aggregation induced by thrombin after O^6-methylguanine repair by hAGT was followed by UV–visible spectroscopy, dynamic light scattering (DLS) and magnetic resonance imaging (MRI) measurements [28]. MRI technique was only used in the case of gold-coated iron oxide supermagnetic nanoparticles.

The direct removal of the alkyl group at position 6 of guanine by hAGT inspired alternative methods for the measurement of hAGT activity. In these methods a fluorescent or a quencher tag is covalently linked to the alkyl group, which will be removed by hAGT, and the complementary Förster resonance energy transfer (FRET)-pair (quencher or fluorophore) is placed nearby. During the repair process, one of the tags is transferred to the protein. The separation of the fluorophore from the quencher results in fluorescence signal light-up, which will allow the quantification of hAGT repair action [29, 30]. Two methods have been described recently using this DNA duplex FRET-pair technology with some differences. The fluorescent derivative used and the location of the tag molecules, in the same strand or in an opposite one, are the main differences [29, 30]. These DNA probes may prove valuable in assessing potential resistance arising from cancer treatment with alkylating agents [30] and for the analysis of hAGT inhibitors [29].

9.3 The Effect of Chemical Modifications in Non-Canonical DNA Structures

In this part we cover some of the synthetic nucleotides designed to modulate the properties observed in non-canonical DNA structures such as triplex,

G-quadruplex, and i-motif. For a long time, these structures were mainly studied from a structural or theoretical point of view. However, since the demonstration of the *in vivo* presence of G-quadruplexes [31] and their importance in several biological processes, the interest in these structures has increased. Synthetic nucleotides may alter the stability and geometry of these non-canonical structures, as well as the interaction between nucleic acids and proteins. There are several reviews describing the impact of modified nucleotides in DNA/RNA structures [32, 33]. Here, we will describe the recent advances that have focused on the introduction of synthetic nucleotides such as conformationally restricted nucleosides or acyclic nucleotides to enhance some of the properties observed in oligonucleotides forming triplex, G-quadruplex, or i-motif DNA structures (Figure 9.2).

9.3.1 Triplex-Forming Oligonucleotides

Triplexes are formed when a duplex containing a polypurine/polypyrimidine track interacts with a third strand by means of specific hydrogen bonds in the major groove of the duplex. The experimental existence of triplexes was demonstrated for the first time in 1957 [34]. Depending on the orientation of the third strand, triplexes are classified into two main categories: parallel triplex, in which a triplex forming pyrimidine strand binds to the purine strand of the duplex in parallel fashion and antiparallel triplex in which a purine-rich strand binds to the purine strand of duplex in an antiparallel orientation. In the first class, the third strand binds through Hoogsteen hydrogen bonds to duplex, forming d(T-A·T) and protonated d(C-G·C+) triads (Figure 9.2). This type of triplex is more stable at low pH values. In the antiparallel triplexes, purine-rich strands bind to the duplex using reverse Hoogsteen base-pairing forming d(T-A·T), d(T-A·A), and d(C-G·G) triads [35]. The specificity and selectivity of the third strand to the duplex makes this structure very attractive in molecular biology, diagnostics, and therapeutics. Triplex-forming oligonucleotides (TFOs) may regulate gene expression and induce site-directed recombination as well as site-specific mutations [36].

A number of modifications have been introduced in the backbone, in the bases or in the 3′ or 5′-end of TFO's to improve selectivity and stability [32, 37, 38]. This includes phosphate modifications, such as phosphorothioate linkages, methylphosphonate or N3′-NP5′ phosphoramidate. The nucleobase modification has mainly focused on the design of cytidine and guanine derivatives to obtain parallel triplexes at neutral pH values [36, 38]. The introduction of psolaren groups to TFOs have also been pursued in order to form adducts and interstrand cross-links to study processes of DNA replication of transcription [36]. In addition, bulge insertions of twisted intercalating nucleic acids (TINAs) have resulted in triplexes and quadruplex with an extraordinarily high thermal stability [39]. Less attention has been paid to the introduction of modifications in the duplex moiety of a triplex structure. However, 8-aminopurines (8-aminoadenine and 8-aminoguanine) are proven to be excellent modifications for parallel- and antiparallel-triplex stabilization [40, 41]. This stabilization has

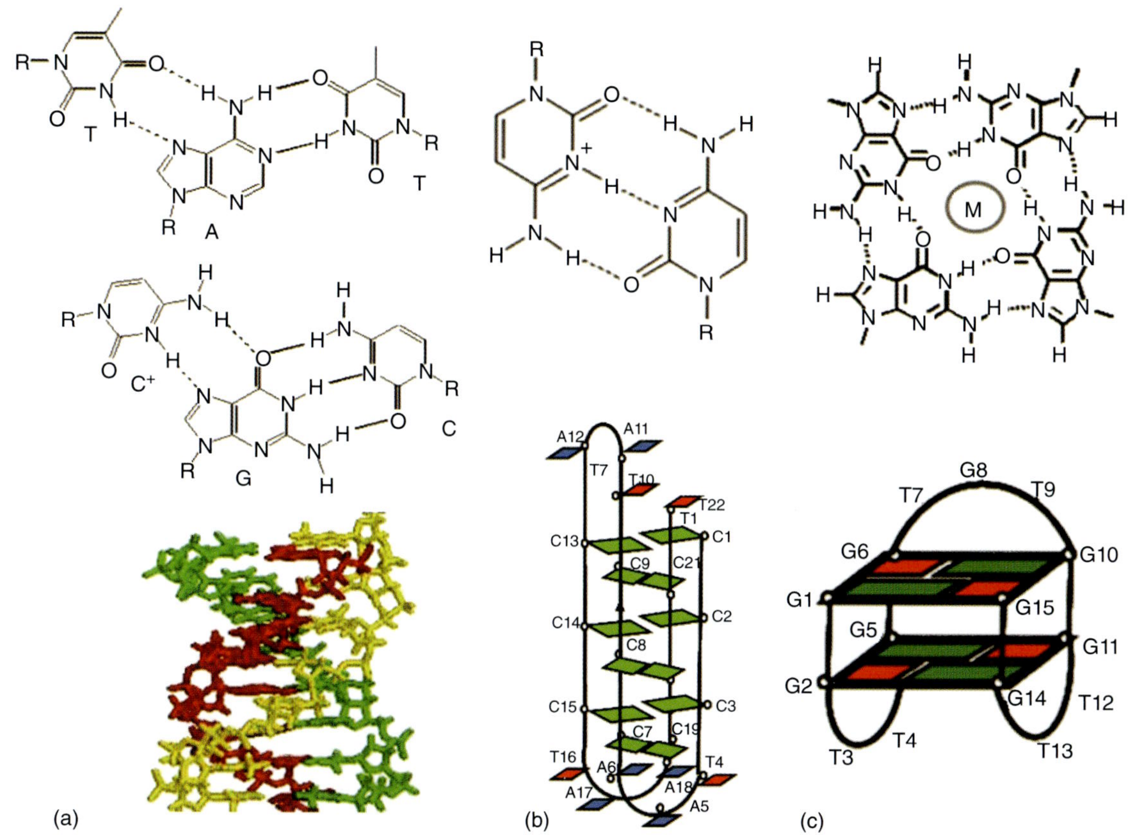

Figure 9.2 Non-canonical nucleic acids structures. (a) the parallel triplex (b) the i-motif, and (C) the G-quadruplex. In the top, is shown a scheme of base-pairing units, T-A·T and C-G·C+ triads is present in parallel triplex, C:C+ in i-motif and G-tetrads in quadruplex. In the bottom are depicted schemes of the corresponding tridimensional structures.

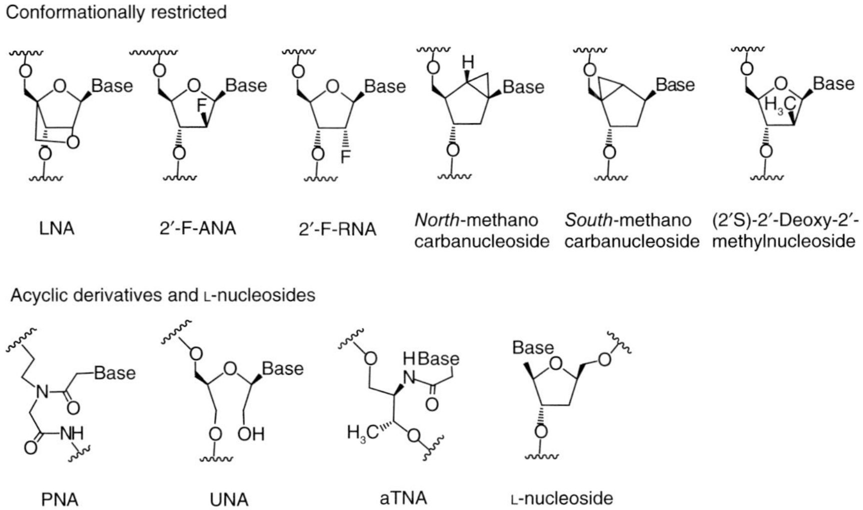

Figure 9.3 Scheme of modified nucleoside backbones used in the study of non-canonical DNA structures.

been critical for several recent applications based on triplex-affinity capture [42–44].

Importantly, the sugar backbone modifications have been the most studied modifications in triple strand formation. In that sense, PNAs (peptide nucleic acids) formed by a 2-aminoethylglycine units (Figure 9.3) is an uncharged backbone with a strong affinity to complementary DNA or RNA strands. Moreover, it is able to form stable parallel triplex [45]. In addition, several conformationally restricted ribose modifications have also been considered as good candidates in TFOs. Specifically, LNAs (Locked Nucleic Acids, Figure 9.3) and its derivatives have been extensively studied in triplex. LNA is a 2′,4′-bridged nucleic acid locked in an 3′-endo conformation. The restriction of the sugar-phosphate backbone facilitates a more efficient stacking of the nucleobases. These derivatives improve affinity toward DNA and RNA complements and the stability against nucleases when incorporated into oligonucleotides [46]. Several variants of LNAs have also been reported during the last decade. LNA is a remarkable nucleic acid modification that has been employed in therapeutic strategies both *in vitro* and *in vivo*. Excellent applications have been found in controlling gene expression and in the modulation of the formation of other nucleic acids structures, rather than duplex. It was reported that the incorporation of LNA derivatives in the homo-pyrimidine strand increased the binding affinity of TFOs in both types of triplexes [47–49]. However, position-dependent stability was observed when they are introduced in the different triads.

9.3.2 G-quadruplex-Forming Oligonucleotides

G-rich sequences may form G-quadruplex structures where there is a cyclic arrangement of four guanines (G-tetrads) through Hoogsteen hydrogen bonding

(Figure 9.2). These sequences have been extensively studied as they have been found in telomeric and promoter regions of certain oncogenes. They are very polymorphic and may form intramolecular/intermolecular arrangements. G-quadruplex may be classified according to the orientation of the strands as parallel or antiparallel. Owing to their biological relevance, this structure has been studied in detail and several modifications have been described [50].

Modifications of the nucleobases have been reported for G-quadruplex [51]. Especially the bases that form the tetrads are very important in the formation and the stability of this structure. Importantly, modifications of 8-position of guanine in G-quadruplex have been studied [52]. In the G-quadruplex, the sugar can assume both C2′-and C3′-endo conformations and the guanine can be found either in *syn* or *anti* conformations. The limitation on the conformational equilibrium by the addition of restricted nucleosides affects the stability of this structure [53]. In antiparallel quadruplexes, the conformation of 2′-deoxyguanosines (dGs) alternate between *anti* and *syn* conformers. In positions *anti*, LNA are compatible and produce stable G-quadruplex structures [53]. A detailed work of the influence of nucleobase orientation and restricted conformation using methanocarba-(MC)-nucleosides *N*-(*anti*)-MCdG and *S*-(*syn*) MC-dG (Figure 9.3) was accomplished using the TBA [25, 54] as a model compound of antiparallel quadruplex [55]. MC-nucleosides have been studied as substrates for kinases and for their inhibitory effects on cell growth [56]. These compounds are pseudonucleosides in which the sugar conformation is restricted. Both *North*- and *South*-locked derivatives are prepared by shifting the position of the fused cyclopropane ring in these constrained nucleosides [56]. This study shows that having the correct nucleobase orientation is more important for the stability of G-quadruplex than having the correct sugar puckering [55]. In addition, the substitution of thymidines by *North*-MC-dT at the TGT loop of TBA was studied. These substitutions produce a destabilizing effect in the antiparallel quadruplex [57]. In addition, it has been reported that 2′-fluoroarabino nucleic acids-(FANAs)-G and 2′-fluororibo-G (F-RNA) (Figure 9.3) can be used for the manipulation of G-quadruplex polymorphism [58]. 2′-Fluoro nucleosides are RNA derivatives that contain a fluorine atom at 2′ position. 2′-F-RNA derivatives contain a fluorine atom in α adopting a sugar 3′-endo conformation whereas in 2′-FANA, the fluoride is located in β position adopting a 2′-endo conformation. Both types of derivatives increase oligonucleotide serum stability and binding affinity to their complementary sequence. Similarly, oligonucleotides modified with 2′-fluoronucleosides provide also a sensitive probe for structural study of nucleic acids and their interactions with other biological moieties [59].

Other classes of modifications were introduced in nucleic acids to confer more flexibility in the ribose phosphate backbone. Specifically, a detailed study of the effect of unlocked nucleic acid (UNA, Figure 9.2) derivatives in the stability of TBA G-quadruplex has been reported [60]. The introduction of these derivatives may control the affinity of the aptamer for the protein in a position-depending manner. UNAs are RNA mimics that lack the bond between the C2′ and C3′ atoms of the ribose ring. This derivative is more flexible than an unmodified RNA monomer. The introduction of UNA monomers may increase or decrease mismatch discrimination against DNA/RNA target strands when it is incorporated in a duplex [61]. Other acyclic nucleoside mimetics studied in G-quadruplex are the acyclic threoninol nucleic acids (aTNAs, Figure 9.3) [62]. This mimetic is constituted by threoninol (2-amino-1,3-butanediol) tethered to one of the natural nucleobases A, C, G, and T by a carboxymethyl linker [63]. Although the aTNA skeleton is characterized by a more flexible backbone than the natural DNA/RNA, it has been demonstrated that it forms a very stable homoduplex with a right-handed structure in an antiparallel orientation [63]. In addition, it has been reported that a fully modified aTNA strand cannot hybridize with the complementary DNA/RNA strand [64]. However, the incorporation of a few aTNA residues in the sequence TG_4T is compatible with the formation of the tetrameric parallel quadruplex [62].

L-nucleosides are opposite enantiomers to D-nucleosides and they have also been introduced in oligonucleotides (Figure 9.3). Their increased nuclease resistance compared to the natural ones has attracted the interest of the groups working in therapeutics. These mirror images have been introduced in aptamers, siRNAs, or ribozymes producing long-lasting structures. L-nucleotides have also been incorporated to G-quadruplex. It has been observed that when all D-DNA nucleotides are replaced by L-nucleotides the resulting structure preserves the same biological properties [65]. Recently, L-RNA nucleotides have been also studied. Specifically, it has been demonstrated that L-RNA aptamer is able to bind the natural D-form of HIV-1 trans-activation responsive (TAR) RNA [66]. More recently, a thermodynamic study regarding the characterization of structural motifs formed by β-L-RNA has been reported [67].

9.3.3 Oligonucleotides Forming i-Motif Structures

I-motif is another secondary structure formed by a C—C+ base pair, in which two parallel cytosine-rich strands forming duplex are intercalated in antiparallel orientation (Figure 9.2). This structure is found in the complementary regions of G-rich sequences of telomeric regions and protooncogenes [68, 69]. These are more stable at low pH values and may adopt several conformations that are sensitive to pH. Based on these versatile features, i-motif oligonucleotides are attractive compounds in the fabrication of nanodevices or nanostructures [70] and as a potential scaffold for drug delivery [71, 72]. Synthetic nucleic acid analogues that can modulate the stability of the i-motif structures are needed in order to control their properties. In this context, several modified nucleotides have been introduced such as 5-methylcytosines and other epigenetic derivatives [73], PNA [74] or aTNA nucleotides [75]. Recently, 2′-C-methyl-2′-deoxycytidine has shown to stabilize i-motif (Figure 9.2) [76]. 2′-C-methyl-2′-deoxynucleosides are interesting nucleosides because they show differential preferred sugar conformations depending on the absolute configuration at the 2′-carbon. The (2′S)-2′-deoxy-2′-C-methylnucleosides mainly adopt the C3′-endo puckering while the 2′R isomers prefer the C2′-endo state. These derivatives are nuclease resistant and have been introduced in therapeutic antisense and DNAzymes oligonucleotides [77]. The limited formation of i-motif at neutral conditions is overcome by the use of 2′-deoxy-2-fluoroarabinose. 2′-F-AraC stabilizes i-motif formation over a wide pH range, including pH 7 [78]. The availability of these oligonucleotides to form i-motif at physiological conditions may find several applications in the study of biological i-motif sequences. In addition, applications in the nanotechnology field based on i-motif conformational changes have been also described [78].

9.4 Modified siRNAs for Gene Silencing

The discovery of RNA interference and microRNA has triggered an intense research activity in the development of modified RNA as drugs [79]. Short-interfering double-stranded RNAs (Figure 9.4) have been demonstrated to be efficient for downregulation of gene expression by targeting a selected mRNA to degradation by the RNA-interfering silencing complex (RISC) [79]. Nowadays, there is no doubt that nucleic acids can be used to interfere with cellular metabolism, in a way that can be developed as novel medicines with potentially more specificity and less toxicity than classical small molecule drugs [80].

However, nucleic acids face multiple obstacles in order to be developed as therapeutic agents. Extensive research effort has been carried out in order to overcome such important barriers [79]. Some of these accomplishments have been carried out by chemically modifying the carbohydrate moiety [81, 82], nucleobases [83], and/or phosphate backbones [84]. Furthermore, the presence of such chemical modifications and others introduced at specific positions of the siRNA guide strand, such as the two-base overhangs at the 3′-termini or at the siRNA 5′-termini [85], are important parameters that govern the efficacy and activity of RNAi [81, 86]. In the following section we will focus on the chemical modifications introduced into the siRNA 3′/5′-overhangs (PIWI/Argonaut/Zwille (PAZ) and middle (MID) domains). Historically, chemical modifications at the 3′, 5′-ends have been considered to be the simplest and safest way to enhance nuclease resistance of siRNAs. But recently, it has been proved that, in addition, these modifications can modulate siRNA selectivity and siRNA potency decreasing the amount of potential off-target effects.

9.4.1 Modifications of the 3′-Overhangs

One of the strategies carried out to enhance siRNA potency and thereby reduce off-target effects relates to modifying the 3′-overhangs of siRNA oligonucleotides. This has resulted in increasing the strength of the interaction between the siRNA 3′-termini and the PAZ domain of Ago2 [87]. In this regard, Somoza et al. proposed the introduction of L-threoninol derivatives containing several aromatic residues (e.g. pyrene (PYR), anthracene (ANTH), naphthalene (NAPH), trifluoromethylbenzene (TFMB) and fluorobenzene (FB)) at the 3′-termini of both siRNA strands [88]. The introduction of such modifications of distinct sizes and electronic properties allowed the authors to gain insight into the binding effects among 3′-overhangs and the PAZ domain. The presence of some aromatic

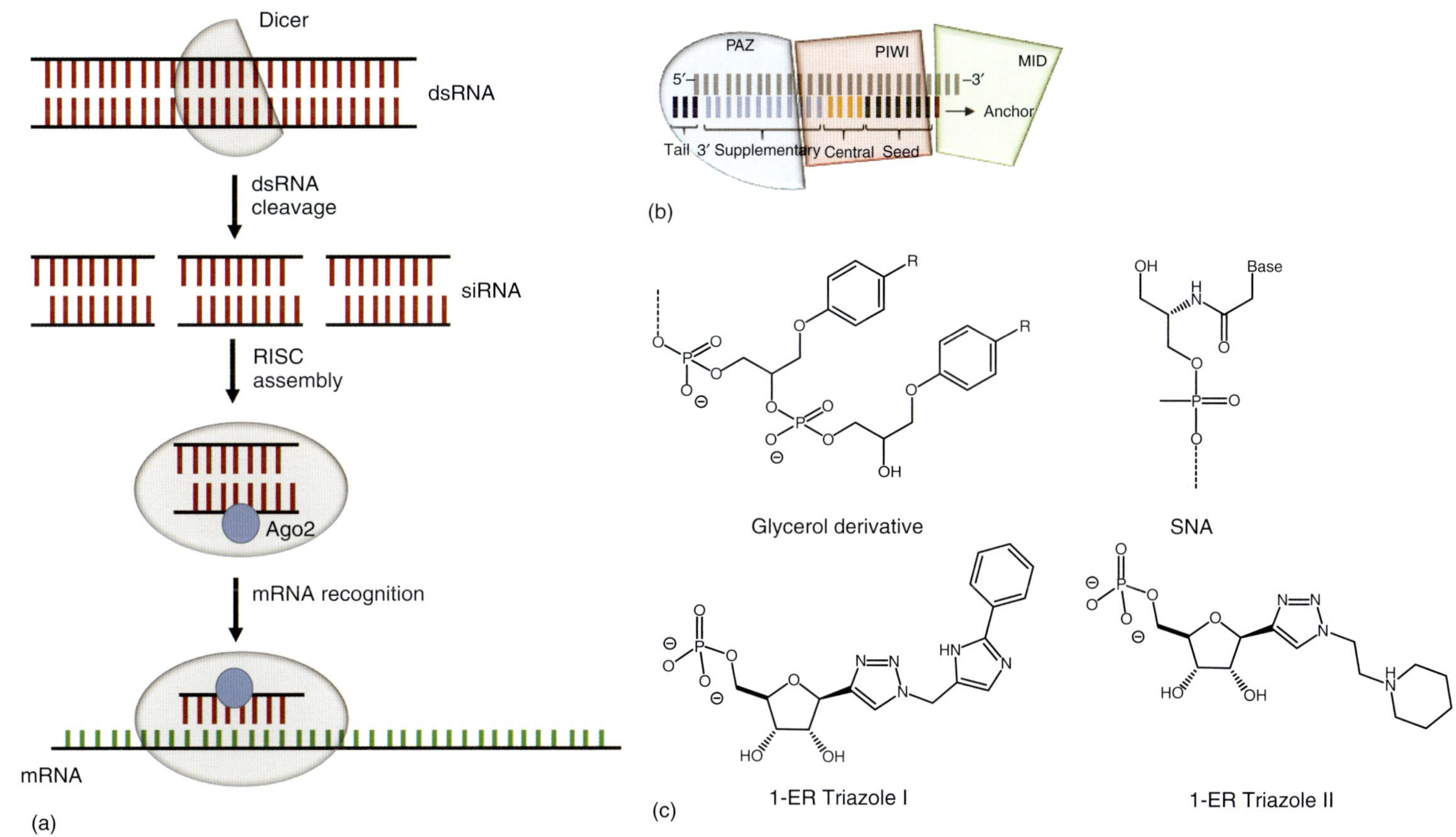

Figure 9.4 (a) Scheme of the RNA interference pathway. (b) Scheme of Ago2 and small interfering RNA (siRNA) molecule. Top strand is the passenger or sense and bottom strand is the guide or antisense. (c) Chemical structure of some of the modifications described for the 3′-overhang and 5′-end.

residues at the guide strand was able to stabilize and modulate the interactions with the PAZ pocket. The authors hypothesize that those aromatic modifications capable of promoting stabilization through electrostatic interactions with the PAZ domain, such as PYR and ANTH, resulted in lowest RNAi activities. However, the lower binding affinities of siRNAs modified with TFMB and FB in the PAZ domain resulted in potent RNAi activities comparable to unmodified siRNA at 32 pM [88].

Other aromatic groups like tyramine (TYR), diphenylpropylamine (DPP) and tryptamine (TRYP) have been also introduced in siRNA in order to study the PAZ domain recognition with the modified 3′-overhangs [89]. Gaglione et al. prepared a series of siRNAs containing such aromatic residues at the 3′-termini of the guide and sense strands as well as simultaneously in both strands. The authors found that all single and double modifications were accepted by the RNAi machinery and afforded higher nuclease stabilities in serum than unmodified siRNA. Interestingly, siRNA conjugates only modified with TRYP at the 3′-termini of the guide strand and double modifications containing either DPP or TRYP at both guide and passenger strands conferred an increased silencing effect according to time-course experiments. The 3′-overhang interaction of the modified siRNAs with PAZ domain was further corroborated by computational studies.

Similarly, structure–activity relationship (SAR) studies carried out by Xu et al. unveiled the effect produced by glycerol-based aromatic compounds (Figure 9.4) on the correct strand selection for improving siRNA efficacy [90]. The authors prepared a series of siRNAs in which the first two positions of the guide and the passenger strands were modified at the 3′-termini with glycerol aromatic derivatives. This strategy allowed the authors to measure the affinity of modified siRNAs with the PAZ domain of Ago2. Computational and gene silencing studies suggested that modified siRNAs with two-nucleotide 3′-overhangs containing glycerol derivatives, in the guide strand, favored their ability to interact with the PAZ domain. Further computational studies confirmed these findings [91, 92]. Lee et al. found different interaction profiles of siRNA duplexes with the PAZ domain of Ago2 depending on the nature of the 3′-nucleotide overhangs of the guide strand [91]. This favored interaction was more pronounced in the case of 3′-UU overhangs than in the case of other nucleotide combinations [91, 92]. Another parameter to be borne in mind toward increasing this strand selection is the pivotal role of the modified 3′-overhangs of the siRNA sense strand, which might promote the recognition of the 5′-termini of the siRNA antisense strand by MID domain thereby improving RNAi activity [90].

Other small molecules have been used to modify either guide or passenger strands of siRNA oligonucleotides. Kamiya et al. replaced the ribose backbone by serinol-based artificial nucleic acids (SNAs) (Figure 9.4) [93]. SNAs are efficient DNA mimetics containing phosphodiester linkages that have been able to interact with DNA and RNA [63]. The authors prepared a series of siRNA molecules containing preferentially SNA residues at the 3′-termini of passenger and guide strands. These modifications exhibited a greater stability enhancement against exonucleases, when SNAs were introduced at the termini of siRNAs. Furthermore, the presence of the SNAs at the 3′-overhangs of both siRNA strands resulted in an improvement of the antisense selection versus sense strand with optimal RNAi silencing activities.

Alagia et al. observed a similar behavior with siRNA oligonucleotides 3′-modified with L-threoninol backbone (Figure 9.3) [94]. Importantly, this acyclic modification was recognized by the RNAi machinery and also conferred good stabilities in the presence of serum and 3′/5′-exonucleases. As described in other studies containing acyclic modifications, the presence of L-threoninol introduced at the 3′-overhang of the antisense strand, as well as the presence of phosphorothioate linkages, promoted potent silencing activities when compared to unmodified siRNAs. These results were hypothesized to be due to the greater affinity of these modifications to the PAZ domain of Ago2, as observed in other studies [95, 96]. Time-course silencing experiments up to five days confirmed that siRNA duplex containing the 3′-overhangs of the passenger and guide strands modified with L-threoninol remarkably retained the RNAi activity and therefore inhibited both luciferase production and ApoB mRNA expression.

9.4.2 Modifications of the 5′-End

The 5′-ends of siRNAs have been also modified with small molecules and the effect on RNAi activity has been evaluated. Great efforts in this field have been mainly carried out by Kubo et al. modifying several parameters such as siRNA length and nature of the chemical entities. Lipids and aromatic molecules have been the two major modifications selected by Kubo et al. introduced at the 5′-termini [97, 98]. The synthetic strategy to incorporate such modifications was carried out by introducing covalently a 5′-amino-modifier into the siRNA sense strand. The final 5′-conjugation with the corresponding hydrocarbonated alkyl chains (palmitic acid, cholesterol, lauric acid) [97] and aromatic pendent groups (phenyl, hydroxyphenyl, naphthyl, and pyrenyl) were carried out obtaining the corresponding 21-mer siRNA conjugates modified with lipids and aromatic molecules, respectively [98].

In the case of lipid–siRNA conjugates, the authors prepared a series of siRNA conjugates targeting *Renilla* luciferase activity and endogenous vascular endothelial growth factor (VEGF) containing 2-nt DNA (dT) as well as the corresponding inverted thymidine (idT) at the 3′-overhang. This idT residue has proved to be resistant to nuclease degradation [99]. Formulated lipid–siRNA conjugates modified with palmitic acid at the 5′-sense strand and the 3′-overhangs containing idT modifications confirmed long-term gene silencing inhibition activities up to five days and exhibited superior silencing efficiencies than Cholesterol (Chol) derivatives. Despite these interesting results at the nanomolar level, these transfection experiments were tarnished in the absence of cationic lipids, which showed a reduction in VEGF levels with moderate efficiencies (~25%) after 48 hours incubation using 1 μM of palmitic–siRNA conjugate.

Synthetic dsRNAs of 27-mer in length have revealed to be Dicer-substrates with the ability to activate RNAi process and consequently promote gene silencing with improved efficiencies and longer duration than classical 21-mer siRNAs [100]. These findings inspired Kubo et al. to modify symmetrical 27-mer Dicer-substrate dsRNA either with amino derivatives [101] or palmitic acid at the 5′-termini of the sense strand [102], according to the same synthetic strategy described previously [97]. The best silencing results were achieved when authors modified 27-mer dsRNA with palmitic acid. The authors showed that Dicer was able to cleave dsRNA to generate the corresponding 21-mer lipid–siRNA conjugates exhibiting resistance to nucleases in cell culture medium. Interestingly, 27-mer dsRNA conjugated with palmitic acid afforded better efficiencies and RNAi potencies in the presence of lipofectamine than unmodified siRNAs and 21-mer dsRNA conjugates modified with palmitic acid, Chol, and lauric acid at 0.2 nM. Transfection experiments realized without using lipofectamine confirmed the suitability of 27-mer dsRNAs modified with palmitic acid to inhibit luciferase and VEGF production at 600 nM and 5 μM, respectively.

Computational studies have proved to be an important tool in the design of modifications at the 3′-overhangs to discern conformational changes of Ago2 like the two-state model or lodging/dislodging motion. As previously described, these parameters have been directly involved in RNAi efficacy [89, 96, 103, 104]. In addition to studying 3′-overhangs of the guide strand, Beal and coworkers performed studies of human-Ago2 (hAgo2) binding modes by computational screenings of purine analogues introduced at the 5′-termini of the guide strand [105]. This computational analysis allowed the authors to engage a rational design for effective MID and P-element-induced wimpy testes (PIWI) domain binders to enhance RNAi activity [106]. The authors synthesized three purine derivatives: 7-ethyyl-8-aza-7-deazaadenosine (7-EAA), 7-EAA-triazol and 2-propargylaminopurine triazol derivative (2-AP-triazol) (Figure 9.4) [107]. All of them were introduced at the 5′-termini of a siRNA guide strand targeting PIK3CB gene [108] and were able to promote gene silencing except for those purine derivatives containing nitrogen substituents or modified Hoogsteen faces. Interestingly, the authors also found new modifications for nucleosides that remarkably improved the silencing effect, when they were introduced at the inner positions (either position 12 or 19) of the siRNA guide strand.

The inner positions of the siRNA guide strand have also been the focus of research in order to determine the efficacy and potency of siRNA oligonucleotides. Many groups have modified such positions in order to obtain stable and more effective siRNA molecules against nucleases with longer half-lives. The correct designing of such siRNAs focusing on improving stability and reducing off-target effects has allowed to launch novel modified siRNA conjugates for *in vivo* applications [109]. Such modifications can also involve blunt 19-mer duplexes containing multiple modifications [110]. In particular, Kaufmann and coworkers found that modifying the 2′-position of the ribose with a 2′-*O*-methyl pendent group greatly increased the melting denaturation of the siRNA duplex and also enhanced its nuclease resistance. Significant contributions in the RNAi activity were revealed depending on the 2′-*O*-methyl ribose position in the siRNA duplex. Thus, while fully modified siRNAs completely lost siRNA inhibition activity, alternating 2′-*O*-methyl RNA bases in the guide strand improved the stability in the serum and showed a good tolerance and potency when compared with unmodified siRNAs.

Similarly, other kinds of 2′-RNA nucleobase modifications have been established and evaluated. Prakash et al. studied the effect of 2′-F, 2′-OMe and

2′-OMOE RNA bases introduced at several positions of the siRNA guide strand [111]. While 2′-F RNA bases did not affect the inhibition levels of phosphatase and tensin homolog (PTEN) mRNA regardless of the strand position, the position of OMe and OMOE substituents remarkably affected the RNAi activity. Thus, the authors found that these bulky modifications actively preferred the 3′-termini strand than the inner positions of the siRNA guide. Overall, the correct design of siRNAs containing 2′-modifications in the ribose such as F, OMe, and/or OMOE groups have proved to be promising siRNA modifications for the validation in animal models [112].

In addition to modifying nucleobases, other chemical modifications have been proposed, which can be introduced either into the guide or the passenger strand without altering the inhibitory properties of natural siRNAs. Alagia et al. put their emphasis on positions 9 and 10 of the siRNA guide strand, positions which are directly involved in the cleavage of the mRNA target by Ago2 [113]. The authors prepared siRNA oligonucleotides containing acyclic nucleic acids based on L-threoninol-thymine and their results were compared to those siRNAs containing a natural uridine at the same scissile phosphate positions in the guide strand. A rational study carried out by the authors allowed them to show that introducing the thymidine modification at position 10 strongly affected the RNAi activity (10 times less) whereas position 11 produced a slight decrease in the siRNA potency when compared with unmodified siRNAs. Further off-target studies involving L-threoninol-thymine building block were conducted by the same authors. The sense/antisense ratio was evaluated by introducing the modification at position 2 of the passenger strand. This produced a total selection of the guide strand versus the passenger strand when L-threoninol-thymine was introduced and therefore produced the expected gene silencing effect with comparable silencing activities to unmodified siRNA.

References

1 Watson, J.D. and Crick, F.H. (1953). Molecular structure of nucleic acids; a structure for deoxyribose nucleic acid. *Nature* 171 (4356): 737–738.

2 Wing, R., Drew, H., Takano, T. et al. (1980). Crystal structure analysis of a complete turn of B-DNA. *Nature* 287 (5784): 755–758.

3 Stein, C.A. and Castanotto, D. (2017). FDA-approved oligonucleotide therapies in 2017. *Mol. Ther.* 25 (5): 1069–1075.

4 Chakraborty, C., Sharma, A.R., Sharma, G. et al. (2017). Therapeutic miRNA and siRNA: moving from bench to clinic as next generation medicine. *Mol. Ther. Nucleic Acids* 8: 132–143.

5 Bennett, C.F., Baker, B.F., Pham, N. et al. (2017). Pharmacology of antisense drugs. *Annu. Rev. Pharmacol. Toxicol.* 57: 81–105.

6 Zhou, J. and Rossi, J. (2017). Aptamers as targeted therapeutics: current potential and challenges. *Nat. Rev. Drug Discovery* 16 (3): 181–202.

7 Merkel, M., Peewasan, K., Arndt, S. et al. (2015). Copper-free postsynthetic labeling of nucleic acids by means of bioorthogonal reactions. *ChemBioChem* 16 (11): 1541–1553.

8 Egli, M. and Pallan, P.S. (2010). Crystallographic studies of chemically modified nucleic acids: a backward glance. *Chem. Biodivers.* 7 (1): 60–89.

9 Tintoré, M., Eritja, R., and Fàbrega, C. (2014). DNA nanoarchitectures: steps towards biological applications. *ChemBioChem* 15 (10): 1374–1390.

10 Iyama, T., David, M., and Wilson, D.M. (2013). DNA repair mechanisms in dividing and non-dividing cells. *DNA Repair* 12 (8): 620–636.

11 Middleton, M.R. and Margison, G.P. (2003). Improvement of chemotherapy efficacy by inactivation of a DNA-repair pathway. *Lancet Oncol.* 4 (1): 37–44.

12 Kawanishi, S., Hiraku, Y., and Oikawa, S. (2001). Mechanism of guanine-specific DNA damage by oxidative stress and its role in carcinogenesis and aging. *Mutat. Res.* 488 (1): 65–76.

13 Cadet, J., Wagner, J.R., Shafirovich, V., and Geacintov, N.E. (2014). One-electron oxidation reactions of purine and pyrimidine bases in cellular DNA. *Int. J. Radiat. Biol.* 90: 423–432.

14 Sabharwal, A. and Middleton, M.R. (2006). Exploiting the role of O^6-methylguanine-DNA-methyltransferase (MGMT) in cancer therapy. *Curr. Opin. Pharmacol.* 6 (4): 355–363.

15 Pepponi, R., Marra, G., Fuggetta, M.P. et al. (2003). The effect of O^6-alkylguanine-DNA alkyltransferase and mismatch repair activities on the sensitivity of human melanoma cells to temozolomide, 1,3-bis(2-chloroethyl)1-nitrosourea, and cisplatin. *J. Pharmacol. Exp. Ther.* 304 (2): 661–668.

16 Pegg, A.E., Swenn, K., Chae, M.Y. et al. (1995). Increased killing of prostate, breast, colon, and lung tumor cells by the combination of inactivators of O6-alkylguanine-DNA alkyltransferase and N,N′-bis(2-chloroethyl)-N-nitrosourea. *Biochem. Pharmacol.* 50 (8): 1141–1148.

17 Wilson, B.D., Strauss, M., Stickells, B.J. et al. (1994). An assay for O6-alkylguanine-DNA alkyltransferase based on restriction endonuclease inhibition and magnetic bead separation of products. *Carcinogenesis* 15 (10): 2143–2148.

18 Dolan, M.E., Schicchitiano, D., and Pegg, A.E. (1988). Use of oligodeoxynucleotides containing O^6-alkylguanine for the assay of O^6-alkylguanine-DNA-alkyltransferase activity. *Cancer Res.* 48 (5): 1184–1188.

19 Moser, A.M., Patel, M., Yoo, H. et al. (2000). Real-time fluorescence assay for O^6-alkylguanine-DNA alkyltransferase. *Anal. Biochem.* 281 (2): 216–222.

20 Wu, R.S., Hurst-Calderone, S., and Kohn, K.W. (1987). Measurement of O6-alkylguanine-DNA alkyltransferase activity in human cells and tumor tissues by restriction endonuclease inhibition. *Cancer Res.* 47 (23): 6229–6235.

21 Juillerat, A., Heinis, C., Sielaff, I. et al. (2005). Engineering substrate specificity of O^6-alkylguanine-DNA alkyltransferase for specific protein labeling in living cells. *ChemBioChem* 6 (7): 1263–1269.

22 Juillerat, A., Gronemeyer, T., Keppler, A. et al. (2003). Directed evolution of O^6-alkylguanine-DNA alkyltransferase for efficient labeling of fusion proteins with small molecules in vivo. *Chem. Biol.* 10 (4): 313–317.

23 Keppler, A., Gendreizig, S., Gronemeyer, T. et al. (2003). A general method for the covalent labeling of fusion proteins with small molecules in vivo. *Nat. Biotechnol.* 21 (1): 86–89.

24 Li, X., Qian, S., Zheng, L. et al. (2012). A mechanism-based fluorescent probe for labeling O^6-methylguanine-DNA methyltransferase in live cells. *Org. Biomol. Chem.* 10 (16): 3189–3191.

25 Macaya, R.F., Schultze, P., Smith, F.W. et al. (1993). Thrombin-binding DNA aptamer forms a unimolecular quadruplex structure in solution. *Proc. Natl. Acad. Sci. U.S.A* 90 (8): 3745–3749.

26 Tintoré, M., Aviñó, A., Ruiz, F.M. et al. (2010). Development of a novel fluorescence assay based on the use of the thrombin binding aptamer for the detection of O6-alkylguanine-DNA alkyltransferase activity. *J. Nucleic Acids* 2010: 632041, 9 pages. doi: 10.4061/2010/632041.

27 Tintoré, M., Gállego, I., Manning, B. et al. (2013). DNA Origami as DNA repair nanosensor at the single-molecule level. *Angew. Chem. Int. Ed.* 52 (30): 7747–7750.

28 Tintoré, M., Mazzini, S., Polito, L. et al. (2015). Gold-coated superparamagnetic nanoparticles for single methyl discrimination in DNA aptamers. *Int. J. Mol. Sci.* 16 (11): 27625–27639.

29 Tintoré, M., Grijalvo, S., Fàbrega, C., and Eritja, R. (2015). Synthesis of oligonucleotides carrying fluorescently labelled O^6-alkylguanine for measuring hAGT activity. *Bioorg. Med. Chem. Lett.* 25 (22): 5208–5211.

30 Beharry, A.A., Nagel, Z.D., Samson, L.D., and Kool, E.T. (2016). Fluorogenic real-time reporters of DNA repair by MGMT, a clinical predictor of antitumor drug response. *PLoS One* 11 (4): e0152684.

31 Lam, E.Y., Beraldi, D., Tannahill, D., and Balasubramanian, S. (2013). G-quadruplex structures are stable and detectable in human genomic DNA. *Nat. Commun.* 4: 1796.

32 Bell, N.M. and Micklefield, J. (2009). Chemical modification of oligonucleotides for therapeutic, bioanalytical and other applications. *ChemBioChem* 10 (17): 2691–2703.

33 Doluca, O., Withers, J.M., and Filichev, V.V. (2013). Molecular engineering of guanine-rich sequences: Z-DNA, DNA triplexes, and G-quadruplexes. *Chem. Rev.* 113 (5): 3044–3083.

34 Felsenfeld, G., Davies, D.R., and Rich, A. (1957). Formation of a three-stranded polynucleotide molecule. *J. Am. Chem. Soc.* 79 (8): 2023–2024.

35 Soyfer, V.N. and Potaman, V.N. (2012). *Triple-helical nucleic acids.* New York: Springer-Verlag.

36 Duca, M., Vekhoff, P., Oussedik, K. et al. (2008). The triple helix: 50 years later, the outcome. *Nucleic Acids Res.* 36 (16): 5123–5138.

37 Simon, P., Cannata, F., Concordet, J.P., and Giovannangeli, C. (2008). Targeting DNA with triplex-forming oligonucleotides to modify gene sequence. *Biochimie* 90 (8): 1109–1116.

38 Robles, J., Grandas, A., Pedroso, E. et al. (2002). Nucleic acid triple helices: stability effects of nucleobase modifications. *Curr. Org. Chem.* 6 (14): 1333–1368.

39 Doluca, O., Boutorine, A.S., and Filichev, V.V. (2011). Triplex-forming twisted intercalating nucleic acids (TINAs): design rules, stabilization of antiparallel DNA triplexes and inhibition of G-quartet-dependent self-association. *ChemBioChem* 12 (15): 2365–2374.

40 Aviñó, A., Frieden, M., Morales, J.C. et al. (2002). Properties of triple helices formed by parallel-stranded hairpins containing 8-aminopurines. *Nucleic Acids Res.* 30 (12): 2609–2619.

41 Avinó, A., Cubero, E., González, C. et al. (2003). Antiparallel triple helices. Structural characteristics and stabilization by 8-amino derivatives. *J. Am. Chem. Soc.* 125 (51): 16127–16138.

42 Carrascosa, L.G., Gómez-Montes, S., Aviñó, A. et al. (2012). Sensitive and label-free biosensing of RNA with predicted secondary structures by a triplex affinity capture method. *Nucleic Acids Res.* 40 (8): e56.

43 Aviñó, A., Huertas, C.S., Lechuga, L.M., and Eritja, R. (2016). Sensitive and label-free detection of miRNA-145 by triplex formation. *Anal. Bioanal. Chem.* 408 (3): 885–893.

44 Huertas, C.S., Aviñó, A., Kurachi, C. et al. (2018). Label-free DNA-methylation detection by direct ds-DNA fragment screening using poly-purine hairpins. *Biosens. Bioelectron.* 120: 47–54.

45 Nielsen, P.E. (2010). Peptide nucleic acids (PNA) in chemical biology and drug discovery. *Chem. Biodivers.* 7 (4): 786–804.

46 Vester, B. and Wengel, J. (2004). LNA (locked nucleic acid): high-affinity targeting of complementary RNA and DNA. *Biochemistry* 43 (42): 13233–13241.

47 Obika, S., Uneda, T., Sugimoto, T. et al. (2001). 2′-O, 4′-C-methylene bridged nucleic acid (2′, 4′-BNA): synthesis and triplex-forming properties. *Bioorg. Med. Chem.* 9 (4): 1001–1011.

48 Alvira, M. and Eritja, R. (2010). Triplex-stabilizing properties of parallel clamps carrying LNA derivatives at the Hoogsteen strand. *Chem. Biodivers.* 7 (2): 376–382.

49 Kosbar, T.R., Sofan, M.A., Abou-Zeid, L., and Pedersen, E.B. (2015). Thermal stability of G-rich anti-parallel DNA triplexes upon insertion of LNA and α-L-LNA. *Org. Biomol. Chem.* 13 (18): 5115–5121.

50 Collie, G.W. and Parkinson, G.N. (2011). The application of DNA and RNA G-quadruplexes to therapeutic medicines. *Chem. Soc. Rev.* 40 (12): 5867–5892.

51 Sagi, J. (2014). G-quadruplexes incorporating modified constituents: a review. *J. Biomol. Struct. Dyn.* 32 (3): 477–511.

52 Lech, C.J., Lim, J.K.C., Lim, J.M.W. et al. (2011). Effects of site-specific guanine C8-modifications on an intramolecular DNA G-quadruplex. *Biophys. J.* 101 (8): 1987–1998.

53 Li, Z., Lech, C.J., and Phan, A.T. (2013). Sugar-modified G-quadruplexes: effects of LNA-, 2′ F-RNA–and 2′ F-ANA-guanosine chemistries on G-quadruplex structure and stability. *Nucleic Acids Res.* 42 (6): 4068–4079.

54 Aviñó, A., Fàbrega, C., Tintoré, M., and Eritja, R. (2012). Thrombin binding aptamer, more than a simple aptamer: chemically modified derivatives and biomedical applications. *Curr. Pharm. Des.* 18 (14): 2036–2047.
55 Saneyoshi, H., Mazzini, S., Avino, A. et al. (2009). Conformationally rigid nucleoside probes help understand the role of sugar pucker and nucleobase orientation in the thrombin-binding aptamer. *Nucleic Acids Res.* 37 (17): 5589–5601.
56 Sjuvarsson, E., Marquez, V.E., and Eriksson, S. (2015). Selective phosphorylation of south and north-cytidine and adenosine methanocarba-nucleosides by human nucleoside and nucleotide kinases correlates with their growth inhibitory effects on cultured cells. *Nucleosides Nucleotides Nucleic Acids* 34 (8): 544–564.
57 Aviñó, A., Mazzini, S., Ferreira, R. et al. (2012). The effect on quadruplex stability of North-nucleoside derivatives in the loops of the thrombin-binding aptamer. *Bioorg. Med. Chem.* 20 (14): 4186–4193.
58 Lech, C.J., Li, Z., Heddi, B., and Phan, A.T. (2012). 2′-F-ANA-guanosine and 2′-F-guanosine as powerful tools for structural manipulation of G-quadruplexes. *Chem. Commun.* 48 (93): 11425–11427.
59 Martín-Pintado, N., Yahyaee-Anzahaee, M., Campos-Olivas, R. et al. The solution structure of double helical arabino nucleic acids (ANA and 2′F-ANA): effect of arabinoses in duplex-hairpin interconversion. *Nucleic Acids Res.* 40 (18): 9329–9339.
60 Pasternak, A., Hernandez, F.J., Rasmussen, L.M. et al. (2010). Improved thrombin binding aptamer by incorporation of a single unlocked nucleic acid monomer. *Nucleic Acids Res.* 39 (3): 1155–1164.
61 Langkjær, N., Pasternak, A., and Wengel, J. (2009). UNA (unlocked nucleic acid): a flexible RNA mimic that allows engineering of nucleic acid duplex stability. *Bioorg. Med. Chem.* 17 (15): 5420–5425.
62 Zhou, J., Murayama, K., Amrane, S. et al. (2013). A "sugar-deficient" G-quadruplex: incorporation of aTNA in G4 structures. *Chem. Sci.* 4 (9): 3693–3698.
63 Asanuma, H., Toda, T., Murayama, K. et al. (2010). Unexpectedly stable artificial duplex from flexible acyclic threoninol. *J. Am. Chem. Soc.* 132 (42): 14702–14703.
64 Murayama, K., Tanaka, Y., Toda, T. et al. (2013). Highly stable duplex formation by artificial nucleic acids acyclic threoninol nucleic acid (aTNA) and serinol nucleic acid (SNA) with acyclic scaffolds. *Chemistry* 19 (42): 14151–14158.
65 Urata, H., Kumashiro, T., Kawahata, T. et al. (2004). Anti-HIV-1 activity and mode of action of mirror image oligodeoxynucleotide analogue of Zintevir. *Biochem. Biophys. Res. Commun.* 313 (1): 55–61.
66 Sczepanski, J.T. and Joyce, G.F. (2013). Binding of a structured D-RNA molecule by an L-RNA aptamer. *J. Am. Chem. Soc.* 135 (36): 13290–13293.
67 Szabat, M., Gudanis, D., Kotkowiak, W. et al. (2016). Thermodynamic features of structural motifs formed by β-L-RNA. *PLoS One* 11 (2): e0149478.
68 Benabou, S., Avino, A., Eritja, R. et al. (2014). Fundamental aspects of the nucleic acid i-motif structures. *RSC Adv.* 4 (51): 26956–26980.

69 Day, H.A., Pavlou, P., and Waller, Z.A. (2014). i-Motif DNA: structure, stability and targeting with ligands. *Bioorg. Med. Chem.* 22 (16): 4407–4418.

70 Dong, Y., Yang, Z., and Liu, D. (2014). DNA nanotechnology based on i-motif structures. *Acc. Chem. Res.* 47 (6): 1853–1860.

71 Dembska, A. (2016). The analytical and biomedical potential of cytosine-rich oligonucleotides: a review. *Anal. Chim. Acta* 930: 1–12.

72 Xu, C., Zhao, C., Ren, J., and Qu, X. (2011). pH-controlled reversible drug binding and release using a cytosine-rich hairpin DNA. *Chem. Commun.* 47 (28): 8043–8045.

73 Xu, B., Devi, G., and Shao, F. (2015). Regulation of telomeric i-motif stability by 5-methylcytosine and 5-hydroxymethylcytosine modification. *Org. Biomol. Chem.* 13 (20): 5646–5651.

74 Kumar, N., Petersen, M., and Maiti, S. (2009). Tunable c-MYC LNA i-motif. *Chem. Commun.* 45 (12): 1532–1534.

75 Pérez-Rentero, S., Gargallo, R., González, C., and Eritja, R. (2015). Modulation of the stability of i-motif structures using an acyclic threoninol cytidine derivative. *RSC Adv.* 5 (78): 63278–63281.

76 Aviñó, A., Dellafiore, M., Gargallo, R. et al. (2017). Stabilization of telomeric i-motif structures by (2′ S)-2′-deoxy-2′-C-methylcytidine residues. *ChemBioChem* 18 (12): 1123–1128.

77 Robaldo, L., Izzo, F., Dellafiore, M. et al. (2012). Influence of conformationally restricted pyrimidines on the activity of 10–23 DNAzymes. *Bioorg. Med. Chem.* 20 (8): 2581–2586.

78 Assi, H.A., Harkness, R.W., Martin-Pintado, N. et al. (2016). Stabilization of i-motif structures by 2′-β-fluorination of DNA. *Nucleic Acids Res.* 44 (11): 4998–5009.

79 Deleavey, G.F. and Damha, M.J. (2012). Designing chemically modified oligonucleotides for targeted gene silencing. *Chem. Biol.* 19 (8): 937–954.

80 Kole, R., Krainer, A.R., and Altman, S. (2012). RNA therapeutics: beyond RNA interference and antisense oligonucleotides. *Nat. Rev. Drug Discovery* 11 (2): 125–140.

81 Chiu, Y.-L. and Rana, T.M. (2003). siRNA function in RNAi: a chemical modification analysis. *RNA* 9 (9): 1034–1038.

82 Terrazas, M., Ocampo, S.M., Perales, J.C. et al. (2011). Effect of North byciclo[3.1.0]hexane-2′-deoxy-pseudosugars on RNA interference: a novel class of siRNA modification. *ChemBioChem* 12 (7): 1056–1065.

83 Terrazas, M., Alagia, A., Faustino, I. et al. (2013). Functionalization of the 3′-ends of DNA and RNA strands with N-ethyl-N-coupled nucleosides: a promising approach to avoid 3′-exonuclease-catalyzed hydrolysis of therapeutic oligonucleotides. *ChemBioChem* 14 (4): 510–520.

84 Eckstein, F. (2002). Developments in RNA chemistry, a personal view. *Biochimie* 84 (9): 841–848.

85 Schwarz, D.S., Hutvagner, G., Du, T. et al. (2003). Asymmetry in the assembly of the RNAi enzyme complex. *Cell* 115 (2): 199–208.

86 Reynolds, A., Leake, D., Boese, Q. et al. (2004). Rational siRNA design for RNA interference. *Nat. Biotechnol.* 22 (3): 326–330.

87 Bramsen, J.B., Laursen, M.B., Nielsen, A.F. et al. (2009). A large-scale chemical modification screen identifies design rules to generate siRNAs with high activity, high stability and low toxicity. *Nucleic Acids Res.* 37 (9): 2867–2881.

88 Somoza, A., Terrazas, M., and Eritja, R. (2010). Modified siRNAs for the study of the PAZ domain. *Chem. Commun.* 46 (24): 4270–4272.

89 Gaglione, M., Potenza, N., Di Fabio, G. et al. (2013). Tuning RNA interference by enhancing siRNA/PAZ recognition. *ACS Med. Chem. Lett.* 4 (1): 75–78.

90 Xu, L., Wang, X., He, H. et al. (2015). Structure-based design of novel chemical modification of the 3′-overhang for optimization of short interfering RNA performance. *Biochemistry* 54 (5): 1268–1277.

91 Lee, H.S., Lee, S.N., Joo, C.H. et al. (2007). Contributions of 3′-overhang to the dissociation of small interfering RNAs from the PAZ domain: molecular dynamics stimulation study. *J. Mol. Graphics Modell.* 25 (6): 784–793.

92 Hutvagner, G. and Simard, M.J. (2008). Argonaute proteins: key players in RNA silencing. *Nat. Rev. Mol. Cell Biol.* 9 (1): 22–32.

93 Kamiya, Y., Takai, J., Ito, H. et al. (2014). Enhancement of stability and activity of siRNA by terminal substitution with serinol nucleic acid (SNA). *ChemBioChem* 15 (17): 2549–2555.

94 Alagia, A., Terrazas, M., and Eritja, R. (2014). RNA/aTNA chimeras: RNAi effects and nucleases resistance of single and double stranded RNAs. *Molecules* 19 (11): 17872–17896.

95 Lingel, A., Simon, B., Izaurralde, E., and Sattler, M. (2004). Nucleic acid 3′-end recognition by the Argonaute2 PAZ domain. *Nat. Struct. Mol. Biol.* 11 (6): 576–577.

96 Alagia, A. and Eritja, R. (2016). SiRNA and RNAi optimization. *WIREs RNA* 7 (3): 316–329.

97 Kubo, T., Yanagihara, K., Takei, Y. et al. (2011). Palmitic acid-conjugated 21-nucleotide siRNA enhances gene-silencing activity. *Mol. Pharmacol.* 8 (6): 2193–2203.

98 Kubo, T., Yanagihara, K., Takei, Y. et al. (2012). SiRNAs conjugated with aromatic compounds induce RISC-mediated antisense strand selection and strong gene-silencing activity. *Biochem. Biophys. Res. Commun.* 426 (4): 571–577.

99 Takei, Y., Kadomatsu, K., Itoh, H. et al. (2002). 5′-3′-Inverted thymidine-modified antisense oligodeoxynucleotide targeting midkine. Its design and application for cancer therapy. *J. Biol. Chem.* 277 (26): 23800–23806.

100 Kim, D.-H., Behlke, M.A., Rose, S.D. et al. (2005). Synthetic dsRNA Dicer substrates enhance RNAi potency and efficacy. *Nat. Biotechnol.* 23 (2): 222–226.

101 Kubo, T., Zhelev, Z., Ohba, H., and Bakalova, R. (2007). Modified 27-nt dsRNAs with dramatically enhanced stability in serum and long-term RNAi activity. *Oligonucleotides* 17 (4): 445–564.

102 Kubo, T., Yanagihara, K., Takei, Y. et al. (2012). Lipid-conjugated 27-nucleotide doublé-stranded RNAs with Dicer-substrate potency enhance RNAi-mediated gene silencing. *Mol. Pharmacol.* 9 (5): 1374–1383.

103 Qiu, S., Adema, C.M., and Lane, T. (2005). A computational study of off-target effects of RNA interference. *Nucleic Acids Res.* 33 (6): 1834–1847.
104 Zander, A., Holzmeister, P., Klose, D. et al. (2014). Single-molecule FRET supports the two-state model of Argonaute action. *RNA Biol.* 11 (1): 45–56.
105 Frank, F., Sonenberg, N., and Nagar, B. (2010). Structural basis for 5′-nucleotide base-specific recognition of guide RNA by human AGO2. *Nature* 465 (7299): 818–822.
106 Onizuka, K., Harrison, J.G., Ball-Jones, A.A. et al. (2013). Short interfering RNA guide strand modifiers from computational screening. *J. Am. Chem. Soc.* 135 (45): 17069–17077.
107 Ibarra-Soza, J.M., Morris, A.A., Jayalath, P. et al. (2012). 7-Substituted 8-aza-7-deazaadenosines for modification of the siRNA major groove. *Org. Biomol. Chem.* 10 (32): 6491–6497.
108 Pu, P., Kang, C., Zhang, Z. et al. (2006). Downregulation of PIK3CB by siRNA suppresses malignant glioma cell growth in vitro and in vivo. *Technol. Cancer Res. Treat.* 5 (3): 271–280.
109 Behlke, M.A. (2008). Chemical modification of siRNAs for in vivo use. *Oligonucleotides* 18 (4): 305–320.
110 Czauderna, F., Fechtner, M., Dames, S. et al. (2003). Structural variations and stabilising modifications of synthetic siRNAs in mammalian cells. *Nucleic Acids Res.* 31 (11): 2705–2716.
111 Prakash, T.P., Allerson, C.R., Dande, P. et al. (2005). Positional effect on chemical modifications on short interference RNA activity in mammalian cells. *J. Med. Chem.* 48 (13): 4247–4253.
112 Paroo, Z. and Corey, D.R. (2004). Challenges for RNAi in vivo. *Trends Biotechnol.* 22 (8): 390–394.
113 Alagia, A., Terrazas, M., and Eritja, R. (2015). Modulation of the RNA interference activity using central mismatched siRNAs and acyclic threoninol nucleic acids (aTNA) units. *Molecules* 20 (5): 7602–7619.

10

Synthesis of Carbohydrate–Oligonucleotide Conjugates and Their Applications

Juan C. Morales

Consejo Superior de Investigaciones Científicas, Instituto de Parasitología y biomedicina "López-Neyra" (IPBLN-CSIC), Avenida del Conocimiento 17, PTS Granada, 18016 Armilla, Granada, Spain

10.1 Introduction

Glycolipids and glycoproteins are biomolecules widespread on the cell surface which are crucial for cell–cell interactions [1] and on the pathogen-cell recognition process [2]. In contrast, glycosylated oligonucleotides (ODNs) are unusual biomolecules that were discovered in 1960 in T-even phages [3], and found later in other bacteriophages [4] and in different parasites such as *Trypanosoma brucei* [5]. The most common glycosylation of DNA involves β-D-glucosylhydroxymethyluracil, names base J **1** (Figure 10.1) [6]. Other carbohydrates such as mannose and galactose have been found modifying nucleosides on tRNAs from rabbit and rat liver [7]. Although the biological role of glycosylated ODN strands is still not clear, these modifications increase DNA resistance to nucleases [8] and they may be involved in transcription regulation [9].

The interest in preparing nonnatural carbohydrate oligonucleotides conjugates (COCs) is basically to take advantage of properties of both biomolecules. On one side, synthetic ODNs are easy to prepare [10], carry a highly specific recognition code through canonical A–T and C–G base pairs, and possess a well-defined 3D structure on different topologies [11]. ODNs have been used as potential therapeutic agents (antisense oligonucleotides (ASOs) [12], triplex forming oligonucleotides (TFOs) [13], aptamers [14], ribozymes [15] and siRNA [16]), supramolecular scaffolds [17], biosensors [18], biochips [19], diagnostic tools [20], etc.

On the other side, carbohydrates are more difficult to synthesize than ODNs [21], possess complex 3D structures that are difficult to predict [22], and participate in crucial biological recognition processes such as pathogen infection, inflammatory pathways, and cancer metastasis [23]. Carbohydrate-binding proteins (also called lectins) [24] usually show low affinity for monosaccharides but tend to bind more strongly to more complex oligosaccharides. When these types of proteins are membrane receptors, they recognize and internalize different glycoproteins efficiently and are very selective for a specific type of oligosaccharide.

Enzymatic and Chemical Synthesis of Nucleic Acid Derivatives, First Edition.
Edited by Jesús Fernández-Lucas and María-José Camarasa Rius.

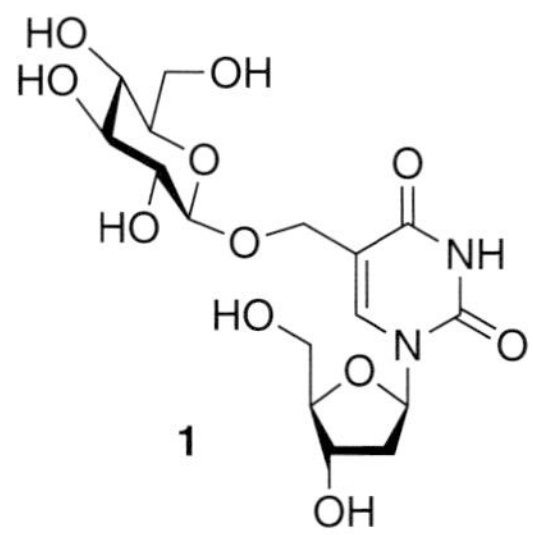

1

Figure 10.1 Nucleoside J.

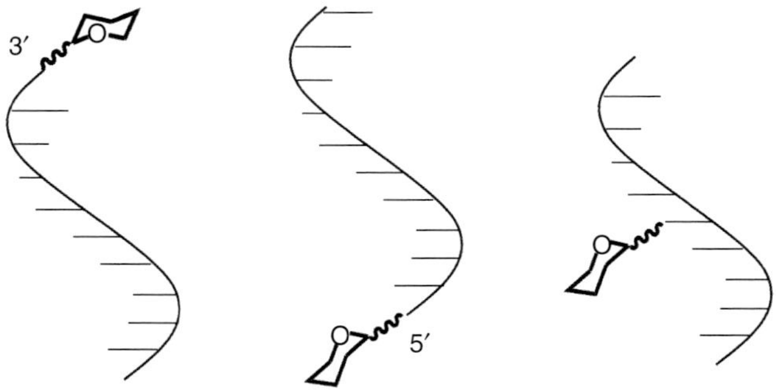

Figure 10.2 Three different types of COCs depending on the location of the carbohydrate on the oligonucleotide.

Since one of the main problems to develop ODNs as drugs is their poor cellular uptake, the preparation of COCs has been a general strategy to improve their cellular delivery. Moreover, selectivity for a specific type of tissue or cell also may be feasible. In fact, COCs started to be designed and synthesized in the mid 1990s for this purpose. In general terms, COCs can be divided into three types depending on where the carbohydrate is linked to the ODNs strand: to the 5′-end, to the 3′-end, or in the middle of the sequence (Figure 10.2). Several reviews have been published on the synthesis and applications of COCs [25]. This chapter aims to give a general view on COCs and cover the progress made on this topic in more detail since 2013.

10.2 Synthesis of COCs

The preparation of COCs is not a simple task since the chemical methods to synthesize oligosaccharides and ODNs are not compatible. Thus, no straightforward methodology has been described. Two main strategies have been used: (i) on-support synthesis, taking advantage of the chemistry developed for ODN preparation; and (ii) solution-phase synthesis, exploiting click chemistry approaches after the ODN has been cleaved from the resin and deprotected and purified.

10.2.1 On-Support Synthesis

The on-support synthesis of COCs can be classified depending on the type of linkage formed between the sugar and the ODNs.

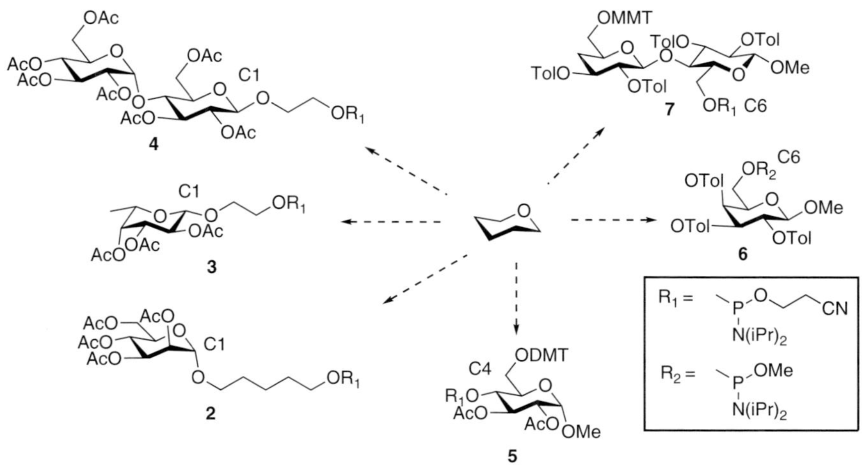

Figure 10.3 Examples of carbohydrate phosphoramidite derivatives showing the different positions of the sugar for the conjugation to the oligonucleotide. C1–C6 indicate the position of the pyranose ring to be linked to the oligonucleotide.

10.2.1.1 Phosphoramidite Chemistry

The most used approach is the one that follows standard ODN elongation on an automated DNA synthesizer using carbohydrate phosphoramidite derivatives for 5′-end or in-sequence conjugation. Actually, the first COC was prepared using this strategy. Akhtar et al. [26] synthesized a mannose phosphoramidite derivative **2** to be attached to a 15-mer ODN sequence through its anomeric position (C1) (Figure 10.3). The rest of the carbohydrate OH groups are normally protected with acyl groups (acetyl, toluoyl, etc.) and get deprotected on the cleavage step with ammonia. Morales et al. [27] prepared mono- and disaccharide phosphoramidite derivatives (e.g. **3**, **4**) using ethylene glycol spacers at C1 (Figure 10.3). The carbohydrate can also be attached through C4 or C6 allowing the exploration of other sugar presentations on the COC such as in compounds **5** and **6** (Figure 10.3). For in-sequence saccharide conjugation, Sheppard et al. [28] used standard carbohydrate protection–deprotection methodology to prepare a toluoyl protected 4′-deoxylactose **7**, with the phosphoramidite group at position C6 and modified with a monomethoxy (MMT) group at C6′ that would allow the elongation of the ODN chain.

For 3′-end conjugation carbohydrates are attached to the support and possess a DMT protected hydroxyl group (Figure 10.4). The first examples used a succinyl spacer to bind monosaccharides to the solid support as for compound **8** [29]. Later, Ikeda et al. [30] linked the anomeric position of a galactose unit to the support through a cyclopentane ring spacer as shown in **9**. Direct binding of the sugar to a Tentagel™ resin was reported by D'Onofrio et al. [31] where the azidoglucose unit was grown through amide chemistry and then modified with a spacer containing a DMT group (**10**).

10.2.1.2 Derivatization of Nucleoside Base Residues

A direct way of glycosylating DNA strands is to attach directly the carbohydrate unit to the nucleoside base and then prepare it for standard

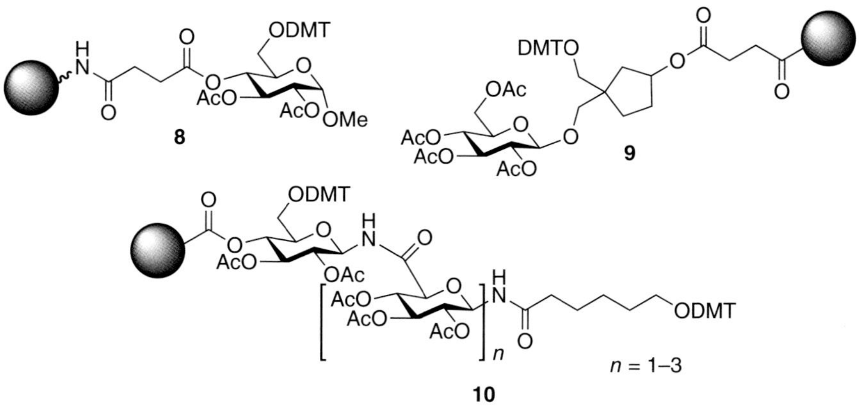

Figure 10.4 Examples of carbohydrates attached to the solid support and ready for oligonucleotide synthesis.

automated DNA synthesis with the corresponding DMT protecting group and phosphoramidite reactive group. Hunziker [32] prepared COCs containing up to five units of a 2-aminoglucose analogue of nucleoside J (Figure 10.5). The DMT carbohydrate-nucleoside phosphoramidite **11** was synthesized from uridine in 11 steps and the sugar was attached to the nucleoside through a glycosylation reaction in good yields. Matsuura et al. [33] attached an alkynyl modified galactose to *t*-butyldimethylsilane (TBDMS) protected 5-iodo deoxyuridine

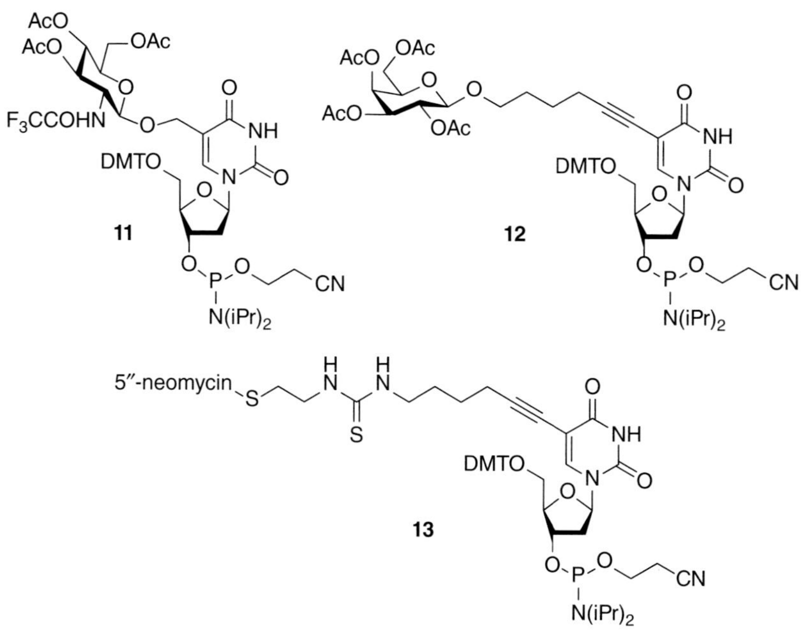

Figure 10.5 Examples of carbohydrates attached to the nucleoside and ready for oligonucleotide synthesis.

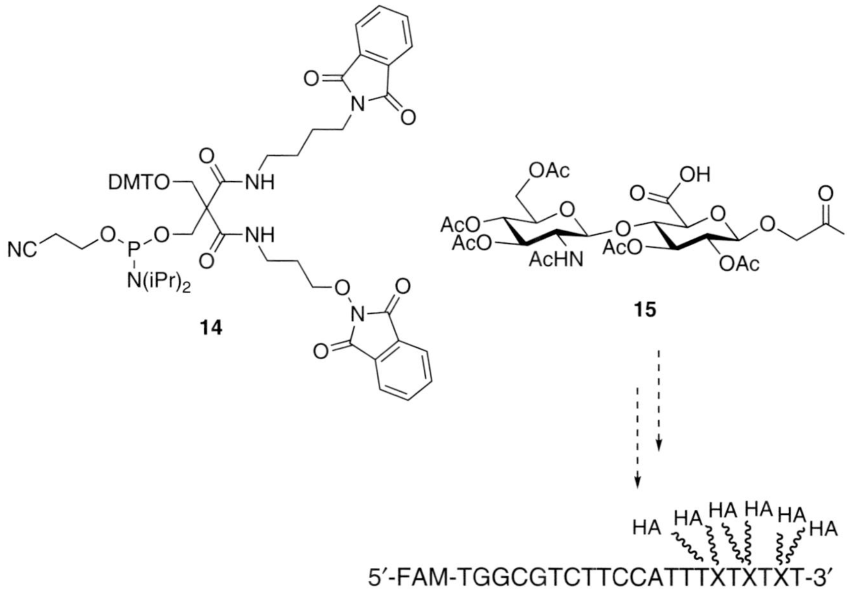

Figure 10.6 Schematic synthesis of HA_6-COC via oxime chemistry.

using a Heck reaction and later added the DMT and phosphoramidite groups to the deoxyribose ring to obtain derivative **12** (Figure 10.5). This methodology allowed to prepare COCs that could be recognized by galactose-binding lectins. With the aim of targeting RNA, Charles et al. [34] prepared an aminoglycoside ODN conjugate. In this case, a 5″-isothiocyanate neomycin derivative was coupled with a 5-amino uridine derivative and later modified with the corresponding DMT and phosphoramidite groups to obtain the thiourea derivative **13** (Figure 10.5). Actually, a 7-mer ODN with just one neomycin modification in the middle of the sequence was capable of increasing Tm by 6 °C when binding to its RNA target in comparison with the unmodified DNA.

10.2.1.3 Oximation Chemistry

Reaction of an aldehyde with an aminooxy group produces an oxime bond. Lonnberg and coworkers [35] used this type of conjugation by preparing a spacer with activated aminooxy groups **14** (Figure 10.6), incorporating it into DNA and later reacting the corresponding aldehyde on the hyaluronic disaccharide unit **15**. With this approach the authors could conjugate up to six hyaluronic disaccharides into DNA. However, the resulting COCs did not improve cellular uptake on cells expressing the hyaluronan (HA) receptor CD44.

10.2.1.4 Amide Chemistry

The amide bond found in peptides and proteins has also been used to conjugate carbohydrates and ODNs. Matulic-Adamic et al. [36] attached *N*-acetylgalactosamine containing a carboxylic acid activated spacer to a 2′-amino-2-deoxyuridine scaffold. However, the corresponding DMT-amidite derivative coupled to the growing ODNs with low yield and only one incorporation could be achieved.

10.2.1.5 Urea Chemistry

Reaction of D-glucosamine or D-galactosamine with an isocyanate alkyl spacer attached to the 5′-end of an anchored ODN produced a COC by formation of a urea bond. The resulting COC showed enhanced stability to endonuclease digestion [37].

10.2.1.6 CuAAC Chemistry

The copper-catalyzed azide alkyne cycloaddition (CuAAC) reaction has been widely employed to prepare COCs. It can be carried out by reaction of alkyne sugars and azide-ODNs or by reaction of azidosugars and alkyne-ODNs. The most frequently used is the latter case since the presence of azido and phosphoramidite groups in the same molecule rapidly decomposes in solution through a Staudinger reaction [38].

One of the few examples where alkynyl carbohydrates were reacted with azide-ODN anchored to the solid support was reported by Lonnberg and coworkers [39] They used *H*-phosphonate of 4′-*C*-azido-methylthymidine **16** to incorporate the azido group into the growing ODN in the resin (Figure 10.7). In this case acyl protected sugars were used and deprotected together with the ODN. On the other side, the use of azidosugars and alkyne-modified ODNs

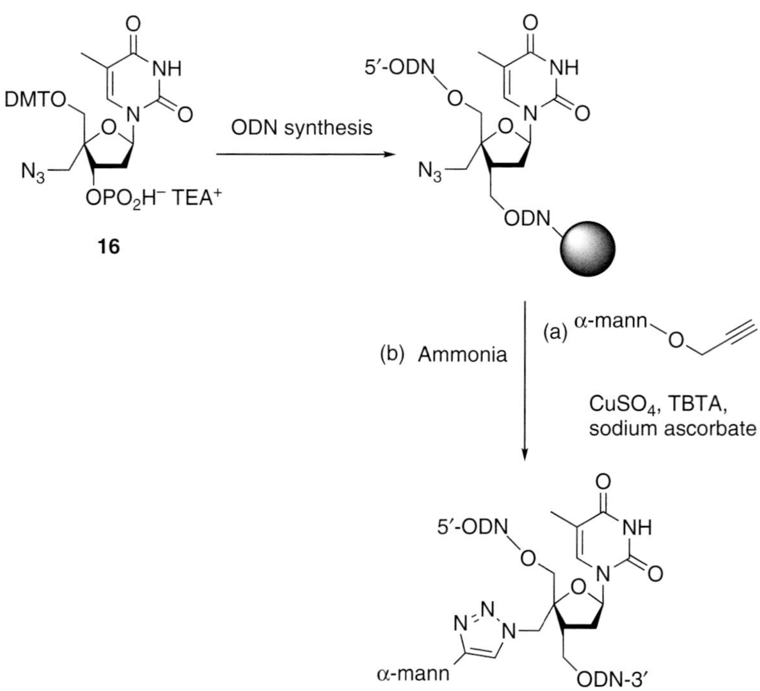

Figure 10.7 Synthesis of COC using alkynyl sugar and azide modified ODN for the coupling CuAAC reaction.

was first described by Morvan and coworkers [40]. The alkyne group was added to the growing ODN by amidative oxidation of propargyl amine into a phosphonate group in the ODN. Finally, the CuAAC reaction was carried out under microwave conditions to shorten reaction times. Later, propargyl groups were attached directly to the support or modified as phosphoramidite derivatives using a 2-(hydroxymethyl)-2-methylpropane-1,3-diol scaffold [41]. These modifications allowed direct growth of the ODN in the resin containing the alkynyl chains. CuAAC reaction with azidosugars right before cleavage from the resin and deprotection resulted in the desired COC. This approach was also used for sequential CuAAC with different azidosugars.

The use of CuAAC and on-support oxime conjugation together has been reported for the synthesis of COCs containing two or more types of carbohydrates [42]. Finally, the use of CuAAC chemistry has been extensively employed to prepare glycocluster–ODN conjugates (see Section 10.3.3) and it has been recently reviewed by Morvan and coworkers [25c]

10.2.2 Solution-Phase Conjugation

Conjugation of two biomolecules in solution needs several requirements such as water solubility, two mutually reactive groups, chemoselectivity, and orthogonality among the potential protecting groups. This concept extends to conjugation between carbohydrates and ODNs. The following reactions have been reported for the synthesis of COCs in solution.

10.2.2.1 Disulfide Formation

Oxidative coupling of two thiols produces a disulfide bond. This type of conjugation was used to prepare COCs by Mahato and coworkers [43] where the binding of the carbohydrate and the ODN is mediated by a protein. Mannose 6-phosphate **17** was conjugated to amino-bovine serum albumin through a thiourea bond. Later, the sugar–bovine serum albumin (BSA) conjugate was reacted with a thiol protected group which after activation was conjugated with a 5′-thiol ODN (Figure 10.8). The resulting COC was used as a TFO with improved delivery to the liver.

10.2.2.2 Nucleophilic Addition on Unsaturated Carbon

Several COCs have been prepared via Michael addition of thiol-substituted ODNs to an unsaturated carbonyl. In the first reported example by Hangeland et al. [44] a tri-GalNAc neoglycopeptide scaffold activated with maleimide was conjugated to a 5′-thiol ODN. The resulting COC showed cell-type specific uptake, with efficient delivery to human hepatocellular carcinoma cells (HepG2) as compared to human fibrosarcoma (HT1080) and human promyleocytic cells (HL-60).

This type of conjugation was also used to prepare a carbohydrate–siRNA conjugate containing an acid labile β-thiopropionate linker. Here, Michael addition reaction of a 5′-thio-modified RNA sequence with a ω-acrylate-PEG-β-lactose derivative produced a sugar-RNA sense sequence. These conjugates were annealed with the corresponding antisense RNA strand and their capacity

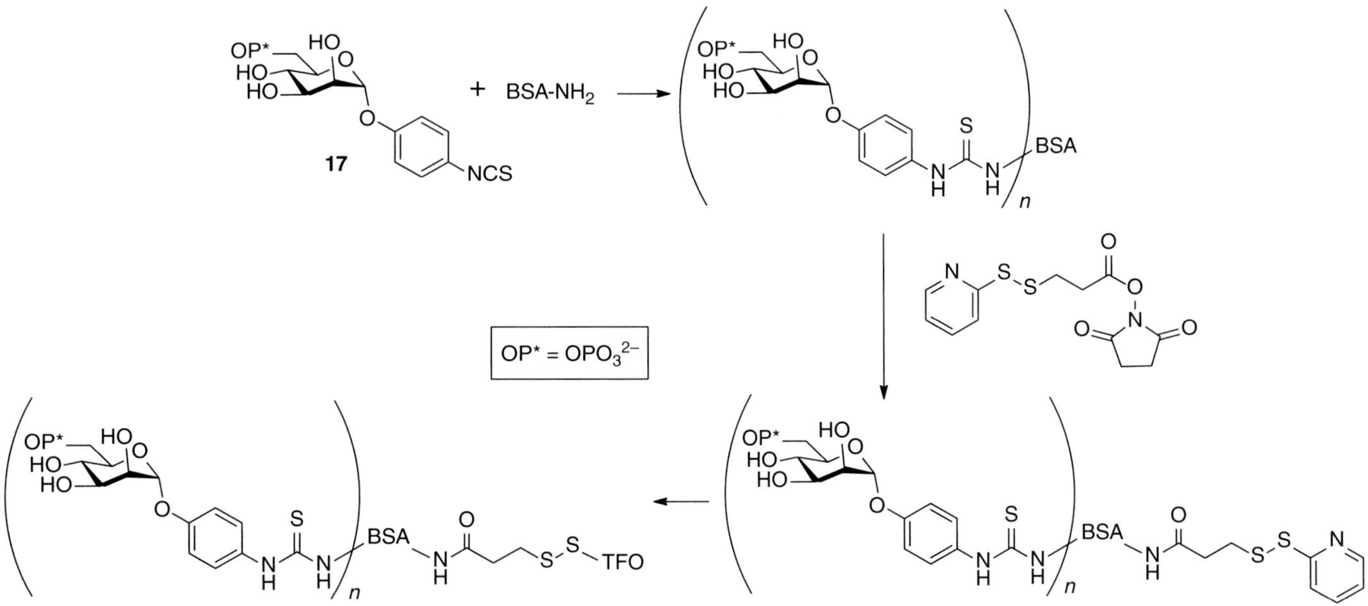

Figure 10.8 Schematic synthesis of mannose-BSA-TFO conjugate.

for gene inhibition was investigated alone or after micelle formation with poly-L-lysine. The authors found that the glycocluster presentation of the lactose–siRNA conjugate within the micelle exhibited higher gene silencing for firefly luciferase expression in HuH-7 cells than the carb–siRNA conjugate alone [45]. The same conjugation strategy was used by Zhu et al. [46] to prepare a galactose–PEG–ODN conjugate. They studied biodistribution of the COC in rats and observed that the galactose-PEG-ODN was rapidly cleared from the circulation and 60.2% accumulated in liver after 30 minutes. They proposed that the COC was being uptaken by hepatocytes through asialoglycoprotein receptor (ASGPR)-mediated endocytosis.

10.2.2.3 Carbonyl Addition–Elimination Reaction

Among this type of reactions, COCs have been prepared using oxime formation and reductive amination. Oxime formation was applied [47] for the conjugation of aminosugars with ODNs containing an aldehyde moiety at the 5′-end. This methodology was carried out without the need of protecting groups and under mild aqueous conditions and produced high yields. The same group used 3′,5′-bifunctionalyzed ODNs with aldehyde groups and could produce COCs with sugars at both ends of the ODNs [48]. Conjugation of unprotected carbohydrate (maltose, lactose, cellobiose, and maltoheptose) with aminoalkylated ODNs under mild reductive amination conditions occurred without notable side reactions as reported by Sando et al. [49]

10.2.2.4 CuAAC Chemistry

COCs have been widely prepared using CuAAC reaction in solution. Most of these examples are glycocluster–ODN conjugates and are discussed in Section 10.3.3. Interesting combinations of CuAAC and other methods have been described. The synthesis of diglycosylated ODNs by sequential oxime and CuAAC was carried out by Morvan and coworkers [50] Either lactose or galactose could be placed at the 5′-end using the azidosugar or at the 3′-end using the aminooxy carbohydrate. Another elegant example of COC synthesis with combined reactions in solution (discussed in Section 10.2.2.7) used first DNA polymerase incorporation of alkynyl-modified nucleosides and then CuAAC reaction with azidosugars [51].

10.2.2.5 Diazocoupling Reaction

Incorporation of multiple carbohydrate moieties in plasmid DNA was achieved by Kobayashi and coworkers [52] by reaction of lactose containing a diazo group with the C-8 position of guanines. The sugar modification in DNA enhanced the resistance to restriction enzymes and showed strong binding to a galactose-specific lectin.

10.2.2.6 Amide Bond Formation

Mahato and coworkers applied standard peptide chemistry to prepare a COC containing mannose-6-phosphate (M6P) attached to a TFO through a tetrapeptide methacrylamide (HMPA) polymer [53]. They first conjugated amine-modified M6P to acyl-activated HMPA, and then to the 3′-amino-TFO via the tetrapeptide linker. The obtained carb–TFO conjugates accumulated in the liver after injection to rats and the authors proposed that they could be potentially used for treating liver fibrosis.

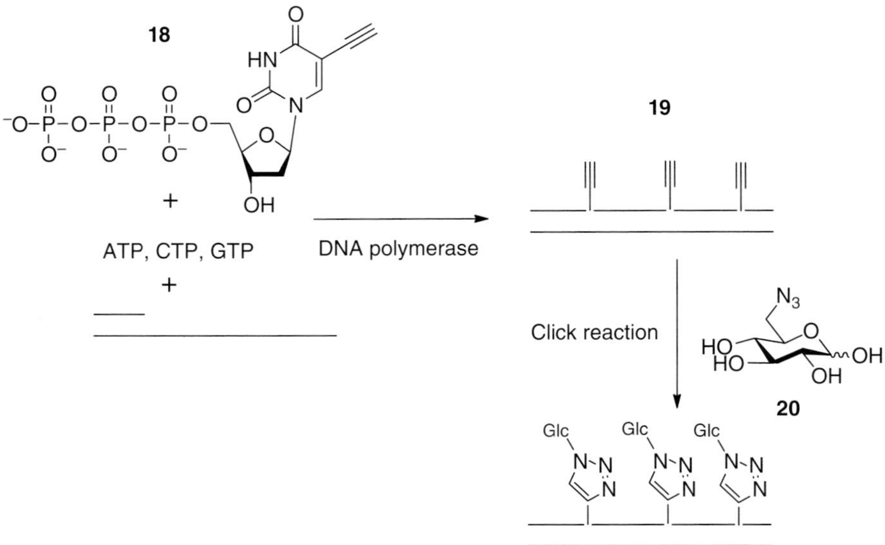

Figure 10.9 Schematic synthesis of COC combining enzymatic incorporation of alkynyl modified nucleosides and posterior CuAAC reaction with azidosugars.

10.2.2.7 Enzymatic Incorporation of Saccharides or Nucleotides

A very elegant way of incorporating extra carbohydrate units to a COC was reported by Wang and Sheppard [54]. After preparing GlcNAc–DNA conjugates by solid-phase synthesis using a GlcNAc phosphoramidite, the resulting COCs were used as substrates for glycosyl transferase enzymes. Thus, successive treatment with β-1,4-galatosyltransferase and α-1,3-fucosyltransferase resulted in a LeXsugar–DNA conjugate.

Burley et al. [51] also employed enzymes to prepare COCs. In this case, they incorporated 5-alkynyl thymine triphosphates **18** into a primer DNA hybridized to a DNA template using a DNA polymerase. Then, the acetylene-decorated ODN **19** was reacted with 6-azidoglucose **20** or other azido sugar constructs under the Cu(I)-catalyzed Huisgen 1,3-cycloaddition reaction conditions (Figure 10.9). The resulting COCs were used as nanoscaffolds to deposit silver and construct a DNA nanowire. The same group also reported the incorporation of alkyne-bearing deazapurine triphosphates by DNA polymerases and follow-up click reaction with galactose azides [55]. A similar approach was applied by Matsui et al. [56] incorporating lactose- and maltose- modified deoxyuridine triphosphates with a DNA polymerase. The authors could introduce up to five sugar units and then prepare a fully sugar-decorated three-way junction (3WJ) (see Section 10.3.2).

10.3 Synthesis of Glycocluster Oligonucleotides

As was mentioned in the introduction, lectins show low binding affinity for monosaccharides. In fact, in nature, this type of recognition processes uses multivalent interactions where receptors bind complex oligosaccharides instead.

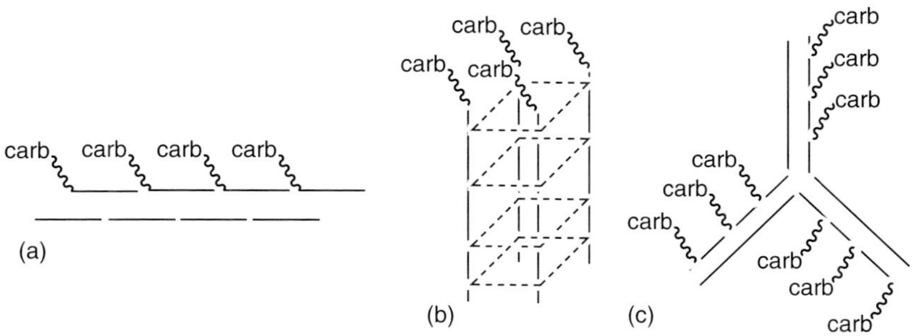

Figure 10.10 (a) COC using a dsDNA scaffold. (b) COC using a G-quadruplex scaffold. (c) COC using a three-way-junction scaffold.

This concept has inspired the design and synthesis of a large number of multivalent glycoclusters attached to ODNs with varied carbohydrate presentation in order to improve lectin binding and ODNs cellular uptake. We present some of the reported glycocluster–ODN conjugates depending on the scaffold used to anchor the corresponding saccharides.

10.3.1 dsDNA Scaffolds

The use of dsDNA as a platform to locate different saccharide units was first described by Kobayashi and coworkers [57]. A COC was synthesized by the reaction of a iodoacetamido oligomeric poly-mannose or poly-galactose with a 5′-thiol ODN. Then, hybridization of the COC with an alternated complementary ssDNA resulted in a fully sugar-decorated dsDNA that could bind Concanavalin A lectin stronger than to the poly-mannose–ssDNA conjugate (Figure 10.10a). The same authors continued this approach by hybridizing ssDNA with COCs where the carbohydrate had been incorporated in the middle of the sequence using a galactose deoxyuridine phosphoramidite and standard DNA automated synthesis [33b].

10.3.2 Non-Canonical DNA Scaffolds (G4 and three-Way Junction)

Several glycoclusters have been assembled taking advantage of the different secondary structures that DNA can adopt. Montesarchio and coworkers [58] exploited the capacity of DNA to form G-quadruplex structures through the stacking of two or more guanine tetrads (Figure 10.10b). They were able to bind mono- and disaccharides at the 3′-end or at the 5′-end of a TGGGAG sequence using carbohydrate phosphoramidite derivatives. These COCs could assemble on a G-quadruplex structure and the 3′-mannose and 3′-glucose tetraplex ODNs showed anti-HIV activity, whereas the corresponding 5′-carbohydrate ODNs were not active.

Matsui and Ebara [59] displayed 3, 6, or 12 maltose units on each branch of a 3WJ (Figure 10.10c). The corresponding COCs were prepared using enzymatic synthesis. When the glycocluster was formed by 3 × 6 maltoses, binding with

ConA lectin was 83-fold higher than for the 1 × 6 maltose array. But more interestingly, it was also better than the 3WJ with 3 × 12 maltoses. The authors argue that the higher flexibility of the glycocluster on the 3 × 6 maltose 3WJ could better fit the binding site of ConA.

10.3.3 Organic Spacer Scaffolds

One of the first approaches to prepare multivalent glycocluster–DNA conjugates was to use small organic scaffolds as those employed for dendrimers that would allow branching from the point of connection between the glycocluster and the ODN. Dubber and Frechet [60] synthesized glycodendrimer–ODN conjugates using carbohydrate phosphoramidites as **21**, spacer phosphoramidites as **22**, and DMT-branched phosphoramidites **23** (Figure 10.11) on a standard automated DNA synthesizer. The glycocluster was formed by three galactose or four mannose units and was linked to the 5′-end of the ODN.

A pentaerythritol-based phosphoramidite, another branching scaffold, was used by Katajisto et al. [61] to synthesize multiantennary ODNs conjugates. This time the branching unit could be placed in the middle of a DNA sequence or at the 5′-end and it was reacted with aldehyde-modified carbohydrates to form the corresponding oxime bonds. Up to six mannose units were bound through three branching units to the 5′-end of a 12-mer ODN. The same authors used a similar scaffold but protected with TBDMS or Lev groups to be cleaved when needed and then connected different monosaccharide phosphoramidite derivatives to these anchoring points. This strategy allowed the attachment of three different glycosyl units, for example, mannose, galactose, and glucose on the same glycocluster–DNA conjugate [62].

The first spatial design of a glycocluster ASO was carried out by Manoharan and coworkers [63] Their design was based on the use of a cholane scaffold and determining where to attach the three galactose units via ε-aminocapramide linkers in order to locate the sugar moieties to the right geometry to bind the binding sites of ASGPR. The carbohydrates were bound to the cholate unit via amide bonds and the resulting triGal-cholate scaffold was attached to the 5′-end of the ODN

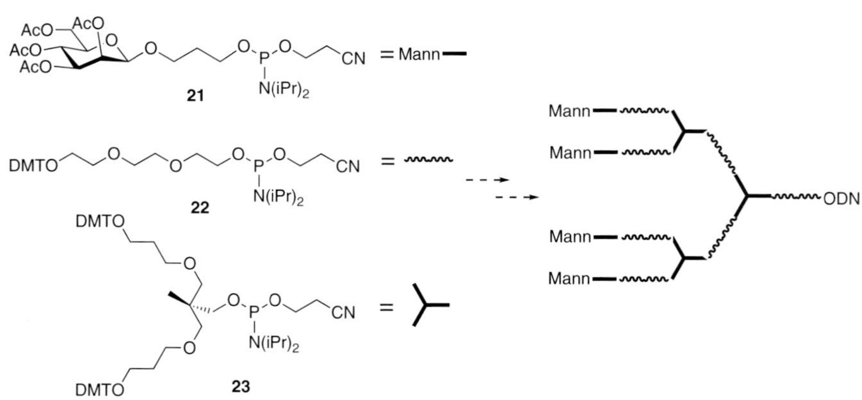

Figure 10.11 Schematic representation of the synthesis of a COC with a dendrimeric scaffold.

using phosphoramidite chemistry and to the 3′-end using a succinic spacer to anchor it to the resin.

More recently, several organic scaffolds have been mixed at once as lego pieces to create diversity on the glycocluster presentation. This is the case where Chevolot et al. [64] used cyclohexane-1,4-diyldimethylene as a semi-rigid spacer, tetraethylene glycol as a flexible spacer and trebler units to create the branching via phosphoramidite chemistry, followed by CuAAC reactions to attach azidogalactose units to produce highly diverse glycocluster–ODN conjugates. The authors hybridized them to DNA chips for lectin recognition. This sort of application is explained in more detailed in Section 10.4.3.

10.3.4 Biomolecules as Scaffolds

In the early days of designing and preparing COCs, Hangeland et al. [65] decided to use a tripeptide as scaffold to spatially locate three GalNAc monosaccharides into a glycocluster–DNA conjugate. The YEE(ahGalNAc)$_3$ moiety was modified with a maleimide group to react with a thiol group at the 5′-end of an ASO (Figure 10.12a). The resulting COC improved cellular uptake in hepatocytes and showed 20-fold better gene inhibition than the unconjugated ODN.

An elegant strategy to present carbohydrates in space was reported by Singh et al. [66]. A cyclodecapeptide scaffold containing five lysine residues was chemically modified to be capable of attaching five carbohydrate units. Serine residues were coupled to the lysine units, oxidized to aldehyde groups, and then reacted with aminooxylated lactose to obtain the corresponding oxime bonds. Then the glycosylated peptide scaffold was bound to the 5′-end of an ODN also via oxime chemistry. The authors found efficient binding of a lectin from *Arachis hypogaea* to the glycocluster–ODN conjugate whereas no binding was observed to a lactose–ODN analogue.

A mannose central core **24** (Figure 10.12b) was employed to attach four extra galactose or fucose units using a combination of phosphoramidite and CuAAC chemistry [67]. This type of glycoclusters could be linked to both the 3′- and 5′-end of an ODN also using CuAAC reactions. These bifunctional COCs were capable of binding to two different lectins: galactophilic lectin PA-IL bound specifically to a ManGal$_4$ cluster and fucose-binding lectin PA-IIL bound specifically to a ManFuc$_4$ cluster. Recently, this type of approach has been used to study the influence of the aromatic aglycon of galactoclusters on the binding of LecA. Docking simulations confirmed that the *O*-biphenyl- and *O*-naphtyl galactoside units participated in extra interactions between the aromatic ring and the histidine 50 and proline 51 in the binding site of LecA that explained the better affinity found for those glycoclusters [68].

A ribonucleoside, uracil nucleoside **25** (Figure 10.12c), has also been used as a central core to attach monosaccharides and prepare new glycoclusters [69]. In fact, ribo-, arabino-, and xylo-furanosyl uracil together with mannitol were modified with alkynyl groups ready to react with azido derivatives of galactose and fucose. Eighteen glycocluster–ODN conjugates with three-, penta-, and decavalent clusters with different topologies were prepared. The authors found that the

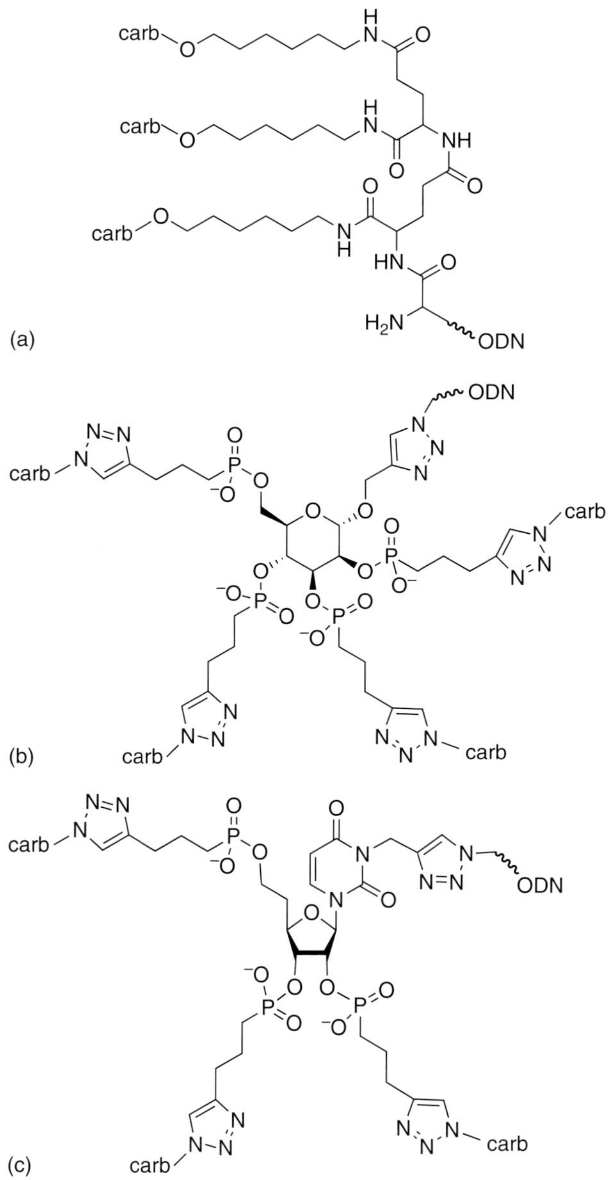

Figure 10.12 (a) COC based on a peptoid scaffold. (b) COC based on a mannose scaffold. (c) COC based on a uracil nucleoside scaffold.

trivalent xylo-centered galactocluster and the ribo-centered fucocluster showed the best affinity for LecA and LecB lectins, respectively.

10.4 Applications of COCs

Several areas of applications have been explored since the discovery of natural COCs and the first nonnatural COCs were prepared. Early on the highly promising therapeutic applications of ODNs clashed into their poor cellular uptake due to their large size and elevated negative electrostatic charge. To use carbohydrates as a vector to deliver ODNs inside cells appeared to be an interesting alternative and researchers started exploring COCs. A useful property of DNA is the well-defined 3D structure of duplex DNA and other DNA secondary structures which has been exploited to study molecular interactions in aqueous media or to prepare sensors. Finally, the advances on DNA immobilization via DNA chips have facilitated the presentation of glycan arrays through the hybridization of COCs. This sort of glycan presentation has allowed investigating lectin binding in a simple and direct way.

10.4.1 Improving Cellular Uptake

Drug delivery has been and is still today one of the major challenges of pharmacology. When the drug is a biomolecule such as an ODN, the difficulty is even bigger. Receptor-mediated drug delivery has been extensively investigated for decades and cell surface carbohydrate receptors started to be explored in the 1990s. The idea is to conjugate a biomolecule to a carbohydrate, then the glycoconjugate will be recognized by the membrane receptor and it will be internalized. This approach has been utilized for COCs taking advantage of different carbohydrate receptors (Table 10.1).

The first target receptor to be explored to improve COCs cellular uptake was the mannose-6-phosphate/insulin-like growth factor 2 receptor (M6P/IGF2R) [87]. It is a multifunctional single transmembrane glycoprotein that is known to regulate diverse biological functions and binds both M6P ligands and IGF2. The receptor cycles continuously between intracellular compartments and the plasma membrane and therefore, it could be a potential gate for biomolecules such as ODNs. Bonfils et al. [70] conjugated a 19-mer ODN to 6-phosphomannosylated serum albumin. They found the COC was more efficiently internalized into J774E cells than the free ODN, by a factor of 20×, and in the case of murine peritoneal macrophages by a factor of 7–12×.

Mahato and coworkers has also prepared several M6P–ODN conjugates for a better delivery to hepatic cells [88]. The conjugation between the carbohydrate and the ODN was mediated by BSA and the sugar multivalent presentation was varied (14, 20, or 27 M6P units). M6P–BSA–TFO conjugates entered rat liver stellate cells (HSC-T6) more efficiently than naked TFO ODNs and the optimal number of M6P for M6P-BSA-TFO was 20 [43]. After intravenous administration in rats, $(M6P)_{20}$-BSA-TFO accumulated in liver to higher extent than naked TFOs [72]. Moreover, in normal rats, 42.3% of total liver uptake was contributed by hepatic stellate cells (HSCs), 34.4% by Kupffer and endothelial cells, and

Table 10.1 Carbohydrate receptors used as potential target for COCs (shown in chronological order).

Target	Carbohydrate	ODN	Cells or animals	Comments	References
M6P/IGF2	Mannose-6-phophate (monovalent)	AS	J774E cells and murine peritoneal macrophages	$(M6P)_{16}$-serum albumin–ODN is internalized 20 times more efficiently than the free ODN	[70]
ASGPR	YEE(ah-GalNAc)3 (trivalent)	AS	Hep G2 cells	[YEE(ah-GalNAc)3]–ODN rapidly uptaken by Hep G2 cells and inhibited by YEE(ah-GalNAc)3	[44]
ASGPR	Galactose (tetravalent)	AS	Parenchymal liver cells	Accumulation of Gal_4–ODN was improved almost 60-fold	[71]
M6P/IGF2	Mannose-6-phophate (20 units)	TFO	Rats (intravenous adm.)	$(M6P)_{20}$-BSA-TFO accumulated in liver to higher extent than TFO	[43]
M6P/IGF2	Mannose-6-phophate (20 units)	TFO	Rats (intravenous adm.)	42.3% of the hepatic uptake of $(M6P)_{20}$-BSA-TFO was in hepatic stellate cells (HSCs), but increased to 82.0% in the HSCs of fibrotic rats	[72]
ASGPR	GalNAc (multivalent)	siRNA	C57BL/6 mice (intravenous adm.)	GalNAc-PBAVE polymer-siRNA inhibited apoB effectively in mice liver	[73]
ASGPR	Galactose	AS	Rats (intravenous adm.)	Gal–PEG–ODN accumulated in the liver at 30 minutes which was significantly higher than ODN.	[46]
M6P/IGF2	Mannose-6-phophate (15 units)	TFO	Rats (intravenous adm.)	Almost 80% of the total liver uptake of $(M6P)_{15}$-GFLG-HPMA-GFLG-TFO in fibrotic rats was contributed by HSCs	[53]
M6P/IGF2 ASGPR	Mannose-6-phophate (monovalent) Galactose	siRNA	HepG2 and HSC-T6 cells	Gal–PEG–siRNA and M6P–PEG–siRNA conjugates could silence luciferase gene expression by about 40%	[74]
GLUT	Glucose (up to tetravalent)	ODN	HeLa and U87.CD4.CXCR4 cells	Glc–ODNs with long spacers were uptaken better than ODNs (twofold)	[75]
GLUT ASGPR	Glucose and galactose (up to tetravalent)	siRNA	HeLa and HuH7 cells	Gal_2 and Gal_4-siRNA inhibited up to 25% TNF-α expression in HuH7 cells	[76]

HA receptor	Hyaluronic acid	siRNA	BALB/c mice (intravenous adm.)	HA – siRNA conjugate + LPEI polymer showed efficient down-regulation of ApoB mRNA levels in the liver	[77]
ASGPR	GalNAc (trivalent)	AS	APOC-III in transgenic mice	TriGalNAc–ODN enhances the potency non-modified ODN 6 to10-fold in mouse liver	[78]
ASGPR	GalNAc (di- and trivalent)	siRNA	C57BL/6 mice (intravenous and subcutaneous adm.)	TriGalNAc-siRNA suppressed >80% of TTR mRNA expression in liver	[79b]
ASGPR	GalNAc (trivalent sequential)	siRNA	C57BL/6 mice (subcutaneous adm.)	TriGalNAc sequential siRNA retained *in vitro* and *in vivo* siRNA activity, similar to the triGalNAc antenna-siRNA	[80]
ASGPR	GalNAc (simplified trivalent)	siRNA	C57BL/6 mice (subcutaneous adm.)	Simplified triGalNAc-siRNA showed vitro and *in vivo* potency similar to the parent trivalent design	[81]
HA receptor	Hyaluronan hexasaccharide	ODN	rats	HA_6–ODN–DOTA conjugate showed the potential of targeting infarcted areas of rats with myocardial infarction	[82]
ASGPR	GalNAc (mono and divalent)	AS	C57Bl-6 mice (subcutaneous adm.)	GalNAc and diGalNAc-AS conjugate can enhance potency 5–10-fold in mice compared to AS (similar to triGalNAc-AS)	[83]
CD22	*N*-acetylneuraminic acid (mono- and trivalent)	AS	B-cells	Neu5Ac–AS conjugates showed reduced activity in CD22 positive B-cells relative to the parent AS	[84]
ASGPR	GalNAc (mono-, di-, and trivalent with diverse presentation)	AS	mouse hepatocytes and C57BL/6NJ mice	GalNAc-AS showed 10-fold reduced ASGR binding affinity relative $(GalNAc)_3$-AS but only twofold reduced activity in mice	[85]
ASGPR	GalNAc (trivalent)	siRNA	C57BL/6 mice (subcutaneous adm.)	Additional protection against 5′-exonuclease attack via additional PS linkages had little impact on plasma stability but resulted in improvements in potency of gene silencing	[86]

Type of carbohydrate and oligonucleotide conjugated on each COC are also indicated.

23.3% by hepatocytes. In contrast, in fibrotic rats which show upregulation of M6P/IGF II receptors, 82.0% of the total hepatic uptake was by HSCs, only 7.4% by Kupffer and endothelial cells, and 10.6% by hepatocytes. However, in the case of free TFO, the difference was much smaller [72]. Later, the same authors conjugated M6P to TFO via *N*-(2-hydroxypropyl)methacrylamide (HPMA) oligomer containing tetrapeptide Gly-Phe-Leu-Gly (GFLG) [53]. Following tail vein injection into rats, $(M6P)_{15}$-GFLG-HPMA-GFLG-TFO accumulated mainly in the liver. Almost 80% of the total liver uptake in fibrotic rats was contributed by HSCs. Mahato and coworker also conjugated a M6P single unit to siRNA via poly(ethylene glycol). In this case, M6P–PEG–siRNA conjugates were able to silence luciferase gene expression by about 40% without any transfection reagents in HepG2 and HSC-T6 cells and could silence endogenous transforming growth factor (TGF)-β1 gene expression as well [74].

The next carbohydrate receptor to be explored was the hepatocyte-specific ASGPR [89]. ASGPR, also named the Ashwell Morell receptor, is a well-characterized membrane-bound lectin receptor, responsible for removing desialylated glycoproteins from circulation through receptor-mediated endocytosis. ASGPR is capable of a receptor recycling time of approximately 10–15 minutes in human cells [90]. This capacity to facilitate multiple rounds of glycoprotein uptake and clearance, together with its ligand specificity, has enabled its use for liver directed delivery of a wide range of small molecules, nanosystems, and biomolecules such as ODNs [91]. Early examples of GalNAc–ODN conjugates already showed the potential of using ASGPR as a gate for ODN delivery into hepatocytes. A trivalent presentation of GalNAc on a peptide conjugated to ASO improved accumulation in HepG2 cells with respect to non-modified ASO [44]. An amphipathic poly(vinyl ether) polymer decorated with multiple GalNAc units was attached to siRNA and showed effective apoB inhibition in mice liver [73].

Galactose is also recognized by ASGPR and therefore it has also been conjugated to ODNs to examine cellular uptake of the corresponding COCs. Monovalent, divalent, and tetravalent galactose presentations were attached to ODNs for their study. Biessen et al. [71] linked four Gal units to an antisense (AS) through a peptide and observed that the COC was more efficiently taken up by parenchymal liver cells than unmodified ASO. Zhu et al. [46] prepared a Gal–PEG–ODN and when administered to rats it accumulated in the liver to a higher extent than the free ODN. Aviñó et al. [76] conjugated 1, 2, or 4 Gal moieties to the 5′-end of the sense strand of a siRNA using a dendrimeric scaffold. They observed up to 25% TNF-α inhibition in HuH7 cells when the COC contained one or four Gal units but no inhibition was attained with two Gal–siRNA conjugates.

More recently, laboratories from two companies (Ionis Pharmaceuticals and Alnylam) have studied meticulously the spatial arrangement of GalNAc units in order to deliver ASO or siRNA, respectively, as drugs for potential treatment of a wide range of diseases involving liver-expressed genes. Both groups reported the use of a triantennary GalNAc arranged on a tris scaffold (Figure 10.13a). This sugar arrangement has the number, distance between the sugar moieties and spatial orientation to show low nanomolar affinity for the ASGPR [79]. When linked to the 3′-end of an ASO, The COC enhances the potency of unconjugated ASOs by 6–10-fold in mouse liver. When combined with modified locked nucleotides such as short S-cEt (*S*-2-*O*-Et-2,4-bridged nucleic acid) ASO, ~60-fold enhancement in potency relative to the parent MOE (2-*O*-methoxyethyl RNA) ASO was observed [78]. If the triantennary GalNAc is conjugated at the 3′-end of the sense strand, the uptake in mouse hepatocytes of triGalNAc-siRNA is significantly higher than that of free siRNA, GalNAc-siRNa or GalNAc$_2$-siRNA [79b]. They also observed a much higher *TTR* gene inhibition in mice when the triGalNAc-siRNA was administered subcutaneously than through intravenous injection. Moreover, addition of extra thiophosphate linkages in the siRNA, which provided enhanced protection against 5′-exonucleases, resulted in a fivefold *in vivo* potency improvement.

Interestingly, Manoharan and coworkers [81] reported triGalNAc-siRNAs with the sugar units linked sequentially at the 3′-end of the sense strand (Figure 10.13b) instead of the previous triantennary, dendrimer-like arrangement. They found that the sequential covalent attachment of the trivalent GalNAc to the 3′-end of the sense strand resulted in a conjugate with *in vitro* and *in vivo* potency similar to that of the parent triantennary GalNAc conjugate [81]. Likewise, GalNAc units resembling a trivalent ligand at or near the 3′ end of the sense strand with the GalNAc units linked to the ribose or the base (Figure 10.13c) retained *in vitro* and *in vivo* siRNA activity, again similar to the parent triantennary GalNAc conjugate [80].

Following a similar strategy, Seth and coworkers [92] investigated different scaffolds to display the three GalNAc units and found only small differences on gene inhibition potency. In fact, they reported later that conjugation of one or two GalNAc sugar units to single stranded chemically modified ASOs can enhance potency 5–10-fold in mice what significantly reduces the synthetic complexity associated with manufacturing such conjugates for clinical applications [83]. At the same time, they observed that the number of phosphorothioate linkages and the chemical composition of the ASOs could assist in the binding and internalization into cells and tissues [85].

Three other membrane proteins have been explored for possible receptor-mediated ODN delivery. Glucose transporters (GLUTs) are found in most mammalian cells. Out of the 14 GLUT described, GLUT4 is internalized via clathrin-mediated and cholesterol-dependent endocytosis [93]. Ugarte-Uribe et al. [75] prepared glucose–ODN conjugates containing 1, 2, or 4 sugar units linked to the 5′-end. They observed that, where the sugar was attached with long spacers (15–18 atoms long), glucose-COCs were more efficiently uptaken in HeLa and U87 cells by 1.5–2 fold than the naked ODN.

The hyaluronan receptor for endocytosis (HARE) clears 14 systemic ligands, including HA, heparin and dermatan sulfate [94]. Park et al. [77] conjugated HA to siRNA via a disulfide bond and later complexed it to linear polyethyleneimine (LPEI). The HA – siRNA/LPEI was specifically delivered to the liver, and showed a significant reduction of ApoB mRNA expression. Recently, direct conjugation of a HA tetra- and hexasaccharide to siRNA modified with NOTA was reported [82]. The authors observed distribution of the COC-NOTA to kidney, bone marrow, salivary gland, and liver of rats. Moreover, the COCs also showed the potential of targeting infarcted areas of rats with myocardial infarction.

CD22, or cluster of differentiation-22 [95], is a sugar-binding transmembrane protein, which specifically binds *N*-acetylneuraminic acid (sialic acid) and is found on the surface of mature B cells. St-Pierre et al. [84] described the

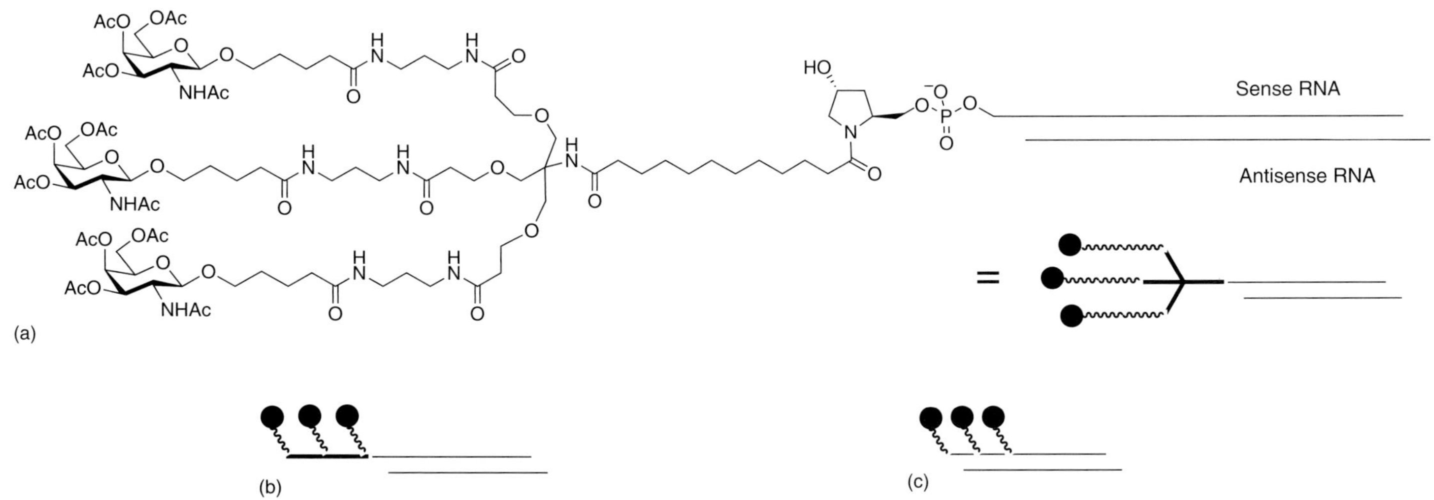

Figure 10.13 (a) Triantennary GalNAc-siRNA conjugate and its schematic representation. (b) Schematic representation of sequentially linked GalNAc. (c) Schematic representation of GalNAc linked to the sense strand.

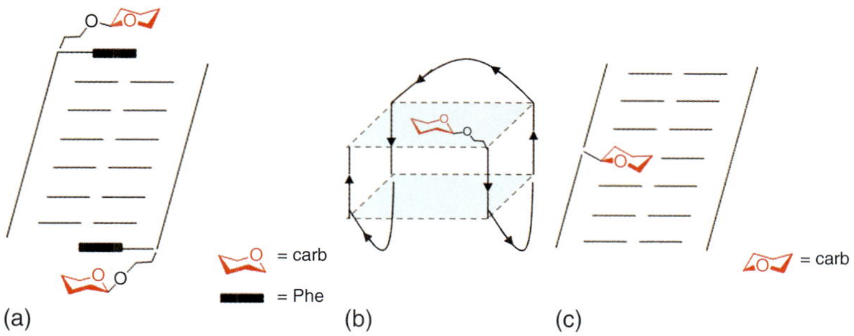

Figure 10.14 Schematic representation of: (a) dangling-end DNA model system to study carbohydrate–aromatic interactions; (b) G-quadruplex carbohydrate conjugate; and (c) DNA double helix containing a carbohydrate–nucleobase pair.

synthesis of monovalent and trivalent sialic acid–ASO conjugates. Surprisingly, The COCs showed reduced activity in CD22 positive B-cells relative to the free ASO perhaps due to the sequestration of ASO by other cell-surface proteins.

10.4.2 Molecular Interactions Probes

Several characteristics of DNA are quite convenient to study molecular interactions in water media. DNA is water-soluble, relatively easy to prepare and folds in well-defined 3D structures. This makes possible to bring together in space two structural motifs and check if they bind to each other. Thus, Morales et al. [27] designed a model system to study carbohydrate–aromatic interactions inspired in a dangling-end DNA duplex previously employed to study π–π stacking interactions in DNA by Turner and coworkers [96]. A carbohydrate-phenyl motif was placed at the 5′-end of a self-complementary DNA duplex using a phenyl nucleoside and a sugar phosphoramidite (Figure 10.14a). The monosaccharide was linked through a flexible spacer to the primary hydroxy group of the benzene nucleoside allowing the pyranose ring to be in contact with the benzene ring or immersed in bulk water. The DNA double helix acts as an energetic probe that allows quantification of the carbohydrate–aromatic interaction when the COC is compared with the control lacking the carbohydrate moiety. The authors found that the fucose-benzene pair was the most stabilizing of the studied series ($-0.4\,\text{kcal mol}^{-1}$) and was in the same range of other interactions with aromatic residues found in proteins, such as Phe–Phe or Phe–His. NMR studies indicated the close proximity between the monosaccharides and the phenyl ring.

The same dangling-end DNA system was used to investigate the interaction between mono- and disaccharides with DNA bases. This monosaccharide–nucleobase interaction had been observed in antibiotic-RNA recognition, where the 2′-amino-2′-deoxyglucose moiety of different aminoglycosides stacks over guanine 1491 of the 16S rRNA A-site [97]. Lucas et al. [98] removed the phenyl nucleoside in the previous model system and observed direct contacts between the sugar moieties and the terminal base pair. The energetics of carbohydrate-nucleobase stacking on the COC system (from −0.5

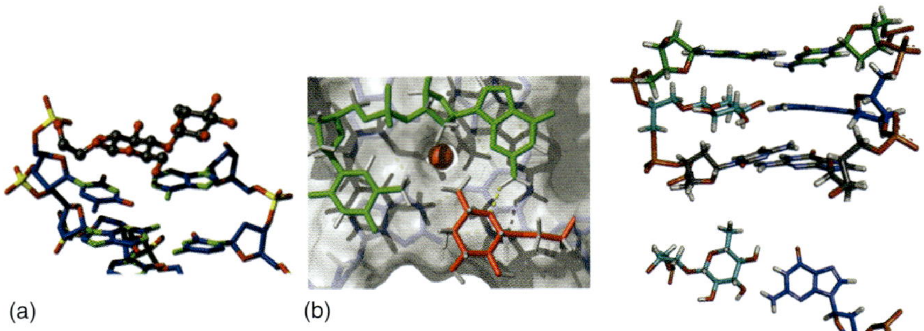

Figure 10.15 (a) Cellobiose stacking on the terminal T–A pair of a DNA duplex. (b) Top view of fucose–thrombin binding aptamer (TBA) quadruplex conjugate. Fucose (red) stacks on the G-tetrad and interacts with the bases of the loop (green). (c) Side view (top) of DNA helix containing a 6-deoxyglucose–guanine pair (bottom).

to $-0.9\,\text{kcal mol}^{-1}$) varied depending on the stereochemistry, polarity, and contact surface of the carbohydrate involved and also on the terminal base pair (Figure 10.15a). They also explored the interaction of fluorosugars with DNA bases and observed similar stabilization to that found for natural saccharides [99]. In contrast, when hydrophobic sugars such as permethylated glucose or cellobiose were linked to the 5′-end of the CGCGCG sequence, stabilization increased two- to threefold with respect to the natural polyhydroxylated mono- and disaccharide DNA conjugates and the sugars stacked on top of the terminal base [100]. Interestingly, the interactions between carbohydrates and DNA bases were also observed in a DNA G-quadruplex context (Figure 10.14b) [101]. Sugars attached to the 5′-end of the thrombin binding aptamers sequence showed favorable interactions with the guanine tetrad and the bases in the loop, not only through stacking but also H-bonding and van der Waals contacts were observed (Figure 10.15b).

Another singular biomolecular interaction found when aminoglycoside paramomycin binds rRNA is a glycoside-adenine pseudo base pair. Morales and coworkers designed a DNA duplex containing a monosaccharide–nucleobase pair in its center to investigate this interaction (Figure 10.14c) [102]. The authors observed glucose and 6-deoxyglucose were able to hybridize with their complementary strand, exhibiting a preference for purine nucleobases. Although the resulting double helices are less stable than the natural ones, they present only minor local distortions with 6-deoxyglucose fully integrated in the helix and its OH groups forming two H-bonds with the opposing guanine (Figure 10.15c). In fact, the 6-deoxyglucose-guanine pair closely resembles a purine-pyrimidine geometry. Quantum chemical calculations indicated that the new glucose–purine pairs are as stable as a natural T–A pair.

10.4.3 Lectin Binding and Glycoarrays

From the first COCs with varied sugar presentation such as those by Kobayashi and coworkers to the synthesis of complex glycocluster–ODN conjugates, all of them have been excellent tools to study lectin binding. Binding of ConA has been studied to galactose- [33b], maltose- [59] and lactose-modified COCs [52a, 66] in order to understand how sugar multivalency was affecting carbohydrate recognition. Nevertheless, the development of glycan arrays using COCs by Chevolot et al. [64] has enormously facilitated the study of lectin binding. These glycocluster arrays are based on hybridization of COCs to DNA chips and posterior study of lectin binding through fluorescence techniques. An excellent recent review summarizes this approach and the binding results with lectins from *Pseudomonas aeuruginosa* [103].

10.5 Outlook

More than two decades later from their first description, the synthesis of COCs has been widely developed. A large variety of reactions have been employed to attach simple carbohydrates or glycoclusters to DNA or RNA strands. Among them, probably phosphoramidite and CuAAC chemistries have been the most used ones, and with the help of different scaffolds complex glycoclusters have been assembled and linked to ODNs. The main application of COCs is still improving ODN delivery, both in the form of ASO or siRNA, although other applications have appeared like the COC model systems to study molecular interactions in aqueous media and the COC glycoarrays to study lectin binding.

COCs are becoming closer to become real drugs and some carb-ASOs and carb-siRNAs are already moving forward to the clinic. The development of COCs as drugs will benefit from the advance on the synthesis of more complex oligosaccharides and varied glycoclusters. For sure, the discovery of new membrane carbohydrate-binding receptors that could internalize COCs will be a key aspect, especially if they are cell-specific or tissue-specific. At the same time, research on the details of receptor-COC molecular recognition and on the mechanism of internalization will give important insights for the advance of COCs as potential drugs.

References

1 (a) Crocker, P.R. (2002). Siglecs: sialic-acid-binding immunoglobulin-like lectins in cell–cell interactions and signalling. *Curr. Opin. Struct. Biol.* 12: 609–615. (b) Sharon, N. and Lis, H. (2004). History of lectins: from hemagglutinins to biological recognition molecules. *Glycobiology* 14: 53R–62R.

2 Bertozzi, C.R. and Kiessling, L.L. (2001). Chemical glycobiology. *Science* 291: 2357–2364.

3 (a) Lichtenstein, J. and Cohen, S.S. (1960). Nucleotides derived from enzymatic digests of nucleic acids of T2, T4, and T6 bacteriophages. *J. Biol. Chem.* 235: 1134–1141. (b) Lehman, I.R. and Pratt, E.A. (1960). On the structure of the glucosylated hydroxymethylcytosine nucleotides of coliphages T2, T4, and T6. *J. Biol. Chem.* 235: 3254–3259. (c) Lunt, M.R.,

Siebke, J.C., and Burton, K. (1964). Glucosylated nucleotide sequences from T2-bacteriophage deoxyribonucleic acid. *Biochem. J* 92: 27–36.

4 Ehrlich, M. and Ehrlich, K.C. (1981). A novel, highly modified, bacteriophage DNA in which thymine is partly replaced by a phosphoglucuronate moiety covalently bound to 5-(4′,5′-dihydroxypentyl)uracil. *J. Biol. Chem.* 256: 9966–9972.

5 (a) van Leeuwen, F., Wijsman, E.R., Kuyl-Yeheskiely, E. et al. (1996). The telomeric GGGTTA repeats of *Trypanosoma brucei* contain the hypermodified base J in both strands. *Nucleic Acids Res.* 24: 2476–2482. (b) van Leeuwen, F., Kieft, R., Cross, M., and Borst, P. (1998). Biosynthesis and function of the modified DNA base beta-D-glucosyl-hydroxymethyluracil in *Trypanosoma brucei. Mol. Cell. Biol.* 18: 5643–5651.

6 Simpson, L. (1998). A base called J. *Proc. Natl. Acad. Sci. U.S.A.* 95: 2037–2038.

7 (a) Okada, N., Shindo-Okada, N., and Nishimura, S. (1977). Isolation of mammalian tRNAAsp and tRNATyr by lectin-Sepharose affinity column chromatography. *Nucleic Acids Res.* 4: 415–423. (b) McCloskey, J.A. and Nishimura, S. (1977). Modified nucleosides in transfer RNA. *Acc. Chem. Res.* 10: 403–410.

8 van Leeuwen, F., de Kort, M., van der Marel, G.A. et al. (1998). The modified DNA base beta-D-glucosylhydroxymethyluracil confers resistance to micrococcal nuclease and is incompletely recovered by 32P-postlabeling. *Anal. Biochem.* 258: 223–229.

9 Borst, P. and van Leeuwen, F. (1997). beta-D-glucosyl-hydroxymethyluracil, a novel base in African trypanosomes and other Kinetoplastida. *Mol. Biochem. Parasitol.* 90: 1–8.

10 Gait, M.J. (1984). *Oligonucleotide Syntesis: A Practical Approach*. Oxford: IRS Press.

11 (a) Bochman, M.L., Paeschke, K., and Zakian, V.A. (2012). DNA secondary structures: stability and function of G-quadruplex structures. *Nat. Rev. Genet.* 13: 770–780. (b) Palecek, E. (1991). Local supercoil-stabilized DNA structures. *Crit. Rev. Biochem. Mol. Biol.* 26: 151–226. (c) Jain, A., Wang, G., and Vasquez, K.M. (2008). DNA triple helices: biological consequences and therapeutic potential. *Biochimie* 90: 1117–1130. (d) Gessner, R.V., Frederick, C.A., Quigley, G.J. et al. (1989). The molecular structure of the left-handed Z-DNA double helix at 1.0-A atomic resolution. Geometry, conformation, and ionic interactions of d(CGCGCG). *J. Biol. Chem.* 264: 7921–7935.

12 Dias, N. and Stein, C.A. (2002). Antisense oligonucleotides: basic concepts and mechanisms. *Mol. Cancer Ther.* 1: 347.

13 Chan, P.P. and Glazer, P.M. (1997). Triplex DNA: fundamentals, advances, and potential applications for gene therapy. *J. Mol. Med.* 75: 267–282.

14 Zhou, J. and Rossi, J. (2016). Aptamers as targeted therapeutics: current potential and challenges. *Nat. Rev. Drug Discovery* 16: 181.

15 Joseph Marr, J. (1996). Ribozymes as therapeutic agents. *Drug Discovery Today* 1: 94–102.

16 (a) Hajeri, P.B. and Singh, S.K. (2009). siRNAs: their potential as therapeutic agents – Part I. Designing of siRNAs. *Drug Discovery Today* 14: 851–858.

(b) Singh, S.K. and Hajeri, P.B. (2009). siRNAs: their potential as therapeutic agents – Part II. Methods of delivery. *Drug Discovery Today* 14: 859–865.

17 Feldkamp, U. and Niemeyer, C.M. (2006). Rational design of DNA nanoarchitectures. *Angew. Chem. Int. Ed.* 45: 1856–1876.

18 Teles, F.R.R. and Fonseca, L.P. (2008). Trends in DNA biosensors. *Talanta* 77: 606–623.

19 Ng, J.H. and Ilag, L.L. (2003). *Biotechnology Annual ReviewElsevier*, vol. 9, 1–149.

20 Erdmann, V.A., Jurga, S., and Barciszewski, J. (2015). *RNA and DNA Diagnostics*. Springer.

21 Binkley, R.W. (1988). Reviews on oligosaccharide synthesis. *J. Carbohydr. Chem.* 7: vii–xi.

22 Duus, J.Ø., Gotfredsen, C.H., and Bock, K. (2000). Carbohydrate structural determination by NMR spectroscopy: modern methods and limitations. *Chem. Rev.* 100: 4589–4614.

23 Varki, A. (2017). Biological roles of glycans. *Glycobiology* 27: 3–49.

24 (a) Ghazarian, H., Idoni, B., and Oppenheimer, S.B. (2011). A glycobiology review: carbohydrates, lectins and implications in cancer therapeutics. *Acta Histochem.* 113: 236–247. (b) Dam, T.K. and Brewer, C.F. (2010). Lectins as pattern recognition molecules: the effects of epitope density in innate immunity. *Glycobiology* 20: 270–279.

25 (a) Zatsepin, T.S. and Oretskaya, T.S. (2004). Synthesis and applications of oligonucleotide-carbohydrate conjugates. *Chem. Biodivers.* 1: 1401–1417. (b) Yan, H. and Tram, K. (2007). Glycotargeting to improve cellular delivery efficiency of nucleic acids. *Glycoconjugate J.* 24: 107–123. (c) Spinelli, N., Defrancq, E., and Morvan, F. (2013). Glycoclusters on oligonucleotide and PNA scaffolds: synthesis and applications. *Chem. Soc. Rev.* 42: 4557–4573.

26 Akhtar, S., Routledge, A., Patel, R., and Gardiner, J.M. (1995). Synthesis of mono- and dimannoside phosphoramidite derivatives for solid-phase conjugation to oligonucleotides. *Tetrahedron Lett.* 36: 7333–7336.

27 Morales, J.C., Reina, J.J., Díaz, I. et al. (2008). Experimental measurement of carbohydrate-aromatic stacking in water by using a dangling-ended DNA model system. *Chem. - Eur. J.* 14: 7828–7835.

28 Sheppard, T.L., Wong, C.H., and Joyce, G.F. (2000). Nucleoglycoconjugates: design and synthesis of a new class of DNA-carbohydrate conjugates. *Angew. Chem. Int. Ed.* 39: 3660–3663.

29 (a) Adinolfi, M., Barone, G., De Napoli, L. et al. (1998). Use of controlled pore glass in solid phase oligosaccharide synthesis. Application to the semi-automated synthesis of a glyconucleotide conjugate. *Tetrahedron Lett.* 39: 1953–1956. (b) Adinolfi, M., Barone, G., Napoli, L.D. et al. (1999). Solid phase glycosidation of oligonucleotides. *Tetrahedron Lett.* 40: 2607–2610. (c) Adinolfi, M., De Napoli, L., Di Fabio, G. et al. (2002). Solid phase synthesis of oligonucleotides tethered to oligo-glucose phosphate tails. *Tetrahedron* 58: 6697–6704. (d) Adinolfi, M., De Napoli, L., Di Fabio, G. et al. (2001). Solid-phase synthesis of glyco-oligonucleotide conjugates. *Synlett* 2001: 0745–0748. (e) Adinolfi, M., De Napoli, L., Di Fabio, G. et al. (2004). Modulating the activity of oligonucleotides by carbohydrate conjugation: solid

phase synthesis of sucrose-oligonucleotide hybrids. *Org. Biomol. Chem.* 2: 1879–1886.

30 Ikeda, Y., Kubota, D., and Nagasaki, Y. (2010). Simple solid-phase synthesis and biological properties of carbohydrate–oligonucleotide conjugates modified at the 3′-terminus. *Bioconjugate Chem.* 21: 1685–1690.

31 D'Onofrio, J., de Champdore, M., De Napoli, L. et al. (2005). Glycomimetics as decorating motifs for oligonucleotides: solid-phase synthesis, stability, and hybridization properties of carbopeptoid-oligonucleotide conjugates. *Bioconjugate Chem.* 16: 1299–1309.

32 Hunziker, J. (1999). Synthesis of 5-(2-amino-2-deoxy-beta-D-glucopyranosyl-oxymethyl)-2′- deoxyuridine and its incorporation into oligothymidylates. *Bioorg. Med. Chem. Lett.* 9: 201–204.

33 (a) Matsuura, K., Hibino, M., Kataoka, M. et al. (2000). Phosphoramidite solid-phase synthesis of site-specifically glycosylated oligodeoxynucleotides. *Tetrahedron Lett.* 41: 7529–7533. (b) Matsuura, K., Hibino, M., Yamada, Y., and Kobayashi, K. (2001). Construction of glyco-clusters by self-organization of site-specifically glycosylated oligonucleotides and their cooperative amplification of lectin-recognition. *J. Am. Chem. Soc.* 123: 357–358.

34 Charles, I., Xi, H., and Arya, D.P. (2007). Sequence-specific targeting of RNA with an oligonucleotide-neomycin conjugate. *Bioconjugate Chem.* 18: 160–169.

35 Karskela, M., Virta, P., Malinen, M. et al. (2008). Synthesis and cellular uptake of fluorescently labeled multivalent hyaluronan disaccharide conjugates of oligonucleotide phosphorothioates. *Bioconjugate Chem.* 19: 2549–2558.

36 Matulic-Adamic, J., Serebryany, V., Haeberli, P. et al. (2002). Synthesis of *N*-acetyl-D-galactosamine and folic acid conjugated ribozymes. *Bioconjugate Chem.* 13: 1071–1078.

37 Yokoyama, K., Kubo, T., Dubey, K.K. et al. (2001). Synthesis and biological properties of DNA-sugar conjugates. *Nucleic Acids Res. Suppl.* 73–74.

38 Jawalekar, A.M., Meeuwenoord, N., Cremers, J.G.O. et al. (2008). Conjugation of nucleosides and oligonucleotides by [3+2] cycloaddition. *J. Org. Chem.* 73: 287–290.

39 Kiviniemi, A., Virta, P., and Lonnberg, H. (2008). Utilization of intrachain 4′-C-azidomethylthymidine for preparation of oligodeoxyribonucleotide conjugates by click chemistry in solution and on a solid support. *Bioconjugate Chem.* 19: 1726–1734.

40 Bouillon, C., Meyer, A., Vidal, S. et al. (2006). Microwave assisted "click" chemistry for the synthesis of multiple labeled-carbohydrate oligonucleotides on solid support. *J. Org. Chem.* 71: 4700–4702.

41 Pourceau, G., Meyer, A., Vasseur, J.J., and Morvan, F. (2009). Azide solid support for 3′-conjugation of oligonucleotides and their circularization by click chemistry. *J. Org. Chem.* 74: 6837–6842.

42 Karskela, M., Helkearo, M., Virta, P., and Lonnberg, H. (2010). Synthesis of oligonucleotide glycoconjugates using sequential click and oximation ligations. *Bioconjugate Chem.* 21: 748–755.

43 Ye, Z., Cheng, K., Guntaka, R.V., and Mahato, R.I. (2005). Targeted delivery of a triplex-forming oligonucleotide to hepatic stellate cells. *Biochemistry* 44: 4466–4476.

44 Hangeland, J.J., Levis, J.T., Lee, Y.C., and Ts'o, P.O. (1995). Cell-type specific and ligand specific enhancement of cellular uptake of oligodeoxynucleoside methylphosphonates covalently linked with a neoglycopeptide, YEE(ah-GalNAc)3. *Bioconjugate Chem.* 6: 695–701.

45 Oishi, M., Nagasaki, Y., Itaka, K. et al. (2005). Lactosylated poly(ethylene glycol)-siRNA conjugate through acid-labile beta-thiopropionate linkage to construct pH-sensitive polyion complex micelles achieving enhanced gene silencing in hepatoma cells. *J. Am. Chem. Soc.* 127: 1624–1625.

46 Zhu, L., Ye, Z., Cheng, K. et al. (2008). Site-specific delivery of oligonucleotides to hepatocytes after systemic administration. *Bioconjugate Chem.* 19: 290–298.

47 Forget, D., Renaudet, O., Defrancq, E., and Dumy, P. (2001). Efficient preparation of carbohydrate–oligonucleotide conjugates (COCs) using oxime bond formation. *Tetrahedron Lett.* 42: 7829–7832.

48 Edupuganti, O.P., Renaudet, O., Defrancq, E., and Dumy, P. (2004). The oxime bond formation as an efficient chemical tool for the preparation of 3′,5′-bifunctionalised oligodeoxyribonucleotides. *Bioorg. Med. Chem. Lett.* 14: 2839–2842.

49 Sando, S., Matsui, K., Niinomi, Y. et al. (2003). Facile preparation of DNA-tagged carbohydrates. *Bioorg. Med. Chem. Lett.* 13: 2633–2636.

50 Meyer, A., Spinelli, N., Dumy, P. et al. (2010). Oligonucleotide sequential bis-conjugation via click-oxime and click-Huisgen procedures. *J. Org. Chem.* 75: 3927–3930.

51 Burley, G.A., Gierlich, J., Mofid, M.R. et al. (2006). Directed DNA metallization. *J. Am. Chem. Soc.* 128: 1398–1399.

52 (a) Akasaka, T., Matsuura, K., Emi, N., and Kobayashi, K. (1999). Conjugation of plasmid DNAs with lactose via diazocoupling enhances resistance to restriction enzymes and acquires binding affinity to galactose-specific lectin. *Biochem. Biophys. Res. Commun.* 260: 323–328. (b) Matsuura, K., Akasaka, T., Hibino, M., and Kobayashi, K. (2000). Facile synthesis of stable and lectin-recognizable DNA-carbohydrate conjugates via diazo coupling. *Bioconjugate Chem.* 11: 202–211.

53 Yang, N., Ye, Z., Li, F., and Mahato, R.I. (2009). HPMA polymer-based site-specific delivery of oligonucleotides to hepatic stellate cells. *Bioconjugate Chem.* 20: 213–221.

54 Wang, Y. and Sheppard, T.L. (2003). Chemoenzymatic synthesis and antibody detection of DNA glycoconjugates. *Bioconjugate Chem.* 14: 1314–1322.

55 Gramlich, P.M., Wirges, C.T., Gierlich, J., and Carell, T. (2008). Synthesis of modified DNA by PCR with alkyne-bearing purines followed by a click reaction. *Org. Lett.* 10: 249–251.

56 Matsui, M., Nishiyama, Y., Ueji, S., and Ebara, Y. (2007). Construction of saccharide-modified DNAs by DNA polymerase. *Bioorg. Med. Chem. Lett.* 17: 456–460.

57 Akasaka, T., Matsuura, K., and Kobayashi, K. (2001). Transformation from block-type to graft-type oligonucleotide-glycopolymer conjugates by self-organization with half-sliding complementary oligonucleotides and their lectin recognition. *Bioconjugate Chem.* 12: 776–785.
58 D'Onofrio, J., Petraccone, L., Martino, L. et al. (2008). Synthesis, biophysical characterization, and anti-HIV activity of glyco-conjugated G-quadruplex-forming oligonucleotides. *Bioconjugate Chem.* 19: 607–616.
59 Matsui, M. and Ebara, Y. (2012). Enhanced binding of trigonal DNA-carbohydrate conjugates to lectin. *Bioorg. Med. Chem. Lett.* 22: 6139–6143.
60 Dubber, M. and Frechet, J.M. (2003). Solid-phase synthesis of multivalent glycoconjugates on a DNA synthesizer. *Bioconjugate Chem.* 14: 239–246.
61 Katajisto, J., Virta, P., and Lonnberg, H. (2004). Solid-phase synthesis of multiantennary oligonucleotide glycoconjugates utilizing on-support oximation. *Bioconjugate Chem.* 15: 890–896.
62 Katajisto, J., Heinonen, P., and Lonnberg, H. (2004). Solid-phase synthesis of oligonucleotide glycoconjugates bearing three different glycosyl groups: orthogonally protected bis(hydroxymethyl)-*N*,*N*′-bis(3-hydroxypropyl)malondiamide phosphoramidite as key building block. *J. Org. Chem.* 69: 7609–7615.
63 Maier, M.A., Yannopoulos, C.G., Mohamed, N. et al. (2003). Synthesis of antisense oligonucleotides conjugated to a multivalent carbohydrate cluster for cellular targeting. *Bioconjugate Chem.* 14: 18–29.
64 Chevolot, Y., Bouillon, C., Vidal, S. et al. (2007). DNA-based carbohydrate biochips: a platform for surface glyco-engineering. *Angew. Chem. Int. Ed.* 46: 2398–2402.
65 Hangeland, J.J., Flesher, J.E., Deamond, S.F. et al. (1997). Tissue distribution and metabolism of the [32P]-labeled oligodeoxynucleoside methylphosphonate-neoglycopeptide conjugate, [YEE(ah-GalNAc)3]-SMCC-AET-pUmpT7, in the mouse. *Antisense Nucleic Acid Drug Dev.* 7: 141–149.
66 Singh, Y., Renaudet, O., Defrancq, E., and Dumy, P. (2005). Preparation of a multitopic glycopeptide–oligonucleotide conjugate. *Org. Lett.* 7: 1359–1362.
67 (a) Pourceau, G., Meyer, A., Chevolot, Y. et al. (2010). Oligonucleotide carbohydrate-centered galactosyl cluster conjugates synthesized by click and phosphoramidite chemistries. *Bioconjugate Chem.* 21: 1520–1529. (b) Gerland, B., Goudot, A., Pourceau, G. et al. (2012). Synthesis of homo- and heterofunctionalized glycoclusters and binding to *Pseudomonas aeruginosa* lectins PA-IL and PA-IIL. *J. Org. Chem.* 77: 7620–7626.
68 Casoni, F., Dupin, L., Vergoten, G. et al. (2014). The influence of the aromatic aglycon of galactoclusters on the binding of LecA: a case study with O-phenyl, S-phenyl, O-benzyl, S-benzyl, O-biphenyl and O-naphthyl aglycons. *Org. Biomol. Chem.* 12: 9166–9179.
69 Ligeour, C., Dupin, L., Angeli, A. et al. (2015). Importance of topology for glycocluster binding to *Pseudomonas aeruginosa* and *Burkholderia ambifaria* bacterial lectins. *Org. Biomol. Chem.* 13: 11244–11254.

70 Bonfils, E., Depierreux, C., Midoux, P. et al. (1992). Drug targeting: synthesis and endocytosis of oligonucleotide-neoglycoprotein conjugates. *Nucleic Acids Res.* 20: 4621–4629.

71 Biessen, E.A., Vietsch, H., Rump, E.T. et al. (1999). Targeted delivery of oligodeoxynucleotides to parenchymal liver cells in vivo. *Biochem. J* 340 (Pt 3: 783–792.

72 Ye, Z., Cheng, K., Guntaka, R.V., and Mahato, R.I. (2006). Receptor-mediated hepatic uptake of M6P–BSA-conjugated triplex-forming oligonucleotides in rats. *Bioconjugate Chem.* 17: 823–830.

73 Rozema, D.B., Lewis, D.L., Wakefield, D.H. et al. (2007). Dynamic polyconjugates for targeted in vivo delivery of siRNA to hepatocytes. *Proc. Natl. Acad. Sci. U.S.A* 104: 12982–12987.

74 Zhu, L. and Mahato, R.I. (2010). Targeted delivery of siRNA to hepatocytes and hepatic stellate cells by bioconjugation. *Bioconjugate Chem.* 21: 2119–2127.

75 Ugarte-Uribe, B., Perez-Rentero, S., Lucas, R. et al. (2010). Synthesis, cell-surface binding, and cellular uptake of fluorescently labeled glucose-DNA conjugates with different carbohydrate presentation. *Bioconjugate Chem.* 21: 1280–1287.

76 Avino, A., Ocampo, S.M., Lucas, R. et al. (2011). Synthesis and in vitro inhibition properties of siRNA conjugates carrying glucose and galactose with different presentations. *Mol. Diversity* 15: 751–757.

77 Park, K., Yang, J.A., Lee, M.Y. et al. (2013). Reducible hyaluronic acid-siRNA conjugate for target specific gene silencing. *Bioconjugate Chem.* 24: 1201–1209.

78 Prakash, T.P., Graham, M.J., Yu, J. et al. (2014). Targeted delivery of antisense oligonucleotides to hepatocytes using triantennary *N*-acetyl galactosamine improves potency 10-fold in mice. *Nucleic Acids Res.* 42: 8796–8807.

79 (a) Meier, M., Bider, M.D., Malashkevich, V.N. et al. (2000). Crystal structure of the carbohydrate recognition domain of the H1 subunit of the asialoglycoprotein receptor. *J. Mol. Biol.* 300: 857–865. (b) Nair, J.K., Willoughby, J.L., Chan, A. et al. (2014). Multivalent *N*-acetylgalactosamine-conjugated siRNA localizes in hepatocytes and elicits robust RNAi-mediated gene silencing. *J. Am. Chem. Soc.* 136: 16958–16961.

80 Matsuda, S., Keiser, K., Nair, J.K. et al. (2015). siRNA conjugates carrying sequentially assembled trivalent *N*-acetylgalactosamine linked through nucleosides elicit robust gene silencing in vivo in hepatocytes. *ACS Chem. Biol.* 10: 1181–1187.

81 Rajeev, K.G., Nair, J.K., Jayaraman, M. et al. (2015). Hepatocyte-specific delivery of siRNAs conjugated to novel non-nucleosidic trivalent *N*-acetylgalactosamine elicits robust gene silencing in vivo. *ChemBioChem* 16: 903–908.

82 Jadhav, S., Kakela, M., Makila, J. et al. (2016). Synthesis and in vivo PET imaging of hyaluronan conjugates of oligonucleotides. *Bioconjugate Chem.* 27: 391–403.

83 Kinberger, G.A., Prakash, T.P., Yu, J. et al. (2016). Conjugation of mono and di-GalNAc sugars enhances the potency of antisense oligonucleotides via ASGR mediated delivery to hepatocytes. *Bioorg. Med. Chem. Lett.* 26: 3690–3693.

84 St-Pierre, G., Pal, S., Ostergaard, M.E. et al. (2016). Synthesis and biological evaluation of sialyl-oligonucleotide conjugates targeting leukocyte B trans-membranal receptor CD22 as delivery agents for nucleic acid drugs. *Bioorg. Med. Chem.* 24: 2397–2409.

85 Schmidt, K., Prakash, T.P., Donner, A.J. et al. (2017). Characterizing the effect of GalNAc and phosphorothioate backbone on binding of antisense oligonucleotides to the asialoglycoprotein receptor. *Nucleic Acids Res.* 45: 2294–2306.

86 Nair, J.K., Attarwala, H., Sehgal, A. et al. (2017). Impact of enhanced metabolic stability on pharmacokinetics and pharmacodynamics of GalNAc–siRNA conjugates. *Nucleic Acids Res.* 45 (19): 10969–10977.

87 (a) Martin-Kleiner, I. and Gall Troselj, K. Mannose-6-phosphate/insulin-like growth factor 2 receptor (M6P/IGF2R) in carcinogenesis. *Cancer Lett.* 289: 11–22. (b) Hawkes, C. and Kar, S. (2004). The insulin-like growth factor-II/mannose-6-phosphate receptor: structure, distribution and function in the central nervous system. *Brain Res. Rev.* 44: 117–140.

88 Yang, N., Singh, S., and Mahato, R.I. (2011). Targeted TFO delivery to hepatic stellate cells. *J. Controlled Release* 155: 326–330.

89 Grewal, P.K. (2010). The Ashwell-Morell receptor. *Methods Enzymol.* 479: 223–241.

90 (a) Schwartz, A.L., Fridovich, S.E., and Lodish, H.F. (1982). Kinetics of internalization and recycling of the asialoglycoprotein receptor in a hepatoma cell line. *J. Biolumin. Chemilumin.* 257: 4230–4237. (b) Spiess, M. (1990). The asialoglycoprotein receptor: a model for endocytic transport receptors. *Biochemistry* 29: 10009–10018.

91 (a) Zimmermann, T.S., Karsten, V., Chan, A. et al. (2017). Clinical proof of concept for a novel hepatocyte-targeting GalNAc-siRNA conjugate. *Mol. Ther.* 25: 71–78. (b) Huang, Y. (2017). Preclinical and clinical advances of GalNAc-decorated nucleic acid therapeutics. *Mol. Ther. Nucleic Acids* 6: 116–132.

92 Prakash, T.P., Yu, J., Migawa, M.T. et al. (2016). Comprehensive structure-activity relationship of triantennary *N*-acetylgalactosamine conjugated antisense oligonucleotides for targeted delivery to hepatocytes. *J. Med. Chem.* 59: 2718–2733.

93 Foley, K., Boguslavsky, S., and Klip, A. (2011). Endocytosis, recycling, and regulated exocytosis of glucose transporter 4. *Biochemistry* 50: 3048–3061.

94 Yan, J., Xin, P., and Guangxi, Z. (2016). Advances in hyaluronic acid-based drug delivery systems. *Curr Drug Targets* 17: 720–730.

95 Tedder, T.F., Tuscano, J., Sato, S., and Kehrl, J.H. (1997). CD22, a B lymphocyte–specific adhesion molecule that regulates antigen receptor signaling. *Annu. Rev. Immunol.* 15: 481–504.

96 (a) Freier, S.M., Alkema, D., Sinclair, A. et al. (1985). Contributions of dangling end stacking and terminal base-pair formation to the stabilities of

XGGCCp, XCCGGp, XGGCCYp, and XCCGGYp helixes. *Biochemistry* 24: 4533–4539. (b) Guckian, K.M., Schweitzer, B.A., Ren, R.X. et al. (1996). Experimental measurement of aromatic stacking affinities in the context of duplex DNA. *J. Am. Chem. Soc.* 118: 8182–8183.

97 Vicens, Q. and Westhof, E. (2003). Molecular recognition of aminoglycoside antibiotics by ribosomal RNA and resistance enzymes: an analysis of x-ray crystal structures. *Biopolymers* 70: 42–57.

98 Lucas, R., Gomez-Pinto, I., Avino, A. et al. (2011). Highly polar carbohydrates stack onto DNA duplexes via CH/pi interactions. *J. Am. Chem. Soc.* 133: 1909–1916.

99 Lucas, R., Penalver, P., Gomez-Pinto, I. et al. (2014). Effects of sugar functional groups, hydrophobicity, and fluorination on carbohydrate–DNA stacking interactions in water. *J. Org. Chem.* 79: 2419–2429.

100 Lucas, R., Vengut-Climent, E., Gomez-Pinto, I. et al. (2012). Apolar carbohydrates as DNA capping agents. *Chem. Commun. (Camb.)* 48: 2991–2993.

101 Gomez-Pinto, I., Vengut-Climent, E., Lucas, R. et al. (2013). Carbohydrate–DNA interactions at G-quadruplexes: folding and stability changes by attaching sugars at the 5′-end. *Chemistry* 19: 1920–1927.

102 Vengut-Climent, E., Gomez-Pinto, I., Lucas, R. et al. (2016). Glucose-nucleobase pseudo base pairs: biomolecular interactions within DNA. *Angew. Chem. Int. Ed.* 55: 8643–8647.

103 Chevolot, Y., Laurenceau, E., Phaner-Goutorbe, M. et al. (2014). DNA directed immobilization glycocluster array: applications and perspectives. *Curr. Opin. Chem. Biol.* 18: 46–54.

11

Advances in Light-Directed Synthesis of High-Density Microarrays and Extension to RNA and 2′F-ANA Chemistries

Jory Lietard[1], Masad J. Damha[2], and Mark M. Somoza[1]

[1] *University of Vienna, Institute of Inorganic Chemistry, Faculty of Chemistry, Althanstraße 14 (UZA II), 1090 Vienna, Austria*
[2] *McGill University, Department of Chemistry, 801 Rue Sherbrooke O, Montréal, Québec H3A 0B8, Canada*

11.1 Introduction

The need for high-throughput techniques is prevalent in current biology, with the advent of new disciplines dedicated to the study of complex mixtures of biomolecules (genomics, proteomics), and as the sensitivity and the computational power of modern instrumentation have tremendously increased over the last three decades. Medicine has also turned its attention toward diagnostics and point-of-care analysis. Directed evolution and artificial gene synthesis have become increasingly popular topics in the field of synthetic biology. Therefore, methods that allow for the preparation of large nucleic acids libraries are highly sought after. DNA microarrays are an attractive and cost-effective approach toward this goal, and one significant asset in microarray fabrication is the ability to perform the synthesis, in parallel, of hundreds of thousands of different oligonucleotide sequences at defined locations on a surface, making them particularly suitable for detection purposes and for the individual study of each sequence variant [1, 2]. Early attempts at arraying oligonucleotides onto surfaces used pre-synthesized or cloned DNA sequences mechanically spotted onto glass microscope slides. While spotting allows for a control over purity and opens the way to chemically modified oligonucleotides, the density (or complexity) of the spotted array is intrinsically limited to the size of the robotic pin. To achieve high density in array fabrication requires either small volumes or equally small reactive surfaces, and both approaches have been developed using *in situ* synthesis and phosphoramidite chemistry. Small volumes are at the core of microarray fabrication via inkjet printing: printer heads have been adapted to contain each of the four DNA phosphoramidites and dispense, as in a standard color printer, microscopic droplets (picolitre range) of a desired phosphoramidite solution onto a silicon wafer decorated with micrometer-sized wells [3]. Surface tension at the interface between the substrate and the droplet keeps the solvent confined within the well and thus prevents mixing of the phosphoramidites. The synthesis of the desired sequence then proceeds using

Enzymatic and Chemical Synthesis of Nucleic Acid Derivatives, First Edition.
Edited by Jesús Fernández-Lucas and María-José Camarasa Rius.

standard solid-phase chemistry and the assembly of all sequences, reaching up to 25 000 [4, 5], is remotely controlled by a computer. Higher density arrays have been obtained using UV light ("photolithography") as a tool to control the space where oligonucleotide synthesis is allowed to happen and was first reported for the preparation of pentapeptides and dinucleotides [6]. Instead of the classical 5′-dimethoxytrityl (DMTr), phosphoramidites carry a photolabile protecting group. To individually address small surfaces for photochemistry, physical masks consisting of a checkerboard pattern of translucent and non-translucent segments were placed over the synthesis substrate. UV light would then only pass through the translucent segments and remove the nitroveratryloxycarbonyl (NVOC) photosensitive protecting group of the surface-bound amino acids or nucleotides directly on the illuminated areas, or "features". This process was further refined in order to synthesize longer oligonucleotides and increase the photolysis rate of the light-sensitive protecting group [7, 8]. However, one major drawback was that each synthesis design and each incremental step in array density/complexity required its own set of physical masks, making it a time-consuming and costly approach. This was solved with the development of maskless photolithography [9]. In Maskless Array Synthesis (MAS), patterned UV light is produced after reflection of 365 nm light onto a Digital Micromirror Device (DMD) and is imaged on the synthesis substrate, triggering the removal of the photolabile protecting group only at desired locations. The DMD consists of a grid of electronically controlled mirrors, initially a 800 × 600 collection of mirrors 16 μm in size. Each mirror can be tilted in either an ON position, whereby UV light is projected onto the surface and leads to photocleavage, or an OFF position reflecting UV light away from the surface (Figure 11.1). Therefore, the pattern of UV light determines the layout of the oligonucleotide array. A DMD of 800 × 600 mirrors allows for the synthesis of 480 000 unique DNA sequences, but higher resolution devices (1024 × 768, 786 432 features and higher) have been introduced to MAS [10].

With oligonucleotides remaining bound to the surface, essentially ruling out any purification, the quality and the fidelity of the *in situ* synthesized oligonucleotides become crucial parameters that dictate the reliability of any result obtained from experiments performed on microarrays. Those issues have been previously explored by cleaving oligonucleotides, which then underwent routine

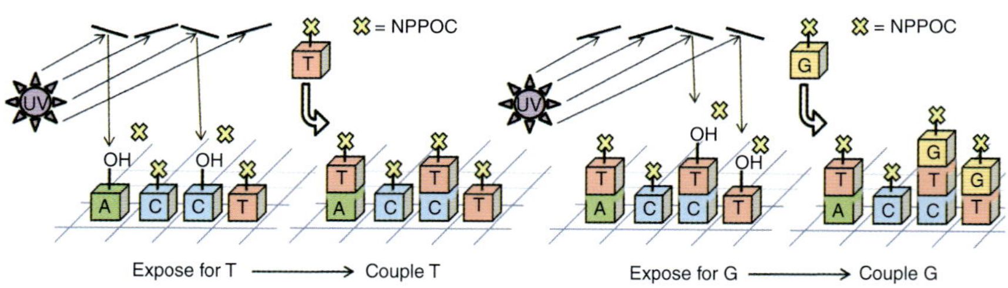

Figure 11.1 Schematic representation of the Maskless Array Synthesis system, where UV light is reflected on the Digital Micromirror Device. "ON" mirrors (black lines tilted toward the surface) reflect light onto the DNA chip. "OFF" mirrors (black lines tilted away from the surface) reflect light away from the chip.

analytical techniques [4, 10, 11]. This allows for fine-tuning of the synthesis parameters. As the efficiency of DNA array fabrication increases, so does the maximal oligonucleotide length that can efficiently be obtained. Combined with the need to prepare arrays of higher complexity, extending oligonucleotide length greatly affects the overall synthesis time. In this chapter, we review the results of our attempts at improving quality, speed, and throughput in the fabrication of high-density DNA microarrays by photolithography.

We will also discuss our recent focus in extending the portfolio of nucleic acids phosphoramidites ready for *in situ* photolithography. Indeed, the overwhelming majority of high-density microarrays are designed for DNA synthesis only. Given the ability of ribonucleotides to convey structure and function to ribonucleic acids and the vast chemical space that nucleoside analogues can probe within DNA and RNA, it appears particularly attractive to design microarrays that can accommodate natural and nonnatural nucleic acids. We present herein our work toward this goal.

11.2 Phosphoramidite Chemistry Applied to the Photolithographic Synthesis of Microarrays

In essence, the *in situ* synthesis of oligonucleotides by photolithography shares all but one aspect of standard solid-phase synthesis: reagents and phosphoramidites are delivered to the substrate using an automated DNA/RNA synthesizer and the synthesis usually proceeds in the 3′ → 5′ direction [12, 13]. Each phosphoramidite is incorporated according to a cycle undergoing the usual four steps: deprotection, coupling, oxidation, and capping (Figure 11.2). However, the fundamental difference with solid-phase synthesis is the substitution of the acid-labile 5′-DMTr for a photolabile protecting group. While early microchips were fabricated using either the NVOC or methylnitropiperonyloxycarbonyl (MeNPOC) as 5′ protecting groups, the attention eventually shifted toward the use of nitrophenylpropoxycarbonyl (NPPOC), a derivative of the *o*-nitrophenyl series with almost quantitative photodeprotection, producing microarrays of higher quality [15, 16]. The mechanism of cleavage has been extensively studied and seems to originate from a photoinduced β-elimination leading to the release of a nitrostyrene product, CO_2, and the 5′-OH oligonucleotide [17–19]. For these reasons, the photocleavage efficiency is optimal when UV illumination at 365 nm is accompanied with immersing the substrate into an appropriate solvent mixed with a small amount of base (usually piperidine or imidazole) to promote the β-elimination reaction. In the MAS optical system, UV light at 365 nm is first emitted from mercury vapor in short arc lamps and then homogenized by passing through a square light-pipe (Figure 11.3). The resulting spatially homogenized UV beam is directed by a series of mirrors to the DMD, where the mirrors have been correctly positioned by an external computer beforehand. The reflected image of the "ON" mirrors is then projected onto an Offner relay optical system which provides a 1 : 1 imaging of the light pattern to the synthesis surface placed in the reaction chamber. The substrate itself is, in most cases, a

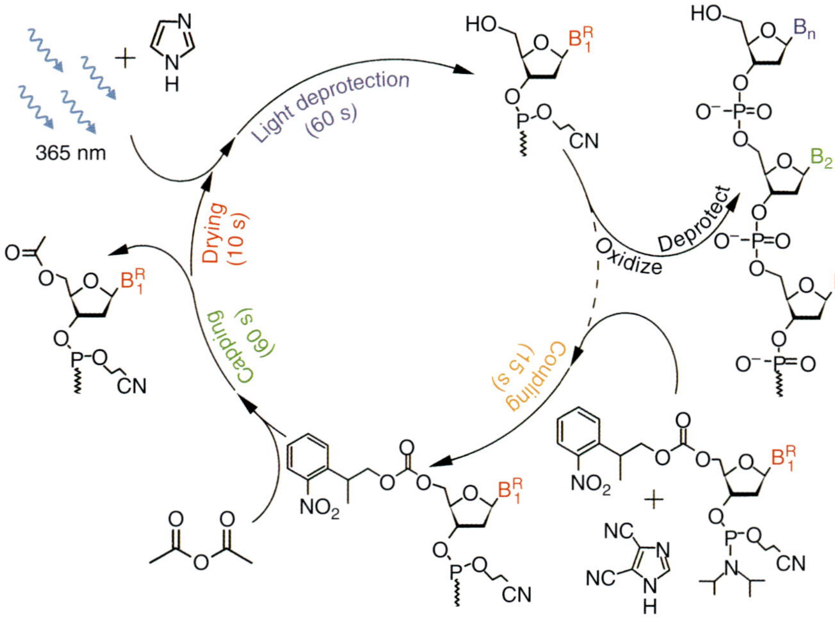

Figure 11.2 Synthetic cycle for the incorporation of an NPPOC-protected phosphoramidite in the *in situ* synthesis of microarrays by photolithography. B = adenine, guanine, cytosine or thymine ($B_1 \rightarrow B_n$ in the 3′ → 5′ direction). R = base protecting group (*tert*-butylPac for adenine, *i*PrPac for guanine, isobutyryl for cytosine). Source: Adapted from Agbavwe et al. [14]. https://creativecommons.org/licenses/by/2.0/.

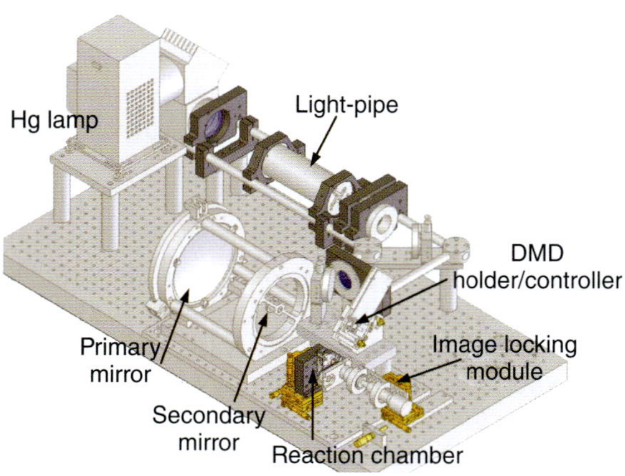

Figure 11.3 3D representation of the MAS optical system. Source: Adapted from Agbavwe et al. [14]. https://creativecommons.org/licenses/by/2.0/.

standard pre-silanized glass microscope slide, but MAS has been performed on glassy carbon [20], carbon-on-metal [21], conductive indium-tin oxide layers [22], and poly(ethylene terephthalate) (PET) substrates [23]. The reaction cell encasing the substrate is connected through delivery lines to the automated synthesizer, which itself is connected to an external computer. Custom-built software allows for control and communication between the DMD and the synthesizer during the UV illumination step.

11.3 Recent Improvements in the Synthesis of DNA Microarrays

High-density microarrays have not only been a laboratory curiosity; they have been promptly brought up to the commercial stage (Affymetrix, NimbleGen, and Agilent microarrays). And although DNA microarray chemistry can be regarded as being analogous to traditional DNA synthesis on solid-phase, its greater complexity as well as the few differences between the two approaches warranted a comprehensive study of the reaction parameters involved in light-directed synthesis (Figure 11.2).

We embarked on this study by tackling the coupling efficiency of DNA phosphoramidites, perhaps the most relevant factor in microarray quality. The first reported coupling efficiencies of MeNPOC and NPPOC amidites fluctuated around 92–94% and 96–99%, respectively [8, 24], which inevitably lowers the final yield of the oligonucleotide probe. While this can partially be compensated for by diminishing probe length or by adding multiple copies of the same probe per array, coupling efficiencies below 99% pose a strict limit in the accessible oligonucleotide length: at 98% coupling efficiency, only 30% of the sequences on a given feature contain a full-length 60mer, and as little as 13% of a 100mer. To accurately measure the coupling efficiency of NPPOC DNA phosphoramidites, a series of 12mers of each homopolymer was synthesized and terminated by a final coupling with a Cy3 phosphoramidite at the 5′ end [14]. The decrease in fluorescence intensity as the sequence length increases illustrates the overall synthesis yield. By fitting an exponential decay curve $f(n) = ae^{-bn}$ (where n is the number of couplings and $1-b$ the stepwise coupling yield) to the fluorescence plot, we were able to measure coupling efficiencies of ~99% for all four phosphoramidites, up to 99.4% and 99.8% for dA and dT, respectively. This corresponds to a 2.5-fold increase in the correct synthesis of a 60mer compared to coupling values previously obtained for NPPOC phosphoramidites. This increase may be the result of several combining factors: improvements in phosphoramidite quality and in coupling protocols (*vide infra*), higher photodeprotection efficiency for NPPOC groups compared to NVOC and MeNPOC, as well as the use of 4,5-dicyanoimidazole (DCI) as the activator. In a mixed-base 40mer, an average stepwise coupling yield of 98.5% was obtained. Interestingly, we found that the stepwise coupling yield does not significantly vary for coupling times ranging between 30 and 150 seconds, but slowly declines for longer coupling times (down to 96% for 600 seconds). A single second, the shortest coupling time mechanically measurable, still yields a 95% coupling efficiency, indicating an extremely fast coupling reaction between the phosphoramidite and the oligonucleotide on a glass substrate.

The absence of DMTr protecting groups in microarray photolithography eliminates the need to use an acidic solution during synthesis, suggesting that the phosphite triester resulting from the coupling reaction could withstand the conditions of oligonucleotide elongation. Indeed, we found that removing the I_2-mediated oxidation step at each cycle in favor of a global oxidation of all phosphite triester linkages at the very end of the synthesis (Figure 11.2) resulted in microarrays of similar and perhaps even higher quality than those produced with a standard oxidation step after each coupling. We also found that the introduction of a drying step after coupling and before exposure to UV resulted in higher coupling efficiencies, particularly with helium. While the exact reason why blowing helium improves the synthesis yield is still elusive, we surmise that clearing out all fluids (mostly acetonitrile (ACN)) before UV exposure helps in submerging the surface of the substrate with the more viscous DMSO as an exposure solvent. Finally, an alternative strategy to the standard acetic anhydride-based capping event was explored with the use of a regular DMTr-dT phosphoramidite. Indeed, given that microarray photolithography bypasses detritylation, a coupling with DMTr-dT can essentially be regarded as capping. Under our experimental conditions, we estimated the capping efficiency of DMTr-dT to be higher than that of acetic anhydride at equal capping times, reaching 97% after 60 seconds compared to 92% for acetic anhydride.

Further improvements in microarray quality must take into consideration unintended photodeprotection as an important source of error in oligonucleotide synthesis. NPPOC removal occurring outside of the intended features is mainly caused by stray light as a result of diffraction and reflection and leads to loss of optical contrast. For example, the area on the microarray surface corresponding to the gap between two "ON" mirrors is particularly prone to receiving stray light originating from the edges of the mirrors. We present several solutions that aim at reducing the amount of synthesis errors. The impact of diffraction can partially be minimized by synthesizing the same oligonucleotide probes within adjacent mirrors while reflection, primarily caused by the quartz block upon which the glass slide is placed, can be mitigated using a coating film absorbing UV light at 365 nm.

We then took an interest in the issue of throughput, a factor directly related to the cost and ease of access to microarrays for the preparation of large nucleic acid libraries. We first envisioned the production of multiple microarrays in a single synthesis, taking advantage of the transparency of glass and exposure solvent. Specifically, a second, identical microscope glass slide was mounted over the first substrate resting on a quartz block (Figure 11.4) [25]. The slides were separated by a 50-μm-thick polytetrafluoroethylene (PTFE) gasket, in the range of the ~60 μm depth-of-focus of the MAS optical system. Under those conditions, the two microarrays are centered around the focal point of the projected UV light and the PTFE gasket allows for a small volume of reactants to flow between both substrates. The resulting pair of arrays is a mirror image of each other and of comparable quality. Not only does this method double the throughput of array fabrication, it offers a simple approach to study array-to-array data reliability by providing two identical microchips originating from the same synthesis.

Directly influencing array throughput, the overall synthesis time was the next parameter that was examined. Singh-Gasson et al. reported synthesis times of 12 hours for simple microarrays [9], while 8–9 hours were necessary to fabricate more complex gene expression arrays using our previously optimized synthetic parameters [14], with the photodeprotection occupying well over a third of the overall fabrication process. Indeed, despite the very high photolysis efficiency of the NPPOC protecting group, its low absorptivity (ε) implies

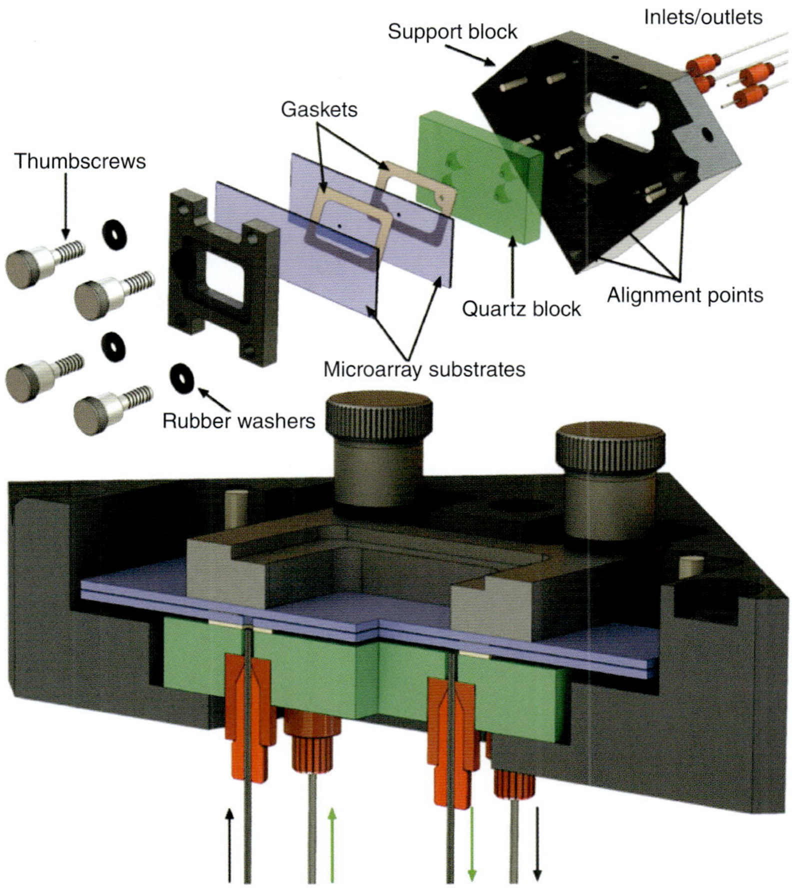

Figure 11.4 Exploded and section view of the new reaction cell allowing for the synthesis of two identical microarrays in parallel. Source: Sack et al. 2013 [25]. Reproduced with permission of ACS.

relatively long photodeprotection times (75 seconds at 80 mW cm^{-2}) to reach a radiant exposure of 6 J cm^{-2}. Alternatives to the NPPOC group with higher light sensitivity have been the subject of meticulous research [18, 19, 26], with however little success in replacing the standard NPPOC. Extended members of the NPPOC family, with aromatic substituents grafted on the nitrophenyl motif – a benzoyl group in *para* position to the nitro function (Bz-NPPOC) and a thiophenyl group in *para* along with an ethyl chain *meta* to the nitro function (SPh-NPPOC) (Figure 11.5) – are significant improvements [27]. The corresponding Bz-NPPOC and SPh-NPPOC protected DNA phosphoramidites were prepared and incorporated into microarrays with similar coupling efficiencies to regular NPPOC amidites (data not published). Moreover, we found that the radiant exposure necessary to reach almost quantitative photolysis is halved for Bz-NPPOC groups (2.7 J cm^{-2}) and tremendously reduced for SPh-NPPOC amidites (0.5 J cm^{-2}). At constant lamp power, this translates into a 2-fold and

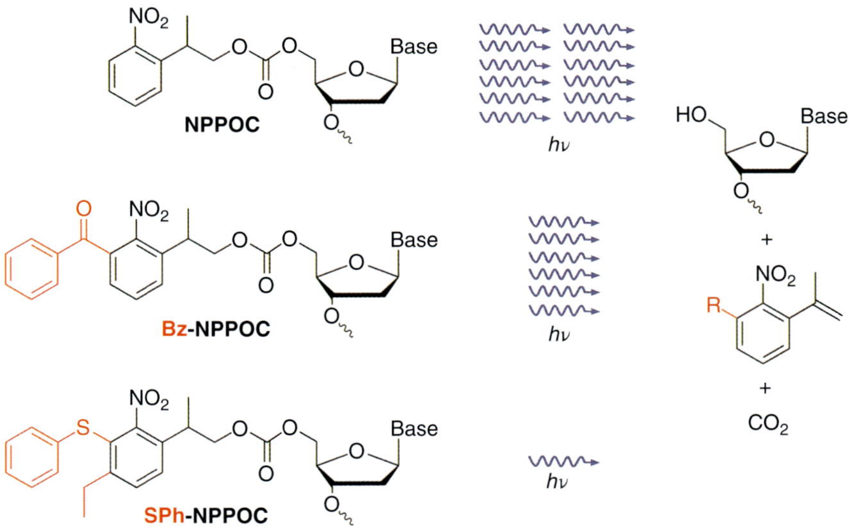

Figure 11.5 Chemical structures of the NPPOC, Bz-NPPOC, and SPh-NPPOC 5′-photoprotected deoxynucleotides and illustrated representation of the amount of light necessary to complete the photocleavage reaction. Source: Taken from Ref. [27].

12-fold reduction in photodeprotection time, down to only seven seconds for the removal of SPh-NPPOC at 80 mW cm^{-2}. Increased photolytic rates for SPh-NPPOC arise from a combination of a higher quantum yield ϕ and a much higher molar absorptivity ε.

Significantly shortening the photodeprotection time was our first step toward the elaboration of an expeditious array fabrication process [28]. We termed "legacy" the original method of gene expression array synthesis by photolithography and "double" the simultaneous fabrication of microarray pairs [25]. Typically, gene expression arrays are genome-wide libraries of 60mers and, as a tool to study the variations in gene expression between two cell populations, are perhaps the most well-known application of DNA microarrays [29]. In our attempts to shorten the fabrication time of complex microchips, we aimed at systematically optimizing each step in the phosphoramidite coupling cycle. With the photodeprotection time now drastically reduced, coupling becomes the most time-consuming step. Previous work on coupling time suggested that coupling of the incoming phosphoramidite to the surface-bound oligonucleotide happens in a timescale of a few seconds only. We revisited this study by considering the use of various amidite activators. We speculated that activators with lower pK_a values could enhance the reaction rate. Three tetrazole-based activators (5-ethylthio-1*H*-tetrazole (ETT), 5-benzylthio-1*H*-tetrazole (BTT), and 5-[3,5-bis(trifluoromethyl)phenyl]-1*H*-tetrazole (Activator 42)) as well as pyridinium chloride and DCI were tested. We found that DCI, despite its relatively higher pK_a compared to tetrazole activators (5.2 versus 4.1–4.8, respectively), yielded consistently high signals in microarrays with coupling

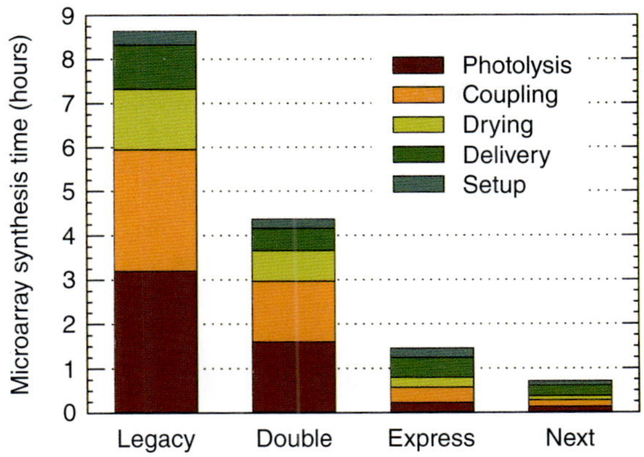

Figure 11.6 Synthesis time for high-density gene expression arrays (60mers) according to various fabrication protocols. Parameters were optimized and resulted in the elaboration of the "express" protocol. Source: Taken from Ref. [28].

times down to 15 seconds. Furthermore, DCI was the only activator to produce uniform feature intensities, as opposed to tetrazole activators that routinely led to the formation of bright spots within the same feature. DCI thus became the activator of choice and allowed for the coupling time to be shortened down to 15 seconds. Other significant adjustments in the synthesis protocol included the introduction of a very short (1 second) oxidation step at each cycle instead of a final oxidation and the reduction of the essential drying step from 30 to 10 seconds. Overall, these changes allowed for the fabrication of gene expression arrays in 90 minutes. Our new process, coined "express", represents a 6-fold decrease in synthesis time compared to the "legacy" method (Figure 11.6).

The latest improvement in DNA microarray photolithography relates to cost and ecological footprint in the process of microarray synthesis. High-pressure short arc mercury lamps are commonplace in photochemistry, as mercury vapor efficiently transforms electricity into UV light. Up until recently, there were few other options for bright and efficient sources of UV. However, the last decade has seen the advent of light emitting diodes (LEDs) as cheap and reliable light sources. UV LEDs emit ultraviolet light with radiance levels that can exceed those of standard mercury lamps. Given the additional polluting and toxic properties of mercury [30], it is not surprising to witness the gradual replacement of mercury lamp for more environmentally friendly UV LEDs. UV light in microarray lithography is traditionally produced with ultra-high-pressure mercury lamps, whose lifetimes do not pass beyond 1000 hours. We have installed a UV LED in our MAS optical system [31] and measured greater amounts of UV light delivered to the synthesis substrate than with mercury lamps (100 mW cm^{-2} versus 80 mW cm^{-2}, respectively). When mounted over a fan-based cooler equipped with a heat sink in order to disperse radiation heat (Figure 11.7), the decrease in light output of the LED was limited to 1% during each exposure cycle. Under those conditions, we estimated the lifetime of the UV LED to reach 15 000 hours. Combined with the lower costs of initial installation, the UV LED appears clearly superior to mercury lamp in light-directed array synthesis.

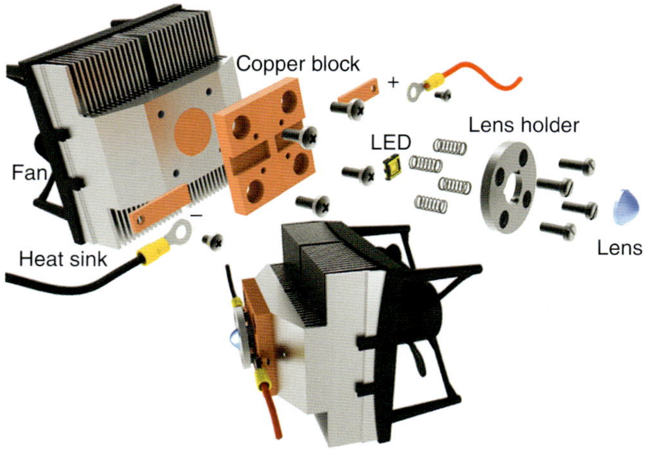

Figure 11.7 Exploded and assembled view of the UV LED as a source of UV light. The LED is soldered through a copper block to a heat sink, which is attached to a standard CPU fan-based cooler. Source: Holz et al. 2017 [31]. https://creativecommons.org/licenses/by/4.0/.

11.4 Synthesis of RNA Microarrays

While DNA microarrays have been available for the past two decades from various fabrication routes and the underlying chemistry was improved and refined, adapting those protocols to the synthesis of RNA arrays has been a very challenging endeavor. Perhaps the most fundamental problem in adapting solid-phase RNA phosphoramidite chemistry to microarrays is the unsuitable method for the removal of the *tert*-butyldimethylsilyl (TBDMS) protection of the 2′-OH group. Indeed, the use of fluoride-containing reagents for TBDMS deprotection is detrimental to the integrity of the glass surface of a microarray. Therefore, the compatibility of RNA phosphoramidites for microarray photolithography depended on the development of a novel, base-labile protecting group at the 2′-OH position. The acyl group is a simple example of a base-labile protecting group, but 2′-acyl groups in ribonucleotides have a tendency to migrate over to the 3′-OH position [32, 33]. Migration can however be prevented by separating the acyl group from the 2′-hydroxyl via an acetal moiety, a strategy that has been successfully applied to the preparation of acetal-based protecting groups for RNA synthesis on solid-phase [34, 35]. The levulinyl chain is a convenient protecting strategy for RNA phosphoramidites as it can be cleaved under mild conditions, and we had previously reported on the synthesis of 2′ and 3′-levulinylated RNA synthons [36]. In order to hinder 2′,3′ migration of the levulinyl and, in so doing, facilitating the synthesis of the corresponding phosphoramidites, we installed an acetal function to act as a chemical spacer between the 2′-hydroxyl and the levulinyl chain (acetal levulinyl ester, ALE). With the introduction of a photolabile protecting group at the 5′-OH position, we isolated the first set of RNA phosphoramidites compatible with *in situ* microarray photolithography [37]. The synthesis of the RNA monomers commences at the free nucleoside stage (Scheme 11.1), which are directly protected with a bridged silyl group at the 3′ and 5′ position using

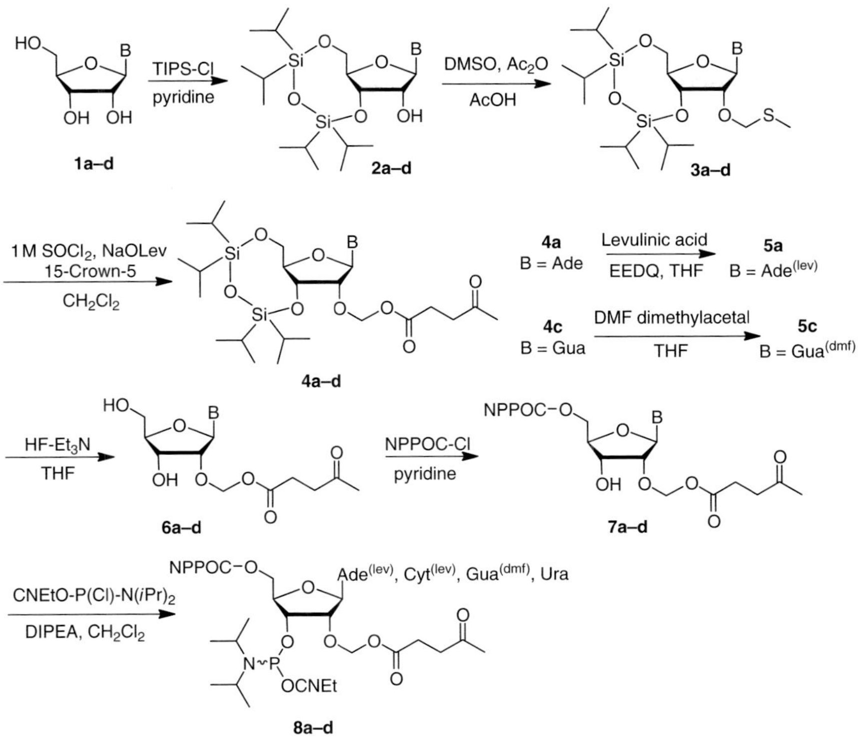

Scheme 11.1 Synthesis of the 5′-NPPOC 2′-*O*-ALE RNA phosphoramidites. Ade, adenine; Cyt, cytosine; Gua, guanine; Ura, uracil; lev, levulinyl; dmf, dimethylformamidine; CNEt, cyanoethyl.

Markiewicz's reagent [38]. The 2′-OH function of the silyl ethers were transiently protected with a thioacetal, subsequently displaced with sodium levulinate for rA, rC, and rU monomers, and with cesium chloride and levulinic acid for rG, to afford 2′-*O*-ALE nucleosides **4**. Next, the adenosine and guanine bases were protected with a levulinyl and a dimethylformamidine, respectively. After removal of the 3′,5′-silyl group with HF·Et_3N, the NPPOC was installed at the 5′ position. Finally, phosphitylation under standard conditions yielded the corresponding 5′-NPPOC 2′-*O*-ALE RNA phosphoramidites.

To evaluate their potential in RNA synthesis on microarrays, we first measured the coupling efficiency of our new RNA monomers. Homopolymeric sequences 1–12 nucleotides long were fabricated on microarrays and terminated with a fluorescent dye. The decrease in fluorescence intensity as the sequence length increases was plotted and used to extract a stepwise coupling efficiency (Figure 11.8a). With relatively long coupling times (10–15 minutes), coupling efficiencies approached 95–97%, though rA suffered average coupling yields (86%). We now consistently obtain coupling efficiencies above 99% for all monomers (data not published). The next step was to confirm the complete deprotection of the RNA oligonucleotide synthesized *in situ*. As in regular RNA deprotection after solid-phase, we adopted a two-step strategy for deprotection. The cyanoethyl groups were first removed by treating the microarray with

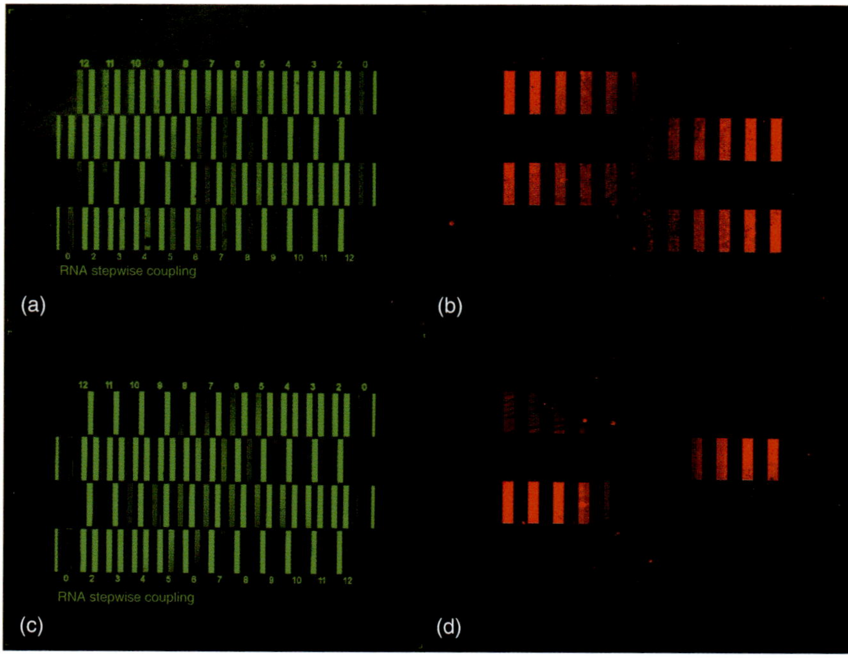

Figure 11.8 (Left) Coupling efficiencies of poly-rU (a) and poly-rA (c) microarrays. For each sequence length (1–12), three features are synthesized: one with a terminal Cy3 label, one without terminal labeling, and one coupling reference. (Right) hybridization of poly-rU (b) and poly-rA (d) sequences (1–12 nt) to their complementary Cy5-dA_{20} and Cy5-dT_{20}, respectively. Source: Lackey et al. 2009 [37]. Reproduced with permission of ACS.

a mixture of Et_3N in ACN, while the ALE and levulinyl were cleaved at a later stage by hydrazinolysis. Hybridization of the resulting deprotected RNA oligonucleotides to their complementary, Cy5-labeled strands, gave brighter signals as the sequence length increased (Figure 11.8b), indicating correct RNA deprotection as well as compatibility of the deprotection cocktails with the microarray substrate.

The efficiency of NPPOC photolysis on RNA oligonucleotides was also investigated by applying a gradient of radiant exposure to the array, from 0 to 14 J cm^{-2}. The deprotected poly-ribouridylate was then hybridized to its complement and the fluorescence intensity served as a measure of photodeprotection efficiency. We found that the fluorescence intensity would plateau at a radiant exposure of approximately 6 J cm^{-2} (Figure 11.9), very similar to DNA oligonucleotides synthesized with NPPOC protecting groups [14].

To further confirm the identity of the synthesized RNA oligonucleotides, we designed a DNA microarray containing a single ribonucleotide insert (rU) and terminally labeled with a fluorescent dye [39]. We then studied the enzymatic cleavage of the DNA–RNA chimera by RNase A, an endoribonuclease that recognizes and cleaves single-stranded RNA preferentially at a pyrimidine residue in 5′-purine-pyrimidine-purine-purine-3′ regions [40]. The substrates (Table 11.1), a 5′-dA-rU-dA-dA-3′ tetranucleotide as well as sequence variants,

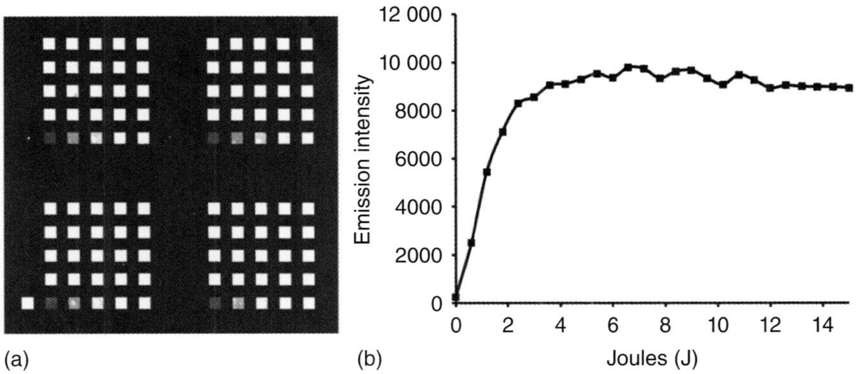

Figure 11.9 Gradient of exposure on a rU_{20} microarray. (a) Cy5 emission scan of the hybridized rU_{20}:dA_{20}-Cy5 duplex. (b) Plot of the measured fluorescence intensity as a function of radiant exposure.

Table 11.1 Sequences synthesized on microarray with single rU inserts and Cy3 terminal labeling.

Sequence name	Microarray sequence 3′ to 5′	Reference substrate 5′ to 3′	Reference RNase activity k_{cat}/K_m (10^7 M^{-1} s^{-1})
A	T_{15}-dArU-Cy3	6-FAM-rUdA-6-TAMRA	2.5 ± 0.3
B	T_{15}-(dA_2)rUdA-Cy3	6-FAM-dArU(dA_2)-6-TAMRA	3.6 ± 0.4
C	T_{15}-(dA_3)rU$(dA)_2$-Cy3	6-FAM-$(dA)_2$rU(dA_3)-6-TAMRA	4.7 ± 0.6
D	T_{15}-(dA_4)rU$(dA)_3$-Cy3	6-FAM-$(dA)_3$rU(dA_4)-6-TAMRA	4.8 ± 0.5

were correctly recognized and cleaved by RNase A (Figure 11.10). As expected, the variant missing a purine 5′ to the rU insert was the slowest to be degraded.

The integrity of the *in situ* synthesized RNA oligonucleotides was independently assessed with mass spectrometry (MS). Microarrays are, by definition, a collection of surface-bound molecules, and the process by which those molecules are synthesized on the substrate inherently restricts the analytical tools that can be used to assess the fabrication quality. The combination of low oligonucleotide density on the surface of glass slides (0.1–1 pmol mm^{-2}) [11, 41] and small synthesis area (1–1.5 cm^2) suggests that, in theory, a maximum of 10–15 pmol of nucleic acids can be synthesized on a microarray, severely limiting the analysis to only the most sensitive methods. Gel electrophoresis of oligonucleotides cleaved from microarrays using ammonia [4] and subsequently radiolabeled with ^{32}P has been successfully shown in a few select cases [5, 10, 11]. Mass spectrometry of cleaved array oligonucleotides had, on the other hand, never been demonstrated. We developed a simple and novel approach for the analysis by mass spectrometry of DNA and RNA oligonucleotides synthesized on microarrays [42]. The sequences are first synthesized over a dT unit containing a succinyl ester at the 3′ position, thus converting the microarray

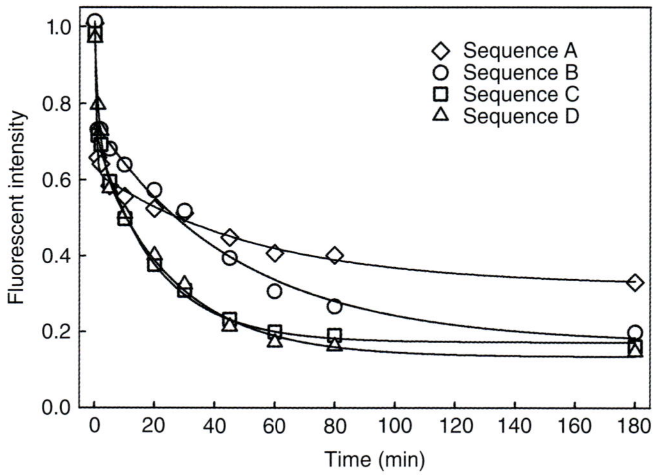

Figure 11.10 Fluorescence intensity of array-bound Cy3-labeled oligonucleotides in contact with RNase A and taken at various points in time. Source: Reproduced from Ref. [39].

into a base-cleavable surface. After synthesis of the remaining sequence, DNA arrays were treated in strictly anhydrous ethylenediamine/toluene 1 : 1 (as opposed to ethylenediamine/ethanol 1 : 1 in standard deprotection) then washed in equally anhydrous ACN. Pleasingly, the deprotected and cleaved DNA oligonucleotides remain electrostatically bound to the microarray surface and can only be eluted by applying a small amount of water over the synthesis area. The collected amount of DNA, oscillating with syntheses around 20 pmol, was injected and analyzed by LC–MS. The resulting MS traces allowed for the characterization of DNA synthesized on microarrays, revealing the presence of the correct full-length sequence as well as crucial information about array quality. Importantly, changes in the chip fabrication parameters had detectable effects on the LC–MS spectra. We applied a similar strategy for the analysis of RNA oligonucleotides synthesized *in situ*. After synthesis, the RNA array was

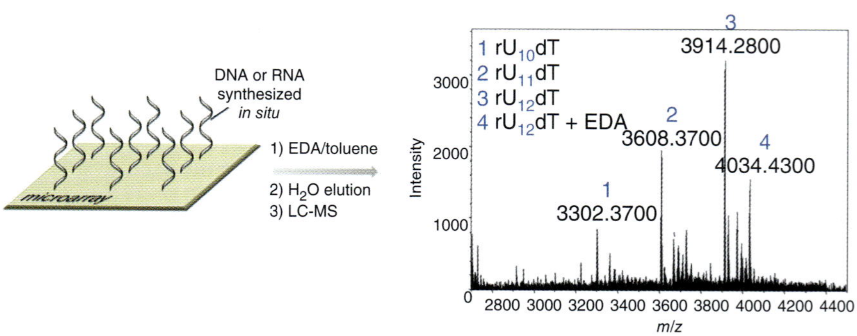

Figure 11.11 Schematic representation of the microarray cleavage procedure leading to LC–MS analysis of the cleaved, deprotected RNA oligonucleotide (rU_{12}). Source: Lietard et al. 2014 [42]. https://creativecommons.org/licenses/by/3.0/.

deprotected in triethylamine followed by hydrazinolysis. To recover the RNA oligonucleotide from the surface, the array was treated with ethylenediamine to promote cleavage of the succinyl ester. The crude chip eluate was then analyzed by LC–MS (Figure 11.11), confirming the synthesis of a full-length, 12-nt-long homopolymer of uridine. As in cleaved DNA, the oligonucleotide contains a dT unit at the 3′ end, a remnant of the cleavage of the dT-succinyl ester. The MS trace also shows the presence of shortmers (rU_{11} and rU_{10}), but the elution protocol does not allow us to conclusively determine whether shorter oligonucleotides are due to coupling failures or base-mediated degradation of the RNA.

11.5 Enzymatic Approaches to RNA Array Synthesis

With the chemical synthesis of RNA arrays still in its infancy long after the rise of high-density DNA microarrays, other methods have been pursued for the preparation of RNA arrays. Given the robustness and cost of DNA chips, the enzymatic transcription of DNA molecules into their RNA counterpart has lately become an attractive alternative to *in situ* synthesis. Enzymatic transcription has the advantage of being able to start from long DNA templates to prepare equally long RNA products, a yet unreachable destination with the phosphoramidite route. In addition, the methodology can be applied to the incorporation of chemically modified nucleoside triphosphates (NTPs), assuming the use of the proper polymerase, to yield RNA analogs with valuable biophysical and structural properties. The challenge however, is to keep the polymerase-extended RNA bound to the array surface. In this section, we will cover strategies that have been developed for the enzymatic synthesis of RNA microarrays.

In a first example, Chen et al. used a flow cell to capture the synthesized RNA. On a gold-coated substrate for Surface plasmon resonance imaging (SPRI) purposes, a pre-synthesized and amino-modified single-stranded DNA (ssDNA) is attached by reacting the 5′-amino function to the surface functionalized with carboxylate groups [43]. The ssDNA is hybridized to its complement and the resulting dsDNA is transcribed into RNA using T7 RNA polymerase [44]. Since the transcription reaction is performed in a microfluidic channel, the ssRNA diffuses along the flow cell and is then captured, about one mm further along the channel, by a short (18 bp) DNA handle complementary to the 5′ end of the RNA in a process called "hybridization–adsorption". The *in situ* transcribed RNA is thus bound to the gold surface through hybridization. Two different RNA aptamers were produced and shown to bind to their protein targets, the human thrombin and the vascular endothelial growth factor (VEGF). A final RNase H treatment confirmed the nature of the captured single-stranded oligonucleotide on the surface.

In a different yet elegant approach, the flow cell of a standard high-throughput sequencing instrument containing clusters of dsDNA at defined locations was reimagined as a DNA microarray. Indeed, the sample DNA is first fragmented and binds to one of the multiple handle sequences covalently attached to the surface of the flow cell. After synthesis of the complementary DNA sequence, the resulting

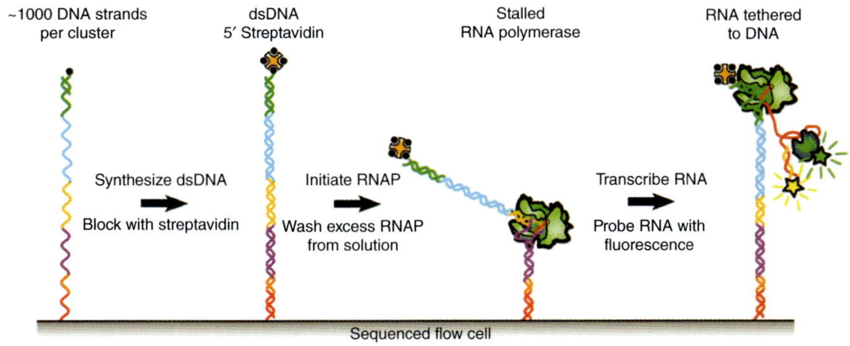

Figure 11.12 Schematic representation of the *in situ* RNA transcription of dsDNA clusters on a sequencing flow cell. The complementary strand of the ssDNA is first polymerized from a biotin-labeled primer. Streptavidin is then added to create a biotin–streptavidin block. In a third step, RNA transcription is initiated with RNAP and proceeds until it reaches the roadblock, where RNAP stalls while still being bound to the dsDNA and to the transcribed RNA (in red). Source: Buenrostro et al. 2014 [46]. Reprinted with permission of Nature Publishing Group.

dsDNA can be used to assess the binding to fluorescently tagged proteins [45]. The procedure was later adapted to the study of RNA-binding proteins [46]. To do so, the dsDNA clusters were transcribed into RNA with an RNA polymerase. A streptavidin-biotin roadblock was introduced at the end of the dsDNA that stalled the RNA polymerase along with the transcribed RNA, essentially tethering the RNA to its dsDNA template (Figure 11.12). The resulting RNA array of 148 000 unique sequences was then probed against a variety of RNA-binding proteins.

High-density DNA microarrays synthesized by photolithography have also been used as templates for transcription into RNA oligonucleotides [47]. In this approach, two different DNA sequences were synthesized within the same feature: a template and its primer for RNA extension. Partial NPPOC removal on the feature allowed for a controlled synthesis of each DNA strand while a long, flexible poly(ethylene glycol) linker was introduced to ensure hybridization between template and primer. The primer was then extended using rNTPs and T7 RNA polymerase. Finally, the DNA template was specifically degraded, yielding a high-density microchip composed of only RNA. This enzymatic approach was found to be applicable to the elongation of 2′-OMe and 2′F-RNA transcripts, making it a convenient method for the fabrication of chemically modified nucleic acids arrays on glass, and also on PET substrates [23].

11.6 Synthesis of 2′F-ANA Microarrays

Chemically modified nucleoside analogues are key elements in DNA/RNA engineering and therapeutics as they allow to tune virtually any biophysical property of nucleic acids. They also play an important role in expanding the structural space of nucleic acids, which is a pivotal aspect of finding and understanding function in DNA and RNA. For those reasons, aptamer and

DNAzyme/RNAzyme development benefits greatly from the incorporation of nucleoside analogues, but the number of combinations of natural and nonnatural nucleic acid within the same sequence quickly becomes unsurmountable for classical solid-phase synthesis. While this can be solved by following a combinatorial approach to the synthesis of nucleic acids libraries, those are typically used to select for the most appropriate candidate and discard poor binders and nonfunctional oligonucleotides. In order to avoid a skewed representation of the chemical space, nucleic acids libraries containing nonnatural nucleoside analogues could be synthesized on high-density microarrays. To this date, there have been few reports on chemically modified DNA/RNA arrays [48].

A prime example of a biologically relevant nucleoside analogue is 2′-deoxy-2′-fluoroarabinonucleic acid (2′F-ANA), a DNA-like derivative of arabinose [49, 50] where the fluorine atom is in "beta" (or "up") configuration (Figure 11.13b). Its resistance against nucleolytic degradation, high base-pairing specificity and its ability to elicit RNase H-mediated cleavage of DNA/2′F-ANA:RNA duplexes has shaped 2′F-ANA into an excellent candidate in antisense and siRNA-based therapeutics [52]. In G-quadruplexes, the introduction of 2′F-ANA nucleotides confers highly unusual stabilization through the formation of internal hydrogen bonds involving the fluorine atom and the H8 of neighboring guanine bases, and between the polarized FC-H2′ hydrogen and the ring oxygen (O4′) of an adjacent nucleotide residue [53, 54]. In addition, 2′F-ANA-modified G-quadruplexes in the thrombin-binding aptamer sequence (TBA1, Figure 11.13a) were found to significantly alter the binding properties to thrombin with, in a few cases, increased binding affinity [55]. We wished to adapt this study to microarrays in order to fully map the binding affinity landscape of 2′F-ANA-modified TBA1 sequences. Indeed, microarrays synthesized by photolithography can

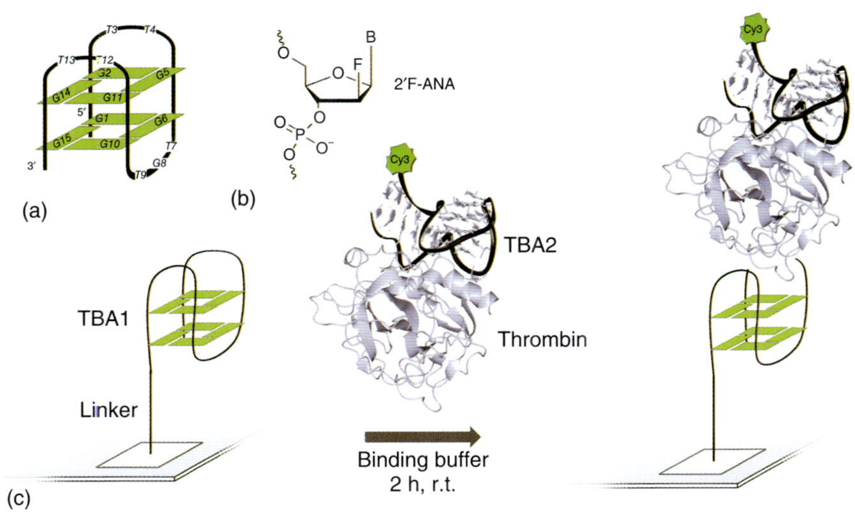

Figure 11.13 (a) Schematic structure of the antiparallel G-quadruplex TBA1 (b) Chemical structure of 2′F-ANA. (c) Overview of the binding assay on G-tetrad TBA1 microarrays to thrombin bound to a secondary, Cy3-labeled aptamer, TBA2. Source: Taken from Ref. [51].

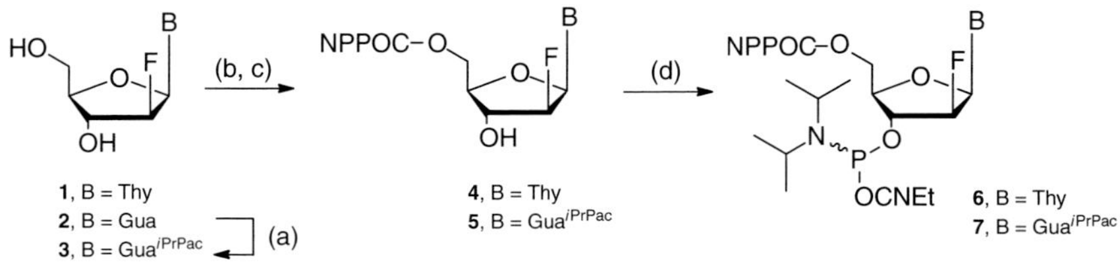

Scheme 11.2 Synthesis of the 2′F-araT and G phosphoramidites. Thy, thymine. Reagents and conditions: (a) TMS-Cl, iPrPac-Cl, pyridine, 5 °C, 45 minutes, 88%; (b) NPPOC-Cl, THF/pyridine 1 : 1 (v/v), 0 °C, 3 hours, 61% (**4**); (c) NPPOC-Cl, pyridine 1 : 1 (v/v), 0 °C, three hours, 45% (**4**); (d) CEP-Cl, DIPEA, THF, 1 hour, r.t., 79% (**6**) and 69% (**7**). Source: Reproduced from Ref. [51].

comfortably accommodate all DNA-to-2′F-ANA permutations in TBA1 (15 nt long; $2^{15} = 32\,768$ combinations), which can then be assayed in parallel against thrombin. The detection method involves the preformation of a complex of thrombin and a secondary, Cy3-labeled thrombin-binding aptamer (TBA2, Figure 11.13c) [51].

Modifying the palindromic TBA1 sequence (GGTTGGTGTGGTTGG) requires the preparation of the corresponding G and T variants of 2′F-ANA. Phosphoramidites of 2′F-araT and 2′F-araG were synthesized from the known 3′,5′-OH 2′F-araT and 2′F-araG nucleosides. After protection of the guanine base with a isopropylphenoxyacetyl group (similar to NPPOC DNA phosphoramidites), the nucleosides were protected at the 5′-OH with NPPOC-Cl then phosphitylated under standard conditions to afford amidites **6** and **7** (Scheme 11.2). The coupling of the photosensitive 2′F-ANA phosphoramidites was found to reach near quantitative levels (>99.9%) for a coupling time of 10 minutes. Expectedly, fully modified 2′F-ANA or mixed 2′F-ANA/DNA oligonucleotides yielded equal or even higher hybridization signals than the corresponding DNA-only sequences. Next, the complete permutation library of the TBA1 sequence was fabricated on a microarray and included multiple replicates of each sequence. After binding the deprotected array to the preformed thrombin·TBA2–Cy3 complex, the recorded fluorescence intensities were used as a measure of binding affinity (Figure 11.14a). We found that the substitution of dT residues for 2′F-araT was generally well accepted except for positions T4 and T13. On the other hand, residues T3 and T12 significantly increased the binding affinity to thrombin when modified with 2′F-araT (Figure 11.14b). Substitution of the dG residues for 2′F-araG was, however, met with decreased binding affinity. The solution structure of the TBA1 sequence modified with 2′F-araT at position T3 was solved by NMR and suggests that the introduction of fluorine at T3 preorganizes the T3–T4 dinucleotide loop, one of the two contact points with thrombin. In parallel, we were able to identify combinations of 2′F-ANA modifications in TBA1 from the microarray library leading to a 3-fold decrease in dissociation constant compared to the original structure. We thus simultaneously showed the attractive properties of high-density microarrays for aptamer discovery and the promising perspective of fabricating arrays containing all four 2′F-ANA nucleotides.

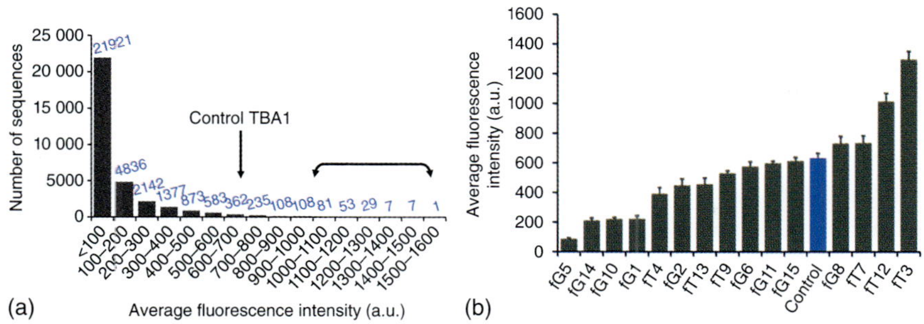

Figure 11.14 Recorded fluorescence intensities of the TBA1 library at a 200 nM thrombin concentration. (a) Higher fluorescence intensities correspond to higher binding affinity. The DNA-only TBA1 is depicted as "control TBA1." The double arrow indicates the subrange of fluorescence intensities that was used for K_d measurements. A range of 178 sequences with high thrombin-binding affinity were identified in this subrange. (b) Average fluorescence intensities for single-modified TBA1 sequences. Source: Taken from Ref. [51].

11.7 Conclusion and Outlook

Phosphoramidite chemistry on microarrays has inherited remarkable efficiency in the preparation of oligonucleotides resulting from decades of optimization in solid-phase synthesis; yet, there remain few differences between microarray and classical solid-phase that justified a deeper investigation of DNA synthesis on glass surfaces. We studied in particular the coupling efficiency, as it dictates the quality of the microarray, perhaps even more so than for solid-phase synthesis. The photodeprotection efficiency was also examined, with novel NPPOC (Bz- and SPh-NPPOC) variants requiring less than a tenth of the light exposure necessary for the removal of the original NPPOC protecting group. This prompted further attempts at shortening the fabrication time of high-density arrays. With high coupling yields (>99%) and a rapid production of complex arrays now available, our attention is turning to improving the quality of high-density DNA arrays for aptamer, gene expression, and gene assembly purposes [56, 57]. In parallel, an additional increase in array complexity and density can be envisaged with higher resolution DMD, as 1920×1080 digital light processors containing two million individually addressable mirrors are becoming commonplace. Similar increase in resolution without changing the number or size of the features could be achieved by decreasing the wavelength of light all the while avoiding absorption of UV by DNA.

Microarray photolithography was shown for the first time to be compatible with the *in situ* synthesis of RNA chips using a new set of base-sensitive NPPOC-protected RNA phosphoramidites. We are now actively involved in the synthesis of complex, mixed-base, high-density RNA arrays and their use as platforms to interrogate RNA-protein interactions. As an alternative, RNA arrays can also be produced with the direct transcription of DNA arrays using an RNA polymerase. The two methods can be seen as complementary; while the phosphoramidite approach is simple to design and a mere adaptation of the *in situ* DNA synthesis, the enzymatic approach profits from the robustness of DNA synthesis to potentially obtain longer transcripts. And while the former may be useful in designing and preparing short RNA probes, the latter may be necessary for the synthesis of structurally and functionally diverse RNA sequences.

Finally, we have presented the recent work toward the diffusion of *in situ* array synthesis to nonnatural nucleic acids, poorly covered in photolithography. We chose to focus on 2′F-ANA, a biologically relevant and DNA-like nucleic acid analogue displaying unusual electronic properties, particularly in the context of G-quadruplexes, i-motifs, and duplex structures [58, 59]. Using photolithography, we have prepared a microarray containing the entire library of DNA-to-2′F-ANA permutations in the 15mer thrombin-binding aptamer (32 768 sequences) and studied the thrombin-binding affinity of each sequence. Having the ability to study each individual 2′F-ANA-modified TBA1 oligomer separately is essential not only for probing the chemical space occupied by 2′F-ANA modifications, but also for mapping the binding landscape of a chemically modified aptamer. We are now aiming toward the incorporation of all 2′F-ANA nucleobases in microarrays, which would result in the extension of our chemical repertoire to three different classes of nucleic acids. In so doing, light-directed array synthesis of DNA/RNA/2′F-ANA chimeras may be possible, thus representing a further step in microchip complexity.

References

1 Trevino, V., Falciani, F., and Barrera-Saldaña, H.A. (2007). *Mol. Med.* 13: 527–541.

2 Hoheisel, J.D. (2006). *Nat. Rev. Genet.* 7: 200–210.

3 Blanchard, A.P., Kaiser, R.J., and Hood, L.E. (1996). *Biosens. Bioelectron.* 11: 687–690.

4 Cleary, M.A., Kilian, K., Wang, Y. et al. (2004). *Nat. Methods* 1: 241–248.

5 LeProust, E.M., Peck, B.J., Spirin, K. et al. (2010). *Nucleic Acids Res.* 38: 2522–2540.

6 Fodor, S.P.A., Read, J.L., Pirrung, M.C. et al. (1991). *Science* 251: 767–773.

7 Pease, A.C., Solas, D., Sullivan, E.J. et al. (1994). *Proc. Natl. Acad. Sci. U.S.A.* 91: 5022–5026.

8 McGall, G.H., Barone, A.D., Diggelmann, M. et al. (1997). *J. Am. Chem. Soc.* 119: 5081–5090.

9 Singh-Gasson, S., Green, R.D., Yue, Y. et al. (1999). *Nat. Biotechnol.* 17: 974–978.

10 Richmond, K.E., Li, M.-H., Rodesch, M.J. et al. (2004). *Nucleic Acids Res.* 32: 5011–5018.

11 LeProust, E., Zhang, H., Yu, P. et al. (2001). *Nucleic Acids Res.* 29: 2171–2180.

12 Matteucci, M.D. and Caruthers, M.H. (1981). *J. Am. Chem. Soc.* 103: 3185–3191.

13 Caruthers, M.H. (1985). *Science* 230: 281–285.

14 Agbavwe, C., Kim, C., Hong, D. et al. (2011). *J. Nanobiotechnol.* 9.

15 Beier, M. and Hoheisel, J.D. (2000). *Nucleic Acids Res.* 28: E11.

16 Pirrung, M.C., Wang, L.X., and Montague-Smith, M.P. (2001). *Org. Lett.* 3: 1105–1108.
17 Walbert, S., Pfleiderer, W., and Steiner, U.E. (2001). *Helv. Chim. Acta* 84: 1601–1611.
18 Buhler, S., Lagoja, I., Giegrich, H. et al. (2004). *Helv. Chim. Acta* 87: 620–659.
19 Woll, D., Smirnova, J., Galetskaya, M. et al. (2008). *Chem. Eur. J.* 14: 6490–6497.
20 Phillips, M.F., Lockett, M.R., Rodesch, M.J. et al. (2008). *Nucleic Acids Res.* 36.
21 Lockett, M.R., Weibel, S.C., Phillips, M.F. et al. (2008). *J. Am. Chem. Soc.* 130: 8611–8613.
22 Mayali, X., Weber, P.K., Brodie, E.L. et al. (2012). *ISME J.* 6: 1210–1221.
23 Holden, M.T., Carter, M.C.D., Wu, C.H. et al. (2015). *Anal. Chem.* 87: 11420–11428.
24 Nuwaysir, E.F., Huang, W., Albert, T.J. et al. (2002). *Genome Res.* 12: 1749–1755.
25 Sack, M., Kretschy, N., Rohm, B. et al. (2013). *Anal. Chem.* 85: 8513–8517.
26 Hasan, A., Stengele, K.P., Giegrich, H. et al. (1997). *Tetrahedron* 53: 4247–4264.
27 Kretschy, N., Holik, A.K., Somoza, V. et al. (2015). *Angew. Chem. Int. Ed.* 54: 8555–8559.
28 Sack, M., Holz, K., Holik, A.K. et al. (2016). *J. Nanobiotechnol.* 14.
29 Bumgarner, R. (2013). *Curr. Protoc. Mol. Biol.* 101: 22.21.1–22.21.11.
30 Park, J.D. and Zheng, W. (2012). *J. Prev. Med. Public Health* 45: 344–352.
31 Holz, K., Lietard, J., and Somoza, M.M. (2017). *ACS Sustain. Chem. Eng.* 5: 828–834.
32 Reese, C.B. and Trentham, D.R. (1965). *Tetrahedron Lett.* 2459–2465.
33 Reese, C.B. and Trentham, D.R. (1965). *Tetrahedron Lett.* 2467–2472.
34 Parey, N., Baraguey, C., Vasseur, J.J., and Debart, F. (2006). *Org. Lett.* 8: 3869–3872.
35 Lavergne, T., Bertrand, J.R., Vasseur, J.J., and Debart, F. (2008). *Chem. Eur. J.* 14: 9135–9138.
36 Lackey, J.G., Sabatino, D., and Damha, M.J. (2007). *Org. Lett.* 9: 789–792.
37 Lackey, J.G., Mitra, D., Somoza, M.M. et al. (2009). *J. Am. Chem. Soc.* 131: 8496–8502.
38 Markiewicz, W.T. (1979). *J. Chem. Res. (S)* 24–25.
39 Lackey, J.G., Somoza, M.M., Mitra, D. et al. (2009). *Chim. Oggi-Chem. Today* 27: 30–33.
40 Kelemen, B.R., Klink, T.A., Behlke, M.A. et al. (1999). *Nucleic Acids Res.* 27: 3696–3701.
41 Beaucage, S.L. (2001). *Curr. Med. Chem.* 8: 1213–1244.
42 Lietard, J., Kretschy, N., Sack, M. et al. (2014). *Chem. Commun.* 50: 12903–12906.
43 Chen, Y.L., Nguyen, A., Niu, L.F., and Corn, R.M. (2009). *Langmuir* 25: 5054–5060.
44 Chen, Y., Nakamoto, K., Niwa, O., and Corn, R.M. (2012). *Langmuir* 28: 8281–8285.

45 Nutiu, R., Friedman, R.C., Luo, S.J. et al. (2011). *Nat. Biotechnol.* 29: 659–U146.
46 Buenrostro, J.D., Araya, C.L., Chircus, L.M. et al. (2014). *Nat. Biotechnol.* 32: 562–568.
47 Wu, C.-H., Holden, M.T., and Smith, L.M. (2014). *Angew. Chem. Int. Ed.* 53: 13514–13517.
48 Yang, F., Dong, B., Nie, K. et al. (2015). *ACS Comb. Sci.* 17: 608–614.
49 Wilds, C.J. and Damha, M.J. (2000). *Nucleic Acids Res.* 28: 3625–3635.
50 Watts, J.K. and Damha, M.J. (2008). *Can. J. Chem.* 86: 641–656.
51 Lietard, J., Abou Assi, H., Gomez-Pinto, I. et al. (2017). *Nucleic Acids Res.* 45: 1619–1632.
52 Deleavey, G.F., Watts, J.K., Alain, T. et al. (2010). *Nucleic Acids Res.* 38: 4547–4557.
53 Anzahaee, M.Y., Watts, J.K., Alla, N.R. et al. (2011). *J. Am. Chem. Soc.* 133: 728–731.
54 Martin-Pintado, N., Yahyaee-Anzahaee, M., Deleavey, G.F. et al. (2013). *J. Am. Chem. Soc.* 135: 5344–5347.
55 Peng, C.G. and Damha, M.J. (2007). *Nucleic Acids Res.* 35: 4977–4988.
56 Franssen-van Hal, N.L.W., van der Putte, P., Hellmuth, K. et al. (2013). *Anal. Chem.* 85: 5950–5957.
57 Mohammadi-Kambs, M., Holz, K., Somoza, M.M., and Ott, A. (2017). *ACS Omega* 2: 1302–1308.
58 Abou Assi, H., Harkness, R.W., Martin-Pintado, N. et al. (2016). *Nucleic Acids Res.* 44: 4998–5009.
59 Abou Assi, H., El-Khoury, R., González, C., and Damha, M.J. (2017). *Nucleic Acids Res.* 45: 11535–11546.

12

SAMHD1-Mediated Negative Regulation of Cellular dNTP Levels: HIV-1, Innate Immunity, and Cancers

Tatsuya Maehigashi[1], *Dong-Hyun Kim*[2], *Raymond F. Schinazi*[1], and *Baek Kim*[1,2,3]

[1] *Emory University, Department of Pediatrics, Center for AIDS Research, School of Medicine, 1760 Haygood Drive Atlanta, GA 30322, USA*
[2] *Kyung-Hee University, College of Pharmacy, 26 Kyungheedae-ro, hoegi-dong, Seoul 130-701, South Korea*
[3] *Children's Healthcare of Atlanta, Center for Drug Discovery, 1760 Haygood Drive Atlanta, GA 30322, USA*

12.1 Cellular dNTP Concentrations

While ribonucleoside 5′-triphosphates (rNTPs) are consumed for multiple biological processes in living cells, including RNA synthesis, metabolic energy carrier (ATP), and cell signaling (ATP and GTP), DNA polymerization is the only cellular event known for the utility of 2′-deoxynucleoside triphosphates (dNTPs). Cellular dNTP concentrations have been extensively investigated, albeit mainly studied in dividing cells. Of note, fast dividing cells such as cancer cells and transformed cells undergo robust cell cycle and chromosomal replication processes, and these cells are known to harbor higher cellular dNTP levels, compared to slow growing or normal cells. In fact, elevated cellular dNTP levels have been considered to be a biochemical marker for cancer cells.

Various quantitative biochemical and chemical assays for cellular dNTP levels, which employ liquid chromatography–mass spectrometry (LC–MS/MS) or DNA polymerase reactions, were previously developed [1, 2]. While these assays were sufficient to determine the amounts of dNTPs extracted from dividing cells where dNTPs are abundant, the dNTP levels of nondividing cells remained undetermined due to the limited dNTP detection sensitivity of these assays. HIV-1 reverse transcriptase (RT) was discovered to be highly efficient for DNA synthesis, even at low dNTP concentrations [3, 4]. Therefore, HIV-1 RT was explored as an enzymatic tool to detect and measure the low levels of dNTPs extracted from not only smaller numbers of dividing cells, but also from nondividing cells. Specifically, dNTP levels in the nondividing human primary monocyte derived macrophages were measured with HIV-1 RT, since they likely harbor low dNTP levels due to their lack of cell division and DNA replication. Indeed, this viral RT-based assay was able to determine the cellular dNTP concentration of human primary macrophages as well as activated/dividing $CD4^+$ T cells [3]. As expected, the dNTP concentration of human primary macrophages was determined to be

Enzymatic and Chemical Synthesis of Nucleic Acid Derivatives, First Edition.
Edited by Jesús Fernández-Lucas and María-José Camarasa Rius.

only 20–40 nM, which is approximately 100 times lower than the average dNTP concentration reported in activated $CD4^+$ T cells (2–5 μM) [3]. In short, this highly sensitive and reliable HIV-1 RT based assay confirmed that dNTP levels in nondividing mammalian cells are significantly low, when compared to levels in dividing and cycling cells.

12.2 SAMHD1 and Negative Regulation of Cellular dNTPs

The regulatory mechanisms of cellular dNTP levels were extensively investigated for many decades, particularly for dNTP biosynthesis. This positive regulation of cellular dNTP levels includes a series of dNTP biosynthesis related enzymes such as ribonucleotide reductase (RNR) and thymidine kinase (TK), which are linked to cell cycles. For example, the activity and/or protein expression of these positive regulators for dNTPs in cells are elevated mainly at the G1/S phase and remain high during the S phase, where dNTPs are consumed for chromosomal DNA replication by cellular replicative DNA polymerases [5].

A novel negative regulatory mechanism of cellular dNTP levels was recently revealed during a series of investigations on a cellular protein, sterile alpha motif (SAM) domain and histidine-aspartate domain (HD) domain containing protein 1, SAMHD1. Human SAMHD1 is a dNTP triphosphohydrorylase that hydrolyzes dNTPs into dNs and triphosphates (Figure 12.1a) [7, 8]. SAMHD1,

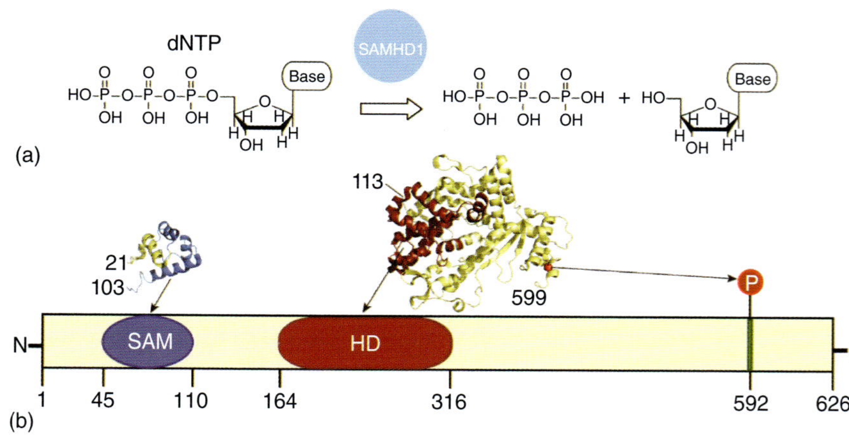

Figure 12.1 Human SAMHD1 as a dNTPase. (a) SAMHD1 is a dNTPase (triphosphohydrolase) that hydrolyzes dNTPs into deoxyribonucleosides (dNs) and inorganic triphosphates. (b) A schematic representation of SAMHD1 protein. Sterile Alpha Motif (SAM) and Histidine-Aspartic (HD) domain are shown in colored oval (blue and red, respectively) with respective residue numbers. An NMR structure of N-terminal construct containing SAM domain is shown on the left (PDB accession number 2E8O) (unpublished) and an example of HD domain containing structure determined through X-ray crystallography is shown on the right (PDB accession number 4mz7) [6]. Both structures are colored according to the schematics below, highlighting corresponding SAM or HD domain.

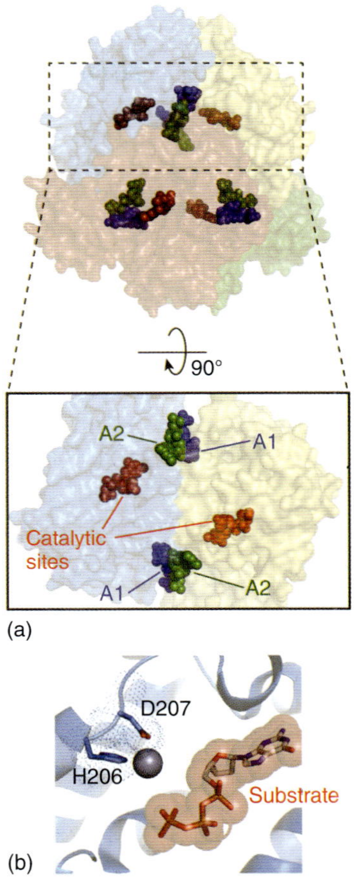

Figure 12.2 Structure of human SAMHD1 tetramer. (a) Active form of SAMHD1 is a tetrameric complex, shown with each monomer represented in a different color in this model (PDB accession number 4mz7) [6]. Allosteric sites (A1 and A2) and catalytic sites are shown with bound nucleotide (dGTP or dATP), colored blue (A1), green (A2), and red (catalytic sites). (b) A detailed view of an active site with bound substrate (dGTP shown) with key catalytic residues His206 and Asp207 shown in sticks. A divalent cation is required for the triphosphohydrolase reaction (Zn shown) and shown in gray sphere. Figures were generated and redered using the program PyMol [21].

a 72 kDa protein, is divided into three domains: N-T SAM domain, middle HD domain, and C-T phosphorylation/regulatory domain (Figure 12.1b). The actual function of the SAM domain of SAMHD1 remains unclear, while the HD domain contains the dNTPase active site that consists of histidine and aspartic acid residues. While the structure of the full length human SAMHD1 protein has not been solved, the structures of the SAM domain (residues 21–103) and the HD domain (residues 113–599) were solved (Figure 12.1b).

Multiple X-ray structures of the HD domain revealed that the enzymatically active form of SAMHD1 is a tetramer, and that the tetramerization of SAMHD1 is facilitated by dGTP/GTP cofactors [6, 9]. As shown in Figure 12.2a, SAMHD1 tetramer harbors two allosteric sites, A1 and A2, and the HD dNTPase catalytic site. The first allosteric site (A1) is for the binding of the dGTP/GTP positive cofactors that induces the initiation of the SAMHD1 tetramerization, whereas the second allosteric site (A2) is for the binding of all four dNTPs. The dNTP hydrolase reaction requires metal ions that are chelated with both histidine (H206) and aspartate (D207) residues to induce the hydrolysis reaction between alpha-phosphate and 5′-carbon of the deoxyribose (Figure 12.2b).

More recently, the C-terminal regulatory domain of SAMHD1 has been extensively investigated. The phosphorylation of T592 residue appears to prevent SAMHD1 monomers from forming enzymatically active tetramers [10, 11]. The negative regulation and phosphorylation of SAMHD1 is catalyzed by the cyclin A/cyclin dependent kinase (CDK) complex mainly at the S-phase in order to block the SAMHD1-mediated dNTP depletion during the S-phase of dividing cells where dNTPs are consumed for DNA replication [12]. Collectively, SAMHD1 dNTPase activity is very similar to that of RNR; both have allosteric regulation and cell cycle-dependent multimerization [13–18], albeit their functional results have opposite outcomes (dNTP degradation versus dNTP synthesis).

12.3 SAMHD1 Substrates, Activators, and Inhibitors

Since SAMHD1 hydrolyzes all four dNTPs (dGTP, dATP, dCTP, and TTP), it was predicted that the dNTPase catalytic site of human SAMHD1 recognizes mainly sugar and 5′-tri-phosphate (TP) moieties of dNTPs, but not the base moiety. However, SAMHD1 does not hydrolyze rNTPs, suggesting that 2′-OH of rNTPs prevents SAMHD1 from hydrolyzing rNTPs [7]. In fact, several residues at the dNTPase active site of SAMHD1 (i.e. L150 and Y374) appear to clash with the 2′-OH of rNTPs ([19]; see Figure 12.3a for a hypothetical GTP binding to the catalytic site). Also, SAMHD1 efficiently hydrolyzes various dNTPs with base modifications in dNTPs such as 2-amino-dNTP, O^6-methyl-dGTP, 5-methyl-dCTP, 2-thio-dTPP, and dUTP, supporting the theory that SAMHD1 does not recognize and interact with the base moiety of dNTP substrates [19]. In addition, nucleotide chain terminators lacking 2′-OH such as 2′3′-dideoxy-NTPs as well as 3′-azide-dideoxythymidine-triphosphate (AZT-TP) containing an azido group at the 3′-position are not hydrolyzed by SAMHD1, supporting the theory that the 2′ OH of dNTPs play a key role in the substrate recognition of SAMHD1 [19, 22, 23]. Since all nucleoside/nucleotide RT inhibitors targeting HIV-1 RT (N/NtRTIs) lack 2′ OH at their sugar moiety, and since N/NtRTIs compete with natural cellular dNTPs during viral reverse transcription, the SAMHD1-mediated dNTP depletion induces more effective anti-HIV effect of the N/NtRTIs, particularly in nondividing macrophages containing low dNTP pools [22]. The moieties of the dNTP substrate recognized by the catalytic site of SAMHD1 are illustrated in Figure 12.3b (yellow bars).

For the allosteric regulators of SAMHD1, both dGTP and GTP work equally as positive SAMHD1 activators that bind to the A1 site [24]. However, since

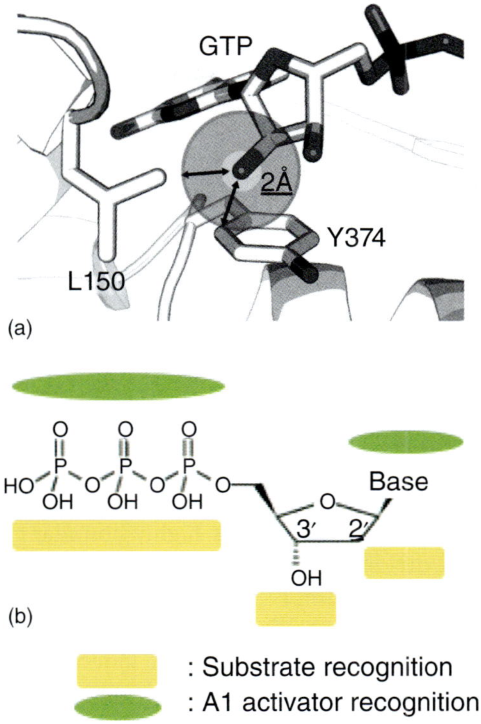

Figure 12.3 Substrates and activators of Human SAMHD1. (a) Close-up view of SAMHD1 catalytic site containing two key residues, L150 and Y374, that select against rNTP by inducing molecular clash with the 2′ OH of rGTP (see arrows). This figure was modeled for the rGTP binding to the SAMHD1 catalytic site. SAMHD1 protein is shown in cartoon, with key amino acid residues (L150 and Y374) shown in sticks. GTP is shown in sticks, with the sphere drawn at the position of 2′-O with a radius of 2 Å, highlighting its proximity to both L150 and Y374 side chains. The measured distance between 2′-O of the modeled GTP to L150 and Y374 side chains are approximately 2.4 Å at the closest point. The GTP molecule was modeled in SAMHD1 active site (PDB accession number 4MZ7) using the program coot [20]. Figure was generated using the program PyMol [21]. (b) Chemical moieties of dNTPs recognized by human SAMHD1 protein. The moieties and chemical positions recognized by the catalytic site and A1 regulatory site of SAMHD1 are marked with rectangular and ovals, respectively.

GTP is much more abundant than dNTP in cells [25], GTP likely works as a primary SAMHD1 activator in cells. Since only dGTP/GTP works as the positive activator of SAMHD1 for the A1 site, it was predicted that the A1 site of SAMHD1 recognizes the base moiety of the activators. Indeed, modified guanine base dGTPs such as O^6-methyl-dGTP fail to fully activate SAMHD1 [24]. In addition, SAMHD1 requires all three phosphates in dGTP and rGTP for its optimal dNTP hydrolysis [24]. The dNTP moieties that are recognized by the A1 site of SAMHD1 is illustrated in Figure 12.3b (green ovals). Unlike the A1 site, all four dNTPs bind to the A2 allosteric site of SAMHD1 [26, 27]. While the mechanistic role of the A2 site remains unclear, the SAMHD1 dNTPase activity requires the full occupancy of both A1 and A2 regulatory sites by dGTP/GTP and dNTPs, respectively.

Finally, a recent study that tested various anticancer dNTP analogues reported that SAMHD1 hydrolyzes araCTP and clofarabine-TP [19, 28], while (2′*S*)-2′-methyl dUTP works as an inhibitor toward the active site of SAMHD1 [19]. The clinical implications of these findings are discussed below.

12.4 SAMHD1 and HIV-1 Reverse Transcription

HIV-1, which belongs to lentiviruses, replicates both dividing/activated $CD4^+$ T cells and nondividing macrophages during the course of its pathogenesis [29]. However, while HIV-1 replicates very rapidly in activated $CD4^+$ T cells, the HIV-1 replication in macrophages is kinetically suppressed [30, 31]. In 2011, SAMHD1 was identified as a restriction host factor that is responsible for the suppressed HIV-1 replication kinetics in nondividing myeloid cells including macrophages [32–34]. In fact, SAMHD1 is highly abundant in macrophages, compared to activated/dividing $CD4^+$ T cells and various cell lines. This abundant SAMHD1 dNTPase, combined with a lack of dNTP biosynthesis, is responsible for the extremely low dNTP concentrations observed in nondividing macrophages, which restricts HIV-1 replication kinetics in macrophages as illustrated in Figure 12.4a for Restrictive Infection.

However, HIV-2 and some simian immunodeficiency viruses (SIV) such as SIV_{sm} encode an accessary protein, called viral protein X (Vpx), and subsequently replicate rapidly even in macrophages, compared to HIV-1 [35, 36]. Vpx counteracts SAMHD1 by inducing the DCAF1/Cul4A/E3 ligase mediated proteosomal degradation of SAMHD1 [37]. As shown in Figure 12.4b, the degradation of SAMHD1 by Vpx elevates the extremely low dNTP levels found in macrophages (see −Vpx, 20–40 nM) close to 1 μM dNTP (see +Vpx), which is close to the dNTP levels found in dividing cells (see T cells). This impact leads to the acceleration of the viral reverse transcription in macrophages [38] (see Figure 12.4a for permissive infection). Cells isolated from SAMHD1 knock-out mice harbor elevated dNTP levels and support more efficient lentivirus replication, compared to the cells from the SAMHD1 wild type mice [39, 40].

In addition to suppressing the replication of lentiviruses, SAMHD1 is able to suppress the replication of other viruses that utilize dNTPs for their viral genomic replication such as vaccinia virus, herpes simplex virus type 1 [41, 42] and hepatitis B virus [43–45]. Overall, it is plausible that SAMHD1, which depletes dNTPs from the cells, is a primitive host-defense mechanism against viral pathogens that consume dNTPs for their replication processes.

12.5 SAMHD1 Mutations and Innate Immunity

SAMHD1 was initially reported as one of the genes that are linked to a rare genetic immune disorder, Aicardi Goutieres Syndrome (AGS). AGS patients

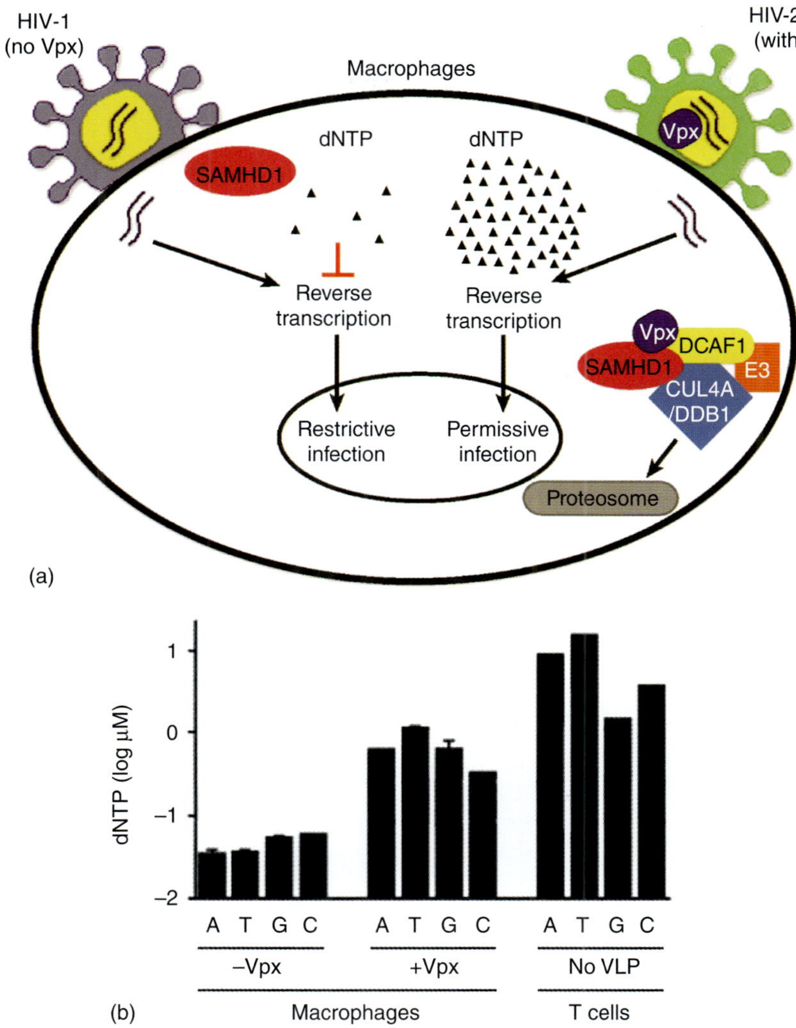

Figure 12.4 Replication of HIV-1, HIV-2, and SIV in macrophages and Vpx-mediated dNTP elevation in macrophages. (a) SAMHD1-mediated low dNTP pools restrict HIV-1 replication in macrophages. Vpx encoded in HIV-2 and SIV enhances viral infectivity in macrophages by proteosomally degrading SAMHD1 through DCAF1/Cul4A/E3 ligase and elevating cellular dNTPs, ultimately leading to the permissive infection. (b) Intracellular dNTP concentrations in human primary monocyte-derived macrophages (−Vpx) and activated $CD4^+$ T cells. Treatment of macrophages with Vpx-containing virus-like particles (+Vpx) elevates dNTP concentration close to that of activated $CD4^+$ T cells.

with mutated SAMHD1 genes display early onset of severe neuro-developmental disorders with lupus-like chronic brain inflammation, which is characterized by hyper-activation and dysregulation of type 1 interferon responses in the absence of any infection of pathogens [46–49]. Other AGS-inducing enzymes are Trex1 nuclease, RNaseH2, adenosine deaminase 1 (ADAR1) and interferon induced RNA helicase C containing protein, which are considered to be involved in

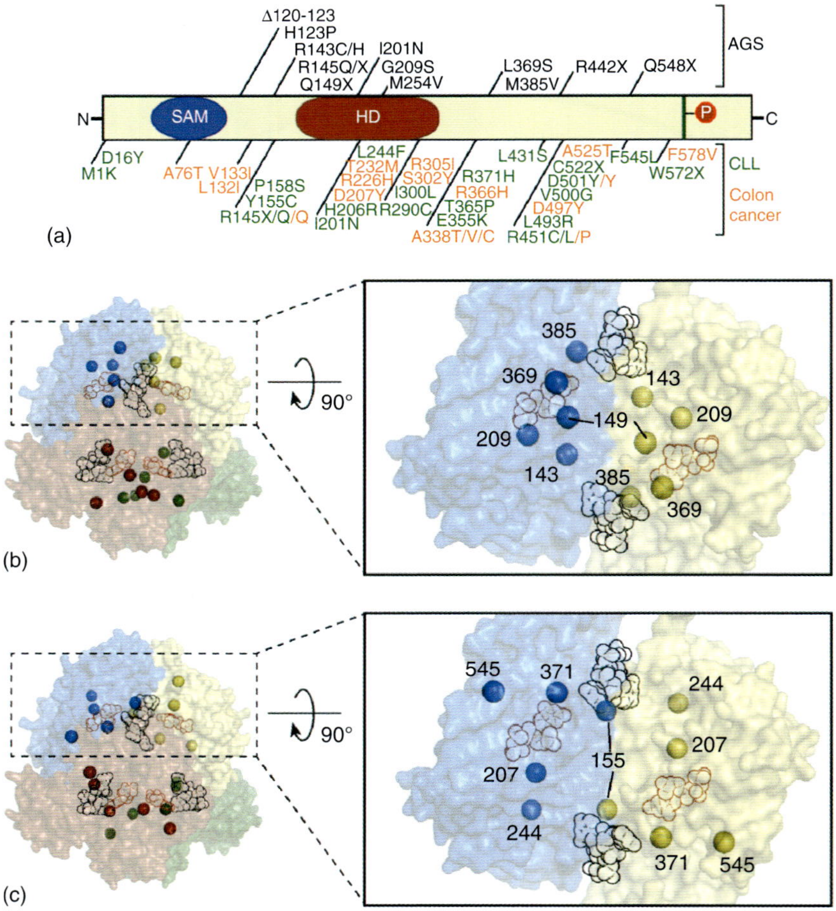

Figure 12.5 Mutations found in AGS and human cancers. (a) Summary of SAMHD1 mutations found in AGS (top) and cancers (bottom) reported previously [48, 52, 53]. (b) A representation of selected AGS-related mutations in SAMHD1. SAMHD1 tetramer (left) and dimer (right) are shown, with locations of mutations indicated by the sphere. Respective positions of allosteric sites (A1 and A2) and catalytic sites are outlined in black and red, respectively. (c) Same as (b), but for cancers. Figures were generated and rendered using the program Pymol [21].

nucleic acid sensing innate immune responses [48, 50, 51]. SAMHD1 mutations known for inducing AGS (Figure 12.5a) are found throughout the genes. Some of these mutations near the catalytic and allosteric sites of SAMHD1 are marked in Figure 12.5b.

Many of these mutated SAMHD1 proteins show various defects in tetramerization, loss of dNTPase activity, and nuclear localization [54]. However, some AGS-related SAMHD1 variants are not degraded by Vpx and maintain anti-HIV restriction capability. Interestingly, G209S SAMHD1 mutant displays all wild type like normal biochemical behaviors and anti-HIV activity, but fails to down-modulate type 1 interferon [54], suggesting the existence of separate domains for AGS and HIV-1 restriction. Collectively, while SAMHD1 is involved in AGS, the mechanism by which the SAMHD1 mutations trigger the AGS phenotype remains unclear and requires further investigations.

12.6 SAMHD1 and Cancers

Recent studies proposed two potential functions of SAMHD1 in cancers: (i) tumor suppressor and (ii) cancer resistance to anticancer nucleoside chemotherapeutic agents, such as araC. The tumor suppressor function of SAMHD1 is based on its dNTPase activity that can negatively suppress the cellular chromosomal DNA replication process and cell division by depleting cellular dNTPs. This putative antitumor activity of SAMHD1 generates a potential anticancer strategy that pharmacologically blocks cyclin A/CDK responsible for the SAMHD1 phosphorylation and dNTPase inactivation.

Secondly, as demonstrated in acute myelogenous leukemia (AML) [55], the high level of SAMHD1 protein in cancer cells reduces the efficacy of araC, an anticancer nucleoside analogue. Indeed, a recent biochemical data demonstrated that SAMHD1 can hydrolyze araC-triphosphate [19], which can lead to the lower cellular araC- triphosphate concentration in cancer cells and ultimately reduce its anticancer efficacy. These findings support the development of SAMHD1 as a clinical oncology biomarker for araC sensitivity in cancers [55].

In addition, a series of the SAMHD1 mutations were found in cancer cells including chronic lymphocytic leukemia [52], lymphoma [56], and colon cancers [53] (Figure 12.5a). These cancer-related SAMHD1 mutations are also found throughout the gene. Some of these mutations near the catalytic and allosteric sites of SAMHD1 are marked in Figure 12.5c. While these findings suggest the role of SAMHD1 as a tumor suppressor, it is also possible that SAMHD1 mutations associated with cancers could be related to lack of DNA repair. Indeed, a recent study reported that SAMHD1 is involved in double-stranded DNA break repair, and this DNA repair related function of SAMHD1 does not require its dNTPase activity [57]. More investigations are needed to elucidate mechanistic interplay among the SAMHD1 functions, tumorigenesis, and cancer cell resistance to araC.

12.7 Summary

SAMHD1 plays multiple roles in various biological and clinical systems including innate immunity, viral replication, and cancers. Owing to recent multidisciplinary studies on SAMHD1, we now know that the regulatory circuit of cellular dNTP metabolism (which has been investigated predominantly through dNTP biosynthesis) harbors both positive and negative biochemical switches in order to maintain optimal dNTP concentrations. Further investigations on SAMHD1 action mechanisms are warranted, and will help to define the role of SAMHD1 as a pharmacological target in various human diseases.

Acknowledgment

We greatly appreciate Dr. James Kohler for the helpful reading. This was supported by NIH grants: R01AI049781 (B.K.), R01GM104198 (B.K.), R01MH100999 and R01MH116695 (R.F.S.), and P30AI050409-19 (C. del Rio for Emory Center for AIDS Research).

References

1 Ferraro, P., Franzolin, E., Pontarin, G., et al. (2010). Quantitation of cellular deoxynucleoside triphosphates. *Nucleic Acids Res.* 38 (6): e85.

2 Wilson, P.M., Labonte, M.J., Russell, J. et al. (2011). A novel fluorescence-based assay for the rapid detection and quantification of cellular deoxyribonucleoside triphosphates. *Nucleic Acids Res.* 39 (17): e112.

3 Diamond, T.L., Roshal, M., Jamburuthugoda, V.K. et al. (2004). Macrophage tropism of HIV-1 depends on efficient cellular dNTP utilization by reverse transcriptase. *J. Biol. Chem.* 279 (49): 51545–51553.

4 Hollenbaugh, J.A. and Kim, B. (2016). HIV-1 reverse transcriptase-based assay to determine cellular dNTP concentrations. *Methods Mol. Biol.* 1354: 61–70.

5 Stillman, B. (2013). Deoxynucleoside triphosphate (dNTP) synthesis and destruction regulate the replication of both cell and virus genomes. *Proc. Natl. Acad. Sci. U.S.A.* 110 (35): 14120–14121.

6 Zhu, C., Gao, W., Zhao, K. et al. (2013). Structural insight into dGTP-dependent activation of tetrameric SAMHD1 deoxynucleoside triphosphate triphosphohydrolase. *Nat. Commun.* 4: 2722.

7 Goldstone, D.C., Ennis-Adeniran, V., Hedden, J.J. et al. (2011). HIV-1 restriction factor SAMHD1 is a deoxynucleoside triphosphate triphosphohydrolase. *Nature* 480 (7377): 379–382.

8 Powell, R.D., Holland, P.J., Hollis, T., and Perrino, F.W. (2011). Aicardi-Goutieres syndrome gene and HIV-1 restriction factor SAMHD1 is a dGTP-regulated deoxynucleotide triphosphohydrolase. *J. Biol. Chem.* 286 (51): 43596–43600.

9 Ji, X., Wu, Y., Yan, J. et al. (2013). Mechanism of allosteric activation of SAMHD1 by dGTP. *Nat. Struct. Mol. Biol.* 20 (11): 1304–1309.

10 Arnold, L.H., Groom, H.C., Kunzelmann, S. et al. (2015). Phospho-dependent regulation of SAMHD1 Oligomerisation couples catalysis and restriction. *PLoS Pathog.* 11 (10): e1005194.

11 Tang, C., Ji, X., Wu, L., and Xiong, Y. (2015). Impaired dNTPase activity of SAMHD1 by phosphomimetic mutation of Thr-592. *J. Biol. Chem.* 290 (44): 26352–26359.

12 Kretschmer, S., Wolf, C., Konig, N. et al. (2015). SAMHD1 prevents autoimmunity by maintaining genome stability. *Ann. Rheum. Dis.* 74 (3): e17.

13 Chimploy, K. and Mathews, C.K. (2001). Mouse ribonucleotide reductase control: influence of substrate binding upon interactions with allosteric effectors. *J. Biol. Chem.* 276 (10): 7093–7100.

14 Guittet, O., Hakansson, P., Voevodskaya, N. et al. (2001). Mammalian p53R2 protein forms an active ribonucleotide reductase in vitro with the R1 protein, which is expressed both in resting cells in response to DNA damage and in proliferating cells. *J. Biol. Chem.* 276 (44): 40647–40651.

15 Hakansson, P., Hofer, A., and Thelander, L. (2006). Regulation of mammalian ribonucleotide reduction and dNTP pools after DNA damage and in resting cells. *J. Biol. Chem.* 281 (12): 7834–7841.

16 Kashlan, O.B., Scott, C.P., Lear, J.D., and Cooperman, B.S. (2002). A comprehensive model for the allosteric regulation of mammalian ribonucleotide reductase. Functional consequences of ATP- and dATP-induced oligomerization of the large subunit. *Biochemistry* 41 (2): 462–474.

17 Nordlund, P. and Reichard, P. (2006). Ribonucleotide reductases. *Annu. Rev. Biochem.* 75: 681–706.

18 Tanaka, H., Arakawa, H., Yamaguchi, T. et al. (2000). A ribonucleotide reductase gene involved in a p53-dependent cell-cycle checkpoint for DNA damage. *Nature* 404 (6773): 42–49.

19 Hollenbaugh, J.A., Shelton, J., Tao, S. et al. (2017). Substrates and inhibitors of SAMHD1. *PLoS One* 12 (1): e0169052.

20 Emsley, P., Lohkamp, B., Scott, W.G., and Cowtan, K. (2010). Features and development of coot. *Acta Crystallogr., Sect. D: Biol. Crystallogr.* 66 (Pt 4): 486–501.

21 Schrodinger, L.L.C. (2015). The PyMOL molecular graphics system, version 1.8.

22 Amie, S.M., Bambara, R.A., and Kim, B. (2013). Anti-HIV host factor SAMHD1 regulates viral sensitivity to nucleoside reverse transcriptase inhibitors via modulation of cellular deoxyribonucleoside triphosphate (dNTP) levels. *J. Biol. Chem.* 288 (28): 20683–20691.

23 Huber, A.D., Michailidis, E., Schultz, M.L. et al. (2014). SAMHD1 has differential impact on the efficacies of HIV nucleoside reverse transcriptase inhibitors. *Antimicrob. Agents Chemother.* 58 (8): 4915–4919.

24 Amie, S.M., Bambara, R.A., and Kim, B. (2013). GTP is the primary activator of the anti-HIV restriction factor SAMHD1. *J. Biol. Chem.* 288 (35): 25001–25006.

25 Kennedy, E.M., Gavegnano, C., Nguyen, L. et al. (2010). Ribonucleoside triphosphates as substrate of human immunodeficiency virus type 1 reverse transcriptase in human macrophages. *J. Biol. Chem.* 285 (50): 39380–39391.

26 Hansen, E.C., Seamon, K.J., Cravens, S.L., and Stivers, J.T. (2014). GTP activator and dNTP substrates of HIV-1 restriction factor SAMHD1 generate a long-lived activated state. *Proc. Natl. Acad. Sci. U.S.A.* 111 (18): E1843–E1851.

27 Ji, X., Tang, C., Zhao, Q. et al. (2014). Structural basis of cellular dNTP regulation by SAMHD1. *Proc. Natl. Acad. Sci. U.S.A.* 111 (41): E4305–E4314.

28 Arnold, L.H. Kunzelmann, S., Webb, M.R., and Taylor, I.A. (2015). A continuous enzyme-coupled assay for triphosphohydrolase activity of HIV-1 restriction factor SAMHD1. *Antimicrob. Agents Chemother.* 59 (1): 186–192.

29 Cunningham, A.L., Donaghy, H., Harman, A.N. et al. (2010). Manipulation of dendritic cell function by viruses. *Curr. Opin. Microbiol.* 13 (4): 524–529.

30 Bergamaschi, A. and Pancino, G. (2010). Host hindrance to HIV-1 replication in monocytes and macrophages. *Retrovirology* 7: 31.

31 Duncan, C.J. and Sattentau, Q.J. (2011). Viral determinants of HIV-1 macrophage tropism. *Viruses* 3 (11): 2255–2279.

32 Berger, A., Sommer, A.F., Zwarg, J. et al. (2011). SAMHD1-deficient CD14+ cells from individuals with Aicardi-Goutieres syndrome are highly susceptible to HIV-1 infection. *PLoS Pathog.* 7 (12): e1002425.

33 Hrecka, K., Hao, C., Gierszewska, M. et al. (2011). Vpx relieves inhibition of HIV-1 infection of macrophages mediated by the SAMHD1 protein. *Nature* 474 (7353): 658–661.

34 Laguette, N., Sobhian, B., Casartelli, N. et al. (2011). SAMHD1 is the dendritic– and myeloid–cell–specific HIV–1 restriction factor counteracted by Vpx. *Nature* 474 (7353): 654–657.

35 Guyader, M., Emerman, M., Montagnier, L., and Peden, K. (1989). Vpx mutants of HIV-2 are infectious in established cell lines but display a severe defect in peripheral blood lymphocytes. *EMBO J.* 8 (4): 1169–1175.

36 Yu, X.F., Yu, Q.C., Essex, M., and Lee, T.H. (1991). The Vpx gene of simian immunodeficiency virus facilitates efficient viral replication in fresh lymphocytes and macrophage. *J. Virol.* 65 (9): 5088–5091.

37 Mashiba, M. and Collins, K.L. (2012). Molecular mechanisms of HIV immune evasion of the innate immune response in myeloid cells. *Viruses* 5 (1): 1–14.

38 Lahouassa, H., Daddacha, W., Hofmann, H. et al. (2012). SAMHD1 restricts the replication of human immunodeficiency virus type 1 by depleting the intracellular pool of deoxynucleoside triphosphates. *Nat. Immunol.* 13: 223–228.

39 Behrendt, R., Schumann, T., Gerbaulet, A. et al. (2013). Mouse SAMHD1 has antiretroviral activity and suppresses a spontaneous cell-intrinsic antiviral response. *Cell Rep.* 4 (4): 689–696.

40 Rehwinkel, J., Maelfait, J., Bridgeman, A. et al. (2013). SAMHD1-dependent retroviral control and escape in mice. *EMBO J.* 32 (18): 2454–2462.

41 Kim, E.T., White, T.E., Brandariz-Nunez, A. et al. (2013). SAMHD1 restricts herpes simplex virus 1 in macrophages by limiting DNA replication. *J. Virol.* 87 (23): 12949–12956.

42 Hollenbaugh, J.A., Gee, P., Baker, J. et al. (2013). Host factor SAMHD1 restricts DNA viruses in non-dividing myeloid cells. *PLoS Pathog.* 9: e1003481.

43 Chen, Z., Zhu, M., Pan, X. et al. (2014). Inhibition of hepatitis B virus replication by SAMHD1. *Biochem. Biophys. Res. Commun.* 450 (4): 1462–1468.

44 Jeong, G.U., Park, I.H., Ahn, K., and Ahn, B.Y. (2016). Inhibition of hepatitis B virus replication by a dNTPase-dependent function of the host restriction factor SAMHD1. *Virology* 495: 71–78.

45 Sommer, A.F., Riviere, L., Qu, B. et al. (2016). Restrictive influence of SAMHD1 on hepatitis B virus life cycle. *Sci. Rep.* 6: 26616.

46 Aicardi, J. and Goutieres, F. (1984). A progressive familial encephalopathy in infancy with calcifications of the basal ganglia and chronic cerebrospinal fluid lymphocytosis. *Ann. Neurol.* 15 (1): 49–54.

47 Crow, Y.J. and Livingston, J.H. (2008). Aicardi-Goutieres syndrome: an important Mendelian mimic of congenital infection. *Dev. Med. Child Neurol.* 50 (6): 410–416.

48 Rice, G.I., Bond, J., Asipu, A. et al. (2009). Mutations involved in Aicardi-Goutieres syndrome implicate SAMHD1 as regulator of the innate immune response. *Nat. Genet.* 41 (7): 829–832.

49 Tolmie, J.L., Shillito, P., Hughes-Benzie, R., and Stephenson, J.B. (1995). The Aicardi-Goutieres syndrome (familial, early onset encephalopathy with calcifications of the basal ganglia and chronic cerebrospinal fluid lymphocytosis). *J. Med. Genet.* 32 (11): 881–884.

50 Crow, Y.J., Hayward, B.E., Parmar, R. et al. (2006). Mutations in the gene encoding the 3′-5′ DNA exonuclease TREX1 cause Aicardi-Goutieres syndrome at the AGS1 locus. *Nat. Genet.* 38 (8): 917–920.

51 Rice, G.I., Kasher, P.R., Forte, G.M. et al. (2012). Mutations in ADAR1 cause Aicardi-Goutieres syndrome associated with a type I interferon signature. *Nat. Genet.* 44 (11): 1243–1248.

52 Clifford, R., Louis, T., Robbe, P. et al. (2014). SAMHD1 is mutated recurrently in chronic lymphocytic leukemia and is involved in response to DNA damage. *Blood* 123 (7): 1021–1031.

53 Rentoft, M., Lindell, K., Tran, P. et al. (2016). Heterozygous colon cancer-associated mutations of SAMHD1 have functional significance. *Proc. Natl. Acad. Sci. U.S.A.* 113 (17): 4723–4728.

54 White, T.E., Brandariz-Nunez, A., Martinez-Lopez, A. et al. (2017). A SAMHD1 mutation associated with Aicardi-Goutieres syndrome uncouples the ability of SAMHD1 to restrict HIV-1 from its ability to downmodulate type I interferon in humans. *Hum. Mutat.* 38 (6): 658–668.

55 Schneider, C., Oellerich, T., and Baldauf, H.M. et al. (2017). SAMHD1 is a biomarker for cytarabine response and a therapeutic target in acute myeloid leukemia. *Nat. Med.* 23 (2): 250–255.

56 Merati, M., Buethe, D.J., Cooper, K.D. et al. (2015). Aggressive CD8(+) epidermotropic cutaneous T-cell lymphoma associated with homozygous mutation in SAMHD1. *JAAD Case Rep.* 1 (4): 227–229.

57 Daddacha, W., Koyen, A.E., Bastien, A.J. et al. (2017). SAMHD1 promotes DNA end resection to facilitate DNA repair by homologous recombination. *Cell Rep.* 20 (8): 1921–1935.

Index

Enzymatic and Chemical Synthesis of Nucleic Acid Derivatives, First Edition.
Edited by Jesús Fernández-Lucas and María-José Camarasa Rius.

d

n